The Atmosphere

THIRD EDITION

RICHARD A. ANTHES

JOHN J. CAHIR

ALISTAIR B. FRASER

HANS A. PANOFSKY

Meteorology Department
The Pennsylvania State University

Charles E. Merrill Publishing Company
A Bell & Howell Company
Columbus   Toronto   London   Sydney

Published by
Charles E. Merrill Publishing Company
A Bell & Howell Company
Columbus, Ohio 43216

This book was set in Optima and Bookman.
Production Editor: Cherlyn B. Paul
Text Designer: Ann Mirels
Cover Design Coordination: Will Chenoweth
Cover Photograph: Four by Five

Library of Congress Catalog Card Number: 81-80372

International Standard Book Number: 0-675-08043-6

Printed in the United States of America

1 2 3 4 5 6 7 8 9 10—85 84 83 82 81

Preface

Although part of modern meteorology is a sophisticated mathematical science, much of the weather around us can be qualitatively understood and appreciated without the use of mathematics. In writing an introductory meteorology book designed to appeal to the reader not mathematically inclined, as well as to the reader who has a strong mathematical and physical science background, we have tried to strike a common responsive chord—curiosity about the atmosphere and its fascinating variety of phenomena.

Rather than emphasize abstract facts or physical concepts, which many people find boring or irrelevant, we have departed somewhat from the traditional introductory textbook format by integrating explanations of basic physical processes into the discussion of atmospheric phenomena or processes that are likely to be of interest to the reader.

The Atmosphere is written for the casual, as well as the sophisticated, observer of the weather to explain those atmospheric phenomena that people are likely to encounter, whether driving across the country on an interstate highway, flying 30,000 feet over the ocean, or skimming across whitecaps in a small sailboat on a summer afternoon. Observation is the key to understanding the atmosphere, and we have tried to show the reader what to observe, and how to interpret these observations in a meaningful way.

The third edition of *The Atmosphere* has been restructured so that the organization is more useful to the student and the instructor. The reorganization was based on detailed comments provided by many users of the earlier editions. The book maintains its nonmathematical approach, which helps students understand the weather as it affects them on a daily and seasonal basis.

The third edition begins with a survey of the development of meteorology as a science. In the second and third chapters, the origin of the atmosphere, the important atmospheric variables, and the horizontal and vertical structure of the atmosphere are introduced to prepare the student for the later chapters treating meteorological phenomena. The basic physics important in meteorology, radiation, and the laws governing the behavior of the atmosphere are treated next, followed by a chapter on moisture and clouds. Synoptic meteorology in extratropical latitudes is discussed in chapters 7 and 8. The discussion of hurricanes, now included as a separate chapter, has been updated and expanded. A new chapter on thunderstorms and tornadoes, which includes an enhanced treatment of lightning, has been added.

A major addition to the third edition is a chapter on climate. In addition to discussing the physical factors affecting climate, numerous maps depict the climate of the earth and North America.

The Atmosphere considers several subjects usually not found in elementary meteorology texts. These include the beautiful and complex world of meteorological optical phenomena, including halos, sun dogs, sun pillars, mirages, and the elusive green flash, as well as the rainbow. One chapter is devoted to air pollution, and covers both the effect of the atmosphere on pollution and the increasingly important effects of pollution on the weather. The controversial, yet vitally important, subject of climate change, both natural and artificial, is discussed in considerable detail. The chapter on hydrology, including the water cycle, has been extensively revised.

As energy supplies become scarce and expensive, we begin to look more closely at the sun and the wind as sources of energy. Some of the meteorological aspects of these alternative sources are discussed. In keeping with the "weather and people" theme, a chapter on biometeorology (the study of interactions of biological life with meteorology) includes human physiological response to the weather.

In addition to explaining how the perceptive observer can infer tomorrow's weather today, we have indicated how weather and climate have shaped human culture throughout history. Included are numerous examples of the direct influence of weather on folklore, literature, music, art, and religion.

Finally, the book is summarized with a review of the annual progression of weather over the United States, illustrating how the various physical processes and atmospheric phenomena unite to provide the ever-repeating, yet never-the-same, weather drama.

We would like to thank Christopher Church, who provided original material on the formation of lightning, Rosa de Pena who carefully reviewed chapter 6, and John Norman who contributed the turbulent wind and temperature data in chapter 2 from his field experiments.

The production editor, Cher Paul cheerfully and competently transformed the manuscript into the finished product. Susan Anthes was very helpful in obtaining reference materials.

We thank Peter Black, Theodore Fujita, Joseph Golden, Ronald Holle, Charlie Hosler, Gustav Lamprecht, Robert McAlister, N. M. Reiss, Frank Schiermeier, Robert Sheets, William Shenk, Dennis Thomson, and Andrew Watson for providing photographs. George Winterling provided material on the humiture index.

Finally, we would like to thank Professors Stephen Berman, Eugene Chermack, Albert Frank, David Marczely, and J. M. Wallace for their helpful reviews of the manuscript, and the following people who gave comments and suggestions for the new edition: L. Dean Bark, Robert Blakely, William Blumen, Edward Brooks, James Carter, Carl Chelius, George M. Hale, Edward L. Janiskee, K. H. Jehn, H. E. Landsberg, John R. Mather, Joseph M. Moran, Frank Nicholas, Joseph Pifer, Allen E. Staver, Charles Stearns, Richard D. Stepp, John Trischka, and Eberhard Wahl.

Richard A. Anthes　　　*Alistair B. Fraser*　　　　　　*December 1980*
John J. Cahir　　　　　*Hans A. Panofsky*

Contents

FOUR   *Radiation*

FIVE   *Behavior of Gases*

SIX   *Moisture and Clouds*

SEVEN   *The Changing Weather Outside the Tropics*

EIGHT   *Watching the Weather*

NINE   *Hurricanes*

TEN   *Thunderstorms and Tornadoes*

ELEVEN   *Climate*

TWELVE   Climate Changes and the Changing Climate

THIRTEEN   Air Pollution Meteorology

ix

FOURTEEN   *Weather and Water*

FIFTEEN   *Meteorological Optics*

SIXTEEN   *Impact of Weather and Climate on People*

SEVENTEEN   *A Year's Weather in the United States*

ONE

Rainbow in Morning

1.1 "SOME ARE WEATHERWISE; SOME ARE OTHERWISE" [Benjamin Franklin, *Poor Richard's Almanac*, 1735]

It is not surprising that people always have been fascinated by the capricious nature of the weather. In their early days as hunters and food gatherers, their comfort and supply of food were dependent upon the weather more than on any other factor. Even as tools and technology have freed us from scurrying about in search of wild animals and fruits and have provided us with warm clothing, houses, and studded snow tires, our dependence on the weather has not ceased, but rather has become more subtle.

Even if people try to ignore the weather, they are likely to fail, for sooner or later the weather will directly affect their lives. The day-to-day effects of weather shape their psychological outlook slowly but surely, and changes in the weather can affect their moods, or even their attitudes toward their business associates, as indicated by the ancient advice "Do business with men when the wind is in the northwest." In his remarkable book *Mainsprings of Civilization*, Ellsworth Huntington argues that climate and weather play a dominant role in determining a nation's history and shaping its culture. Indeed, climate is a major factor in the advancement—or lack of advancement—of civilization.

Besides the subtle molding of character by the climate, misbehavior of the atmosphere can also completely interrupt normal human activities, as anyone whose life has been touched by a flood, hurricane, tornado, or even slippery roads on New Year's Eve can readily attest. Even the most sheltered urban existence is occasionally threatened by varieties of weather, such as traffic-halting blizzards, air pollution events, canceled airplane flights, or power blackouts caused by disruptive weather.

Of course, most of us escape the spectacular weather events that seriously threaten lives. However, we find gray drizzling days depressing, we pay for the corrosion of our cars from road salt used to melt snow, we drag garden hoses around the lawn during droughts to save our grass, and then spend hours behind a mower after the rains resume. On the positive side, we feel exhilarated as we ski down snowy mountain slopes on a bright winter day, or serene as we drowse by the fireside during a cold November rain.

So, the weather is a constant, and occasionally dominating, force in our lives. This inevitable influence, however, is not the only reason for wanting to

2

know how and why the weather acts as it does. Even the casual study of day-to-day weather can be an enjoyable rest from daily pressures. The dramatic evolution of weather events in the middle and high latitudes and the subtle changes in tropical weather patterns can afford the perceptive observer hours of satisfaction that are varied, exciting, and rewarding. Naturally, we may enjoy the bracing north wind of winter or the distant lightning of a faraway summer thunderstorm without understanding the underlying physical processes. As comprehension of the forces that shape these weather events grows, however, so does our enjoyment and appreciation of the weather drama.

Anticipation spices any pleasure, and the weatherwise individual will soon be able to recognize the signs of impending weather changes. For example, great snowstorms in the eastern United States usually give the alert individual warning of a day or two. First, the brilliant cerulean skies of a midwinter cold snap give way to high, delicate cirrus clouds which streak in from the southwest and entangle the sun in a web of icy fibers. The barometer, which has been rising for several days, hesitates, and then slowly begins to fall. Flags which have been rippling from the north now hang limp, and smoke from the fireplace, refusing to rise, drifts slowly westward. The thermometer climbs sluggishly from the teens into the more moderate twenties (°F). Many people might interpret these signs, if they noticed them at all, merely as a welcome respite from the cold, windy weather of the previous two days. However, perceptive observers of the atmosphere will alert themselves to the unmistakable signs of an impending storm.

Twelve hours later, the signs are even more legible. The halo around the moon, caused by a veil of cirrostratus clouds, disappears in a thickening mass of clouds. The barometer, as if making up its mind, falls more rapidly, and bare trees begin to sway in the increasing northeast wind.

In another six hours, the first tiny snowflakes flutter across the landscape. The clouds lower, and soon thick-falling snow reduces the lights from nearby houses and street lamps to hazy patches of luminescence. The snow falls heavily during the day as the barometer continues to fall. Weather observers monitor the falling pressure and the wind direction, looking for the abrupt upward swing of the glass (barometer) and the shift of the wind vane that will signal the end of the storm. At the same time, they watch the rising thermometer, which now indicates $-1°C$ and hints at the possibility of the snow changing to rain.

Suddenly, the steady wailing of the east wind subsides. Snowflakes hesitate in their westward drive before the wind, and a cold gust from the northwest halts the upward progress of the thermometer. The barometer ceases its downward drive and jumps decisively upward. The heavy snowfall is over. Now, stronger northerly (from the north) winds pile the snow into mountains and canyons, sweeping bare exposed areas and piling high drifts behind every obstacle. The mercury in the thermometer contracts hurridly into the teens (°F), ending the possibility of any melting. Now, the snow flurries diminish gradually, and through the scudding clouds we can catch occasional glimpses of the moon. The storm is over.

All the preceding signs were completely missed and unappreciated by many people. To the weather observer, however, with $20 worth of instruments and a pair of open eyes and ears, the signs were far more exciting and revealing than the sterile words of the newspaper forecast: "Cloudy and cold today and to-

3

night with snow likely, possibly accumulating four or more inches. Snow ending tomorrow, followed by clearing weather.''

Most of the daily weather events that touch our lives are not as spectacular, nor the signs as obvious, as those in the preceding example. Nevertheless, a sensitive observer may catch many small clues to the behavior of the atmosphere that nature provides every day. The purpose of this book is to instill in the amateur this interest in observing and interpreting the weather and to describe the physical processes that are responsible for weather events. In this way, we hope to heighten the amateur's appreciation and comprehension of the atmosphere's seemingly mysterious ways.

1.2 THE ANCIENT METEOROLOGIST

We do not know who first peeked out of the cave, saw a halo around the moon, and called off the mammoth hunt for the next day. Yet, it is certain that the earliest people, whose lives were intimately entwined with the weather, would devote some of their intellectual energies trying to comprehend the meaning of the wind and the sky. Indeed, all of the information (except the barometer) that was available to the modern weather observer watching the evolution of the snowstorm in the previous section was available to our ancestors. (Although objections may be raised about the availability of a thermometer, human skin could have served the purpose. Skin is very sensitive to air temperature, and most people can estimate it within 2°C.) Therefore, intelligent and perceptive people would soon notice the signs that portend major weather events, and then verbally pass their weather wisdom to successive generations. In this way, an increasing inventory of weather folklore could be established.

The prehistoric view of meteorology must remain a matter for speculation. Our first real glimpse of the ancient meteorologists begins with the Greeks, who named the study of heavenly phenomena *meteōrología,* and called raindrops, hail, and snowflakes *metéōron,* which means "a thing in the air." In contrast to today's science of meteorology, which includes only atmospheric phenomena, early meteorology was closely associated with astronomy. Thales (640 B.C.) believed in a flat earth, but his views on meteorology were somewhat more sound. He correctly ascribed the four seasons to variations in the position of the sun in the sky. The seasons were separated, as they are today, by the winter and summer solstices (sun-standings), and by the autumn and vernal (spring) equinoxes (dates of equal night and day).

Because of its unmistakable association with human health and psychology, early meteorology was an important part of medicine. Hippocrates (fifth century B.C.) cautioned that physicians, upon reaching a new city, should first study the meteorology of the region, including the prevailing winds, the relative amounts of sunshine and cloudiness, the amount of rainfall, and the exposure of the town in relation to the winter sun. Today we call Hippocrates' type of meteorology biometeorology, and even in a day of technologically controlled human environments, we have essentially confirmed Hippocrates' advice.

1.2.1   ARISTOTLE'S VIEW OF METEOROLOGY

Aristotle was the most famous of the ancient meteorologists. His *Meteorologica*, written about 340 B.C., gives a detailed account of his view of most aspects of weather and climate. We present Aristotle's ideas here in considerable detail, not only because they are interesting, but also because these views dominated meteorological thinking until the Renaissance.

Aristotle was a philosopher, not strictly a meteorologist, and did not differentiate clearly between the sciences of astronomy, geography, geology, and meteorology. In *Meteorologica,* he talks about shooting stars, the aurora borealis, comets, earthquakes, rivers, springs, and the oceans, as well as rain, clouds, mist, dew, snow, hail, winds, thunder, and hurricanes. Aristotle believed that weather phenomena were caused by mutual interactions of the four elements, *fire, air, water,* and *earth,* and the four prime contraries, *hot, cold, dry,* and *moist.* However, he never clearly explained the nature of these interactions.

Aristotle frequently argued against ideas which were closer to the truth than his own. For example, in considering the cause of hail, he presents first the ''wrong'' view of Anaxagoras:

> *Some then think that the cause of the origin of hail is as follows: when a cloud is forced up into the upper region where the temperature is lower . . . the water when it gets there is frozen, and so hailstorms occur more often in summer and in warm districts because the heat forces the clouds up farther from the earth.* [Aristotle, *Meteorologica,* with an English translation by H.D.P. Lee (Cambridge, MA: Harvard University Press, 1952), p. 81]

As will be discussed in a later chapter, Anaxagoras' theory is amazingly correct. Nevertheless, Aristotle denies this view and offers his own theory:

> *The process (hail) is just the opposite of what Anaxagoras says it is. He says it takes place when clouds rise into the cold air: we say it takes place when clouds descend into the warm air, and is most violent when the clouds descend farthest.* [Ibid., p. 85]

After reading the chapters on precipitation and hail, you should be able to point out Aristotle's folly.

A second example of Aristotle's meteorological deductions is illustrated by his concept of the wind:

> *There are some who say that wind is simply a moving current of what we call air . . . and define the wind as air in motion. The unscientific views of ordinary people are preferable to scientific theories of this sort.* [Ibid., p. 89]

In a later chapter, Aristotle presents his belief that winds are simply dry or moist exhalations from a breathing earth. The dry exhalations are the wind; the moist exhalations, the rain.

Aristotle's winds are classified according to points on the compass:

> . . . *westerly winds are counted as northerly, being*
> *colder because they blow from the sunset; easterly*
> *winds are counted as southerly, being warmer*
> *because they blow from the sunrise. Winds from the*
> *sunrise are warmer than winds from the sunset,*
> *because those from the sunrise are exposed to the*
> *sun for longer; while those from the sunset are*
> *reached by the sun later and it soon leaves them.*
> [Ibid., p. 193]

This discourse is typical of some of Aristotle's meteorological observations, because the statement (that east winds are milder than west winds), over much of the middle-latitude regions, is correct, but the interpretation is not.

Aristotle did contribute many accurate explanations for atmospheric phenomena, however. His reasoning for rain could have been lifted from a modern textbook with only a few changes:

> *The earth is at rest, and the moisure about it is*
> *evaporated by the sun's rays and the other heat from*
> *above and rises upwards; but when the heat which*
> *causes it to rise leaves it, . . . the vapor cools and*
> *condenses again as a result of the loss of heat and*
> *the height and turns from air into water. The*
> *exhalation from water is vapor. The formation of*
> *water from air produces clouds.* [Ibid., p. 69]

FIGURE 1.1 *Aristotle's rainbow. "Let B be the outer and* A *the inner, primary rainbows and to symbolize the colors, let us use* Γ *for red,* Δ *for green, and* E *for purple."*

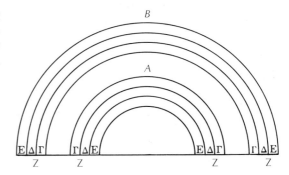

His views on the formation of dew and frost are also essentially valid, and he correctly notes that dew and frost form on calm nights and in the valleys rather than on the mountain peaks. His observations on rainbows, halos, and mock suns (sun dogs) are quite detailed and accurate. For example, figure 1.1 shows Aristotle's diagram of the primary and secondary rainbow. It is interesting that he was well aware of the rare, but possible, night rainbow (rainbow caused by moonlight).

Aristotle noted one of the best-known and most reliable sayings of weather folklore—that halos around the sun or moon are harbingers of rain:

> *This formation is therefore a sign of rain . . . . It is*
> *reasonable to regard it as a sign of rain, since it*
> *shows that a condensation is taking place of the*

> kind, which, if the condensing process continues,
> will necessarily lead to rain. [Ibid., p. 247]

We now know that halos are caused by fine ice crystals high in the sky and are frequently associated with moisture that is racing ahead of an approaching storm.

A concise summary of Aristotle's work in meteorology might be that his observations were mostly correct, but his explanations were frequently erroneous. The chief value of *Meteorologica* is its detailed exposition of the early ideas concerning the causes and effects of meteorological phenomena.

1.2.2   ARCHIMEDES TAKES A BATH

Although best remembered for his ecstatic "Eureka!" as he stepped into his bath, Archimedes (born 287 B.C.) made an important contribution to the future understanding of cumulus cloud formation. (Cumulus clouds are the familiar isolated clouds which develop vertically in the form of rising domes or towers.) At the time of his bath, however, Archimedes was far more interested in a problem about gold than in the formation of clouds. Hiero, King of Syracuse, suspecting that a lump of gold had been diluted with silver by workmen who were making a crown for him, assigned Archimedes the problem of determining whether the fraud had actually occurred. Archimedes debated unsuccessfully with himself until, while stepping into the bath, he realized that any solid body displaces an equal amount of water. This experience gave Archimedes a method for determining the percentage of gold in the crown and led to the recognition of buoyancy as an important force associated with the motions of fluids. This buoyancy principle was the basis for the design of the hot-air balloon, which is sometimes called the Montgolfier balloon after its French inventors. The Montgolfier brothers launched the first balloon on June 5, 1783. Later, hydrogen and helium were used instead of hot air to provide the necessary buoyancy. Weather observations taken from balloons have contributed much to our knowledge of the vertical structure of the atmosphere. Meteorologists today utilize Archimedes' principle as a basis for theoretical investigations of the buoyant rise of cumulus clouds.

RENAISSANCE METEOROLOGY   *1.3*

Science in general, and meteorology in particular, made little progress during the Middle Ages. In meteorology, Aristotle's theories were accepted without reservation, as indicated in the following passages, which were written in the tenth century by an unknown Anglo-Saxon:

> There are four elements in which all earthly bodies
> dwell, which are, aer, ignis, terra, aqua. Aer is a
> very thin corporeal element; it goes over the whole
> world, and extends upward nearly to the moon . . .
> [*Popular Treatises on Science*, Thomas Wright, ed.
> (London: R. and J.E. Taylor, 1841), p. 17]

Concerning the formation of precipitation, the Anglo-Saxons showed little more insight than Aristotle:

> *The atmosphere licks and draws up the moisture of*
> *all the earth, and of the sea, and gathers it into*
> *showers; and when it can bear no more, then it falls*
> *down loosed in rain . . .*
>
> *Snow comes of the thin moisture, which is drawn up*
> *with the air, and is frozen before it be run into drops,*
> *and so immediately falls.* [Ibid., p. 19]

The Renaissance ended the period of blind acceptance of such meteorological dogma. The spirit of healthy doubt was typified by René Descartes' (1596–1650) condemnation of unquestioning assertion:

> *It is not true to say we know a thing simply because*
> *it has been told us . . . to know anything requires*
> *much more than this, and unless the reasons for any*
> *belief are so clear to our minds that we cannot doubt*
> *them, we have no right to say we know it to be true,*
> *but only that we have been told so.* [Arabella B.
> Buckley, *A History of Natural Sciences* (London:
> Edward Stamford, 1894), p. 103]

Readers of this or any other book should take its assertions with a grain of Descartes' salt, until they understand the reasons behind the facts and thereby discover the truth for themselves.

1.3.1 THE AIR HAS WEIGHT . . .

Just as fish are not conscious of weight of water, we do not normally perceive the weight of the atmosphere. Because air seems so nebulous, it is easy to imagine why the ancients considered it weightless. Nevertheless, the atmosphere presses upon the earth's surface at sea level with a weight of approximately 14.7 pounds on every square inch. We do not feel this weight crushing our shoulders because the pressure inside our bodies equals that of the surrounding environment, so there is no net force for us to sense.

Some of the consequences of air pressure were known before it was accepted that air had weight. The siphon and suction tubes operate on the principle that large forces can be achieved if pressure differences are created. For example, if a movable piston is drawn upward in a tube which is set in a tub of water (figure 1.2), water will rise into the tube. Prior to the recognition of air pressure, the popular explanation for this phenomenon was that nature "abhorred a vacuum," and that the rising of the water was nature's way of preventing this crime.

Galileo, however, was puzzled by one aspect of the suction pump. No matter how high the piston was raised, the water could not be made to rise above a height of about 10 meters. It was difficult to explain why nature abhorred a vacuum only to a height of 10 meters, so an alternative explanation was necessary. Torricelli, a friend of Galileo, correctly explained that it is the pressure or

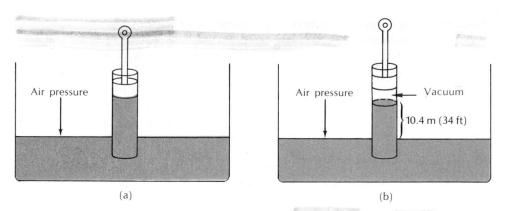

FIGURE 1.2 *Water barometer. Air pressure at sea level can support a column of water approximately 10.4 meters (34 ft) high.*

weight of the atmosphere pressing upon the surface of the water in the tub that causes water to flow into the tube. When the weight of the column of water equals that of the atmosphere, equilibrium is obtained, and no further raising of the water is possible. Torricelli then reasoned that if a denser liquid was used instead of water, the required height for equilibrium would be much less. As predicted, a column of mercury rises only about 760 centimeters (30 inches); hence, the mercury barometer was invented.

Soon after Torricelli correctly explained the behavior of Galileo's water barometer and invented the mercury barometer, seventeenth century scientists were probing the lower atmosphere with this new instrument. At the suggestion of Pascal in 1646, Périer carried a barometer to the top of a mountain, the Puy de Dome in France, and showed that air pressure decreases with increasing elevation. Pascal noted the compressibility of air, and that air in the valleys is more compressed than air on the mountaintops. He likened the atmosphere to a huge pile of feathers; the feathers at the bottom of the pile are more squashed than the ones near the top.

Torricelli, in a letter dated June 11, 1644, indicated that air pressure varies not only with elevation at a given time, but also from time to time at a given location. Soon after these variations in time and place were noted, they were associated with changes in the weather. Early references to the barometer as a weather glass were made by Jean Pecquet in 1651 and by Samuel Pepys in his diary. It was much later, however, that systematic analyses of surface pressure were related to weather events associated with storm systems.

1.3.2 . . . AND BLOWS HOT AND COLD

It was not some sense of modesty that made early humans in the middle latitudes wrap wooly skins around their bodies on frosty autumn mornings. They did not need a thermometer to tell them that some mornings were cold and others hot. However, quantitative measurement of the property of the air that made people shiver or sweat waited until the thermometer was developed, which occurred at about the same time as the development of the barometer. Galileo had a water thermometer, which must have broken with every freeze, and a Dutchman named Cornelis Drebbel combined science with pleasure by making a wine thermometer. By 1670, mercury was being used as the liquid in the hollow glass. Whether mercury, water, or wine, the principle is the same: the liquid expands with heating and contracts with cooling. The scales on thermometers were arbitrary, and several different ones (for example, the Fahrenheit, Celsius, and Reaumur) were proposed by the eighteenth century. These scales were based on one or two reference points, usually the freezing and boiling temperatures of water. The common Fahrenheit and Celsius scales are compared in figure 2.1.

The invention of the barometer and thermometer enabled measurement of pressure and temperature, two of the basic variables important to the scientific study of meteorology. The discovery of the relationship between these two variables and a third, density, was the beginning of a quantitative rather than qualitative description of weather; the door was opened for a more detailed understanding of the weather.

Much of the foundation for later progress in meteorology was laid by the basic advances in chemistry, physics, and mathematics during the seventeenth and eighteenth centuries. In 1661, Robert Boyle discovered the relationship between pressure and density (at a constant temperature, pressure = constant × density), and in 1802, Jacques Charles discovered the realtionship between temperature and density of gases (at a constant pressure, temperature × density = constant). In 1666, Newton, most famous for his discovery of gravity, developed the so-called method of *fluxions,* which is similar to differential calculus and is essential in solving theoretical meteorological problems.

1.3.3 HEAT THAT MELTS THE ICE BUT DOES NOT WARM THE WATER

Latent heat, a form of energy that is very important in meteorological processes, was discovered by a chemist, Dr. Joseph Black, in 1760. We know that when ice is melted, the temperature of the icewater mixture does not change while any ice remains, no matter how much heat is applied. This lost heat, which is required to melt the ice but does not warm the water, is called *latent, or hidden, heat.* This hidden heat is given back to the surrounding air if the water is refrozen. A similar hiding of heat occurs when water is evaporated, with an equal amount of heat being liberated when the water vapor subsequently condenses. Because evaporation, condensation, freezing, and melting occur frequently in the atmosphere, these latent sources of heat greatly influence many meteorological processes. The principle of latent heat was explained by the experiments of Count Rumford in 1798 and Sir Humphry Davy in 1799, which proved that heat consists of molecules in motion. Since molecules locked in the rigid structure of ice crystals have less motion than molecules in the liquid phase at the same temperature, extra heat is required to turn ice to water.

1.3.4 INVISIBLE HEAT RAYS

While chemists were melting ice cubes, the nature of radiation was being investigated by Sir William Herschel. In 1800, Herschel passed a thermometer through the colors that had been separated from light passing through a prism. As expected, the temperature rose as the thermometer passed from the violet into the yellow. Surprisingly, though, the temperature continued to rise as the thermometer moved beyond the red into the darkness. Thus, an important form of energy transfer, the invisible *infrared* ("beyond the red") *rays*, was discovered.

1.3.5 KITE FLYING IN THE EIGHTEENTH CENTURY

Benjamin Franklin was a versatile genius of many passions, meteorology among them. His most dangerous meteorological pursuit was kite flying in thunderstorms, in which he was engaged in 1752 to demonstrate the electrical nature of lightning. By sending his kite into the vicinity of thunderstorms, he induced sparks to jump from a key to his finger. Miraculously, he was not electrocuted; unfortunately, others seeking to duplicate his experiment were not so lucky. A year later, a Russian scientist, Georg Wilhelm Richmann, was struck on the forehead and killed by a lightning stroke while kite flying in a thunderstorm. In spite of this and other accidents, kite flying was a common procedure for exploring the lower atmosphere until the late 1800s.

1.3.6 STORMS THAT MOVE

Prior to the eighteenth century, it was commonly believed that storms were born and died at the same location. With improving communications, however, it gradually became evident that storms move about in a more or less orderly fashion, retaining their identity for several days. Much later, when simultaneous observations could be transmitted rapidly by telegraph, the extrapolation of the storm's motion became an important basis for forecasting. Even so, the realization that storms were only a part of the huge complex weather machine came slowly.

Daniel Defoe, referring to a British storm of November 27, 1703, remarked that the same storm had occurred earlier on the shores of America.* Benjamin Franklin, in his pre-kite-flying days, also deduced that storms move. In 1743, Franklin was ready to observe an eclipse of the moon in Philadelphia when dense clouds moved in just as the eclipse was about to begin. He was probably even more frustrated when he found out that similar clouds in Boston did not obscure the moon until after the eclipse, so that his friends there obtained a splendid view. In a letter to one of these friends, Franklin noted that the storm must

---

*Donald R. Whitnah, *A History of the United States Weather Bureau* (Urbana, IL: University of Illinois Press, 1961), p. 3.

have moved from Virginia to Connecticut, and so reached Philadelphia before Boston.

1.4   NINETEENTH-CENTURY METEOROLOGY

If progress in the seventeenth century could be described by advances in basic science, and in the eighteenth century by a vague awareness of the nature of storms, then the nineteenth century represented the beginnings of our modern understanding of large-scale atmospheric circulations and the typical weather patterns associated with them. During the nineteenth century, controversies on the relationship of the wind, precipitation, and pressure in moving cyclones (large-scale wind circulations about a center of low pressure) arose and were resolved. At the beginning of the century, it was barely recognized that storms move; toward the end of the century, the typical wind, temperature, and rainfall patterns associated with the moving cyclones were documented, and the U.S. Weather Bureau was making forecasts based on simultaneous weather reports. A complete history of meteorological progress during this time is outside the scope of this brief survey. Instead, we mention just a few of the problems that occupied nineteenth-century meteorologists.

1.4.1   THE ROTARY WIND CONTROVERSY

It is now such a well-established fact that hurricanes and the larger, but less intense cyclones associated with our typical rainy days are all whirlwinds that it is difficult to imagine that barely one hundred years ago the question of the wind distribution around low-pressure centers was a lively topic of controversy. Much of the motivation for determining the correct distribution stemmed from marine interests which desired interpretation of the behavior of the barometer in terms of the probable effect on the wind.

In the early 1800s, there were two contradictory theories which described the wind flow around the storms. One theory, with adherents William Redfield and Colonel William Reid, maintained that the northeast storms and hurricanes were giant whirlwinds, with the wind blowing around the center of low pressure (figure 1.3a). A contrasting view was held by James Espy, who believed that the winds associated with these storms blew directly toward the center of low pressure (figure 1.3b). Espy even cautioned mariners against following Redfield's rules on avoiding storms by following the circular wind law:

> *I earnestly recommend to gentlemen who embrace*
> *the whirlwind theory of storms, to abstain from*
> *laying down rules to the practical navigator, founded*
> *on this doctrine, until it is better established than it*
> *is at present.* [*The Philosophy of Storms* (Boston:
> Little and Brown, 1841), p. 253]

Redfield and Reid's theory of the rotary nature of winds around lows was based on hundreds of observations from ships and land stations. Strangely

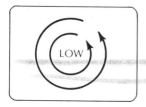
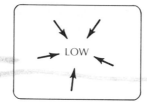
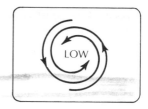

Redfield's storms   Espy's storms   Loomis' storms

FIGURE 1.3   *Three views of storm circulations in the nineteenth century. Arrows show the direction of air movement.*

enough, Espy interpreted many of the same observations as supporting his inflow theory. The reasons for these conflicting interpretations of the same data are understandable when we realize that both theories are partially correct. At the surface of the earth, the wind spirals inward toward the low center and therefore has both a circular and an inflow component (figure 1.3c), as correctly concluded by Professor Elias Loomis in 1859 after a detailed analysis of an 1836 storm:

> *For several hundred miles on each side of the center of a violent storm, the wind inclines toward the area of least pressure, and at the same time circulates around the center contrary to the motion of the hands of a watch.* ["On Certain Storms in Europe and America, December, 1836," *Smithsonian Contributions to Knowledge*, Vol. II (Washington, D.C.: Smithsonian Institution, 1860), p. 25]

This fact, plus the paucity of observations, which made imagination an important ingredient of the analyses, explains the opposite conclusions drawn by the proponents of the two theories.

1.4.2 THE PHILOSOPHY OF STORMS

Although his pure inflow theory was carried a bit too far, Espy made a very important contribution by explaining how the release of latent heat associated with condensing water vapor maintains the necessary warmth of ascending air as it rises to great heights. Without the extra heat of condensation, ascending bubbles of air would soon become colder and denser than their environment and would therefore sink. Because clouds and their offspring—rain, snow, sleet, and other forms of precipitation—depend on deep currents of upward-moving air, it is difficult to overemphasize the importance of latent heat.

Consideration of latent heat led Espy to many correct explanations of phenomena associated with the cumulus-type clouds. These theories, with observations to back them up in many cases, were explained in his book *The Philosophy of Storms*. Espy showed why cumulus clouds have relatively flat roots (bases), all at one level. He described the atmospheric conditions favorable for tornado formation and gave the correct explanation for the formation of the visible funnel of tornados and waterspouts. His theory that rising air is necessary for clouds and precipitation formation solved an old riddle which argued that clouds should be more frequent at night. According to that theory, nights are colder than days, and cold air cannot hold as much water vapor; therefore, condensation should be heaviest at night. Espy showed that although the nocturnal cooling is frequently conducive to fogs, a more efficient way to produce condensation is the lifting of air in cumulus clouds.

1.4.3 EARLY WEATHER MAPS AND FRACTURED FORECASTS

Prior to the nineteenth century, weather records were kept sporadically, and the techniques, time, and frequency of observation were entirely at the whim of the individual recorder. Early observations were taken for Wilmington, Delaware (1644–45) and Boston, Massachusetts (1738–50). Thomas Jefferson (1743–1826) carefully recorded the daily weather at Monticello, in Charlottesville, Virginia, in the late 1700s* However, taking observations on a regular basis did not start in the United States until 1812, when an order from the Surgeon General of the Army was given to hospital surgeons. These army doctors were required to take weather observations and keep climatological records. By 1853, ninety-seven army posts were recording the daily weather.

The mushrooming of weather stations made it possible to study the spatial variation of wind, pressure, and precipitation associated with important storms by plotting the information on a map. Because of slow communications, the first maps were produced long after the storms occurred. Espy postanalyzed many storms during the 1830s and 1840s. One example from *The Philosophy of Storms* is shown in figure 1.4, which summarizes a late-winter storm which brought heavy snow to the northeastern United States.

The map shown in figure 1.4 appears slightly naked to today's meteorologists because of the absence of the familiar web of isobars (lines connecting points of equal pressure) and the fronts (boundaries between cold and warm air masses). In fact, Espy's map is little more than a list of the sequence of weather events at each station during the three-day period of March 16–18, 1838. No attempt was made to analyze the conditions over the entire map at a given time in the storm's history. We may note, however, the typical sequence of events associated with a major snowstorm at many locations. For example, the weather sequence during the three days at station 45 (near Boston) was clear, cloudy, snow, cloudy, and finally clear again.

Espy's map contained too much information to be analyzed in a way that would show significant details of the storm's structure, mainly because the plotted data spanned three days. A much simpler, and more revealing, procedure was to plot all the data for a given time on a separate map. An early set of maps of this type, still barren of fronts, was analyzed in color by Professor Loomis in 1860.** These maps covered the evolution of a major storm in December, 1836.

---

*Ibid., p. 9.

**Elias Loomis, "On Certain Storms in Europe and America, December, 1836," *Smithsonian Contributions to Knowledge*, Vol. II (Washington, D.C.: Smithsonian Institution, 1860), pp. 27–39.

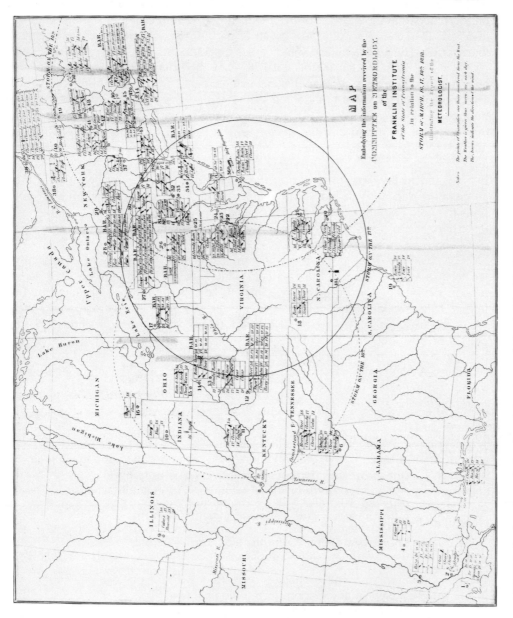

FIGURE 1.4 *Espy's analysis of a winter storm of March 16–18, 1838.*

An example is shown in Plate 2a. Professor Loomis' analysis does not show the pressure and temperature structure associated with the storm in the manner usually seen on today's weather maps. Rather than draw isobars and isotherms (lines connecting points of equal temperature), Loomis analyzed departures from the mean of pressure and temperature at each station. Thus, every point on the +10°F temperature departure line is 10°F above the mean temperature of that station. Loomis also indicated the wind direction by an arrow and depicted the state of the sky (clear or cloudy) and rainy and snowy areas by colors.

Even though Loomis' analyses show no fronts (fronts were not discovered until the twentieth century), distinct evidence of a strong cold front is present. The modern meteorologist, schooled in the frontal theory of storms, would not hesitate in drawing the characteristic heavy blue line (to designate a cold front) in the trough of minimum pressure departures that extends along a north-south line through Michigan, Ohio, Kentucky, and Georgia. Ahead of this front, in the warm air, temperatures are 10°F above the mean, and winds are from the southeast. Behind and to the west of the front, the winds are from the northwest, and the temperature departures fall to −30°F. The cloud and precipitation pattern also supports the cold frontal system, with rain ahead of the front changing to snow behind the front. Far to the southwest, the clear skies are indicative of a high-pressure center moving in from the west.

The maps of Professor Loomis, along with other analyses, gradually led to a qualitative model of the typical middle-latitude cyclone. One such early model, devised by the Reverend W. Clement Ley, is reproduced in Plate 2b. To interpret this model, it is helpful to imagine the sequence of weather events that would be experienced by people at different points along the storm's track. To a person on the southeast side of the storm track, the *cirrostratus* clouds (layer of high, thin clouds) gradually thicken to the *nimbus* (rain) clouds as the storm approaches. As the center passes, there is a sudden clearing as the winds shift around to the northwest. An observer on the northwest side of the track would see a more gradual clearing as the storm passed.

It is interesting that Ley's model also recognizes the existence of the not-yet-discovered cold front. We quote Robert H. Scott's description, written in 1887, of the Ley cyclone model:

> *One of the most striking characteristics of a cyclonic storm is the sudden shift of wind which takes place between S.W. and N.W., accompanied frequently by a heavy squall and a shower, together with an almost instantaneous fall of temperature.* [*Weather Charts and Storm Warnings* (London: Logmans, Green and Co., 1887), p. 65]

This description of the weather changes concurrent with the passage of a cold front would stand up well in any modern meteorological text.

The early weather maps were attempts to summarize in an orderly way the many observations about major storms and served as research tools to determine the relationship of wind and weather to the low-pressure centers. In 1849, however, the first weather observations were transmitted by telegraph, making possible the rapid collection and analysis of data. Then, weather maps could

16

be produced during the storm's lifetime and utilized for forecasting purposes. By 1860, 500 stations were reporting the weather, and forecasting based on more than local observations became possible. The increased data spurred optimism that truly accurate weather forecasting was imminent, and the number of aspiring weather forecasters suddenly multiplied.

Weather forecasting is not an easy job, however, and, unlike other endeavors, it is almost impossible to conceal mistakes. From the beginning and continuing to the present, weather forecasters have been remembered for their failures, or their fractured forecasts. Any forecaster today can sympathize with the poor wretch lambasted by President Lincoln:

> It seems to me Mr. Copen knows nothing about the weather, in advance. He told me three days ago that it would not rain. . . . It is raining now, and has been for 10 hours. I cannot spare any more time to Mr. Copen. [Donald R. Whitnah, A History of the United States Weather Bureau, p. 15]

But weather forecasters, like economic forecasters, do not give up because of such fiascos, for if they did, there would be no forecasting at all. In 1870, a national meteorological service was created by Congress under the direction of the Army Signal Corps. By 1872, national forecasts were being issued on a daily basis. These early forecasters claimed an accuracy of 75 percent, a figure not much below what is claimed today.

Weather forecasting in the late 1800s was based almost entirely on the surface charts, which included pressure, wind direction and velocity, speed of movement of high- and low-pressure systems, and amounts of fallen precipitation. Forecasting consisted of locating the weather-producing systems, determining their direction and speed of movement, and extrapolating to some future time. And even today, although meteorologists are aided by a bountiful supply of upper-air charts, computer forecasts, and satellite information, extrapolation of features on the surface chart is one of the most important methods for short-range forecasting.

EMERGENCE OF MODERN METEOROLOGY 1.5

Progress toward understanding the atmosphere has gained momentum in the twentieth century. Here, we will touch on a few of the major developments during the 1900s. However, first we should update the weather maps discussed in the preceding section by introducing the concepts of cold and warm fronts. A typical modern weather map (see chapters 2 and 7 for examples) shows several fronts, whereas prior to 1930, maps contained no fronts whatsoever. The atmosphere, of course, did not suddenly develop cold fronts in the 1930s, as indicated by Loomis' and Ley's unwitting references. The frontal concept was born of the Norwegians' ideas of the existence of distinct air masses, which were advanced shortly after World War I. Jacob Bjerknes stressed the conflict between masses of hot and cold air, with the boundary separating the two antagonists appropriately called the front. If the cold air was marching southward, routing the warm air, the front was termed a *cold front*. When the forces of the south regrouped and overran the weakened and now retreating cold air, the front was renamed a *warm front*. Stationary fronts marked indecisive boundaries at which neither side was making much progress.

Besides the introduction of the frontal concept in surface analyses, there have been three major technological advances during the 1900s that have made significant impacts on forecasting. The first was the beginning of routine upper-air observations in the late 1930s by balloon-carried instruments called *radiosondes*. These radiosondes radioed the temperature, pressure, and humidity of the upper atmosphere back to the ground stations. With this information, the three-dimensional structure of the atmosphere could be studied, and forecasting became more complicated than simple extrapolation of the movement of surface weather systems.

The second revolutionary advance in weather forecasting came in the 1950s when high-speed computers were developed. These computers are able to solve the equations that describe the wind, temperature, and moisture behavior over the entire atmosphere. The equations are predictive in the sense that they can be solved for a future state of the atmosphere, given the present conditions. This concept of forecasting by numbers will be described in more detail in a later chapter.

The third revolution in weather forecasting began on April 1, 1960, when the first weather satellite, *Tiros 1*, was launched. This satellite provided television coverage of storm systems from an altitude of 725 kilometers and gave meteorologists a fresh look at old weather patterns. Subsequent satellites have been more sophisticated and have been used to deduce the winds, temperature, and humidities of the atmosphere below. One of the major difficulties yet to be solved satisfactorily is the meshing of these abundant data with the forecasting phase of the computer, for no forecast can be better than its initial input. The satellite revolution is still in progress. Pictures from some of the more recent satellites will be used in the discussion of various weather phenomena in later chapters.

Questions

1   What type of cloud often precedes a winter storm and gives warning by producing a halo around the moon?

2   Where does the word *meteorology* come from and what does it mean?

3   Give two examples of Aristotle's views on meteorology that are totally incorrect.

4   Who discovered the principle of buoyancy?

5   Why do we not feel the crushing weight of the atmosphere on our bodies?

6   What is latent heat?

7   Draw diagrams indicating the direction of surface winds around a region of low pressure (storm) according to Redfield, Espy and Loomis. Which do you think is more correct?

8   Why were the boundaries between cold and warm air masses named *fronts?*

Problems

9 Why does atmospheric pressure decrease with increasing height?

10 Where in the United States would Aristotle's view on the temperature of east and west winds be correct, and where would it be incorrect? Consider the following locations: Boston, Washington, New York, and Seattle. Explain whether Aristotle's explanation is correct for those cities where easterly winds are warmer than westerly winds. If not, give a better explanation.

11 Why does a water barometer have to be much taller than a mercury barometer? Would an oil barometer have to be taller or shorter than a water barometer?

12 In which storm circulation theory (Redfield's, Espy's, or Loomis') would you expect more rain, with other factors such as moisture content being equal.

TWO

Atmospheric Variables and Measurements

photosynthesis • temperature • pressure • horizontal wind • vertical
wind • Beaufort • relative humidity • wet-bulb temperature •
dewpoint • isotherms • isobars • mesosphere • cyclone •
anticyclone • fronts • air mass • upper-air analysis • sea-level
pressure • elevation and azimuth angles • radar • sodar • lidar • polar-
orbiting and geostationary weather satellites • visible and infrared
satellite photographs

Observation is at the heart of all science, and this is especially true of meteorology. Before exploring the structure and workings of atmospheric phenomena, we must know what the atmosphere is, the variables used to describe it, and the ways in which these variables are measured.

2.1 ORIGIN AND COMPOSITION OF THE ATMOSPHERE

To understand the probable origin of the atmosphere, we must sift through the geologic history of the earth's 4.5×10^9 years of existence and use our knowledge of chemistry and biology to develop a consistent explanation of how the atmosphere evolved. Although this task may seem hopeless, there is surprising agreement among scientists about how the present atmosphere, consisting mainly of nitrogen and oxygen (see table 2.1), evolved from an infant earth with no atmosphere at all.

TABLE 2.1 *Composition of the lower and middle atmosphere*

Nearly constant constituents	Percent by volume	Highly variable constituents	Percent by volume
Nitrogen (N_2)	78.084	Water vapor (H_2O)	<4.0
Oxygen (O_2)	20.946	Ozone (O_3)	$<7.0 \times 10^{-6}$
Argon (A)	0.934	Sulfur dioxide (SO_2)	$<1.0 \times 10^{-4}$
Carbon dioxide (CO_2)	0.033	Nitrogen dioxide (NO_2)	$<2.0 \times 10^{-6}$
Subtotal	99.997	Carbon monoxide (CO)	$<2.0 \times 10^{-5}$
		Particles (dust, salt)	$<1.0 \times 10^{-5}$
Trace constituents			
Neon (Ne)	18.18×10^{-4}		
Helium (He)	5.24×10^{-4}		
Krypton (Kr)	1.14×10^{-4}		
Xenon (Xe)	0.09×10^{-4}		
Hydrogen (H_2)	0.50×10^{-4}		
Methane (CH_4)	2.0×10^{-4}		
Nitrous Oxide (N_2O)	0.5×10^{-4}		
Radon (Rn)	6.0×10^{-18}		
Subtotal	0.003		
Total	100.0		

Forming from planetary matter with gravitational fields too weak to hold an atmosphere, earth developed its present atmosphere from gases emitted by volcanoes. However, the composition of volcanic gases (table 2.2) is radically different from the present composition of the atmosphere; in particular, volcanoes emit virtually no oxygen.

If we believe the evidence that suggests that the gases emitted by volcanoes in the early years of the earth's history are the same as those emitted by present volcanoes, we must then explain how the present atmosphere evolved.

The volcanic gases in table 2.2 would undergo a number of physical and chemical transformations upon leaving the hot, pressurized interior of the earth. Upon cooling, most of the water vapor would condense, filling the oceans. Much of the light hydrogen would escape the gravitational attraction of the earth. The carbon dioxide would react with surface minerals, producing carbonates. None of the changes, however, would produce oxygen, which of course is essential for the higher forms of life. Instead, the early atmosphere probably consisted mainly of methane (CH_4).

There is considerable evidence for the absence of oxygen during the first quarter of the earth's life. First, the earliest materials show incomplete oxidation; for example, the Blind River uranium deposits in Canada contain uranite, which cannot exist when exposed to air with the present oxygen content. Second, there is no known source of free oxygen. And finally, the generally accepted theories on the origin of life indicate that life formed in an absence of oxygen.

There have been two main theories to explain the production of oxygen. The first mechanism is the photodissociation of water vapor by ultraviolet light and the subsequent escape of the free hydrogen into space, according to the reaction

$$2\ H_2O + \text{ultraviolet light} \rightarrow 2\ H_2 + O_2 \qquad (2.1)$$

For this reaction to be effective in producing free oxygen, however, the hydrogen must escape the earth's gravity before it recombines with the oxygen to produce water. Because the chances of this escape occurring are small, this method does not appear to be the major one for producing the earth's oxygen.

A second, more likely mechanism for generating oxygen is photosynthesis, in which carbon dioxide and water combine to produce carbohydrates and oxygen, according to the reaction

$$6\ CO_2 + 6\ H_2O \rightarrow C_6H_{12}O_6 + 6\ O_2 \qquad (2.2)$$

H_2O	79.31
CO_2	11.61
SO_2	6.48
N_2	1.29
H_2	0.58
Other	0.73
Total	100.00

TABLE 2.2 *Percentage by volume of various gases in Hawaiian volcanoes*

23

When the plant material produced in photosynthesis oxidizes in the decay process, the reverse reaction consumes the same amount of oxygen produced during the original photosynthesis. Therefore, in order to achieve an increase in oxygen, the carbon in the plants must be removed from the possibility of oxidation. The vast reserves of fossil fuel and shale deposits attest to the fact that this removal has occurred. According to some estimates, approximately 99 percent of the total amount of oxygen produced since the earth's creation was produced by photosynthesis, and only 1 percent by photodissociation.

The explanation that life itself was responsible for producing most of the oxygen requires an evolution of life in an extremely hostile environment. Without oxygen, the sun's ultraviolet radiation, which is deadly to all cells, would reach the earth's surface in lethal doses. Currently, a layer of ozone about 40 km above the surface (see section 3.1.2) absorbs most of the ultraviolet radiation, screening surface life on earth.

In a world without oxygen, the first one-celled plants must have inhabited layers of water close enough to the surface to obtain sunlight, yet deep enough to avoid the ultraviolet radiation. In an atmosphere free of oxygen, lethal radiation penetrates to a depth of about 10 m, so the first photosynthesizing plants probably occurred at or slightly below this level. As the plants lived, died, and sank to the bottom where they avoided oxidation, a slow increase in oxygen could occur. As the oxygen and ozone increased in the atmosphere, the ultraviolet radiation reaching the surface would diminish, and the upper layers of the ocean would become more hospitable. The expansion of plants into the layer of water immediately next to the surface, where abundant sunlight exists, would accelerate the photosynthesis process. Finally, the oxygen content in the atmosphere could increase to such an extent that life was able to emerge onto the earth's surface. This migration of life from sea to land began around 420 million years ago.

Although there is no evidence that the current percentage of oxygen is changing on human time scales, it is unlikely that the oxygen increased at a uniform rate to the current levels. Instead, the oxygen content has probably varied with the major climatic fluctuations and the associated variations in photosynthesis. For example, the oxygen content was probably higher in the lush carboniferous period some 300 million years ago. Conversely, during colder conditions, the amount of oxygen, or at least the rate of increase, was probably less than in warmer periods.

In summary, the earth was probably formed without an atmosphere. Volcanic eruptions contributed water vapor, carbon dioxide, sulfur dioxide, nitrogen, hydrogen and minor amounts of other gases (but not oxygen) to form the early atmosphere. The earliest life which formed in this oxygen-free environment was probably confined to subsurface layers of water because of the deadly ultraviolet light at the surface. Gradually the photosynthesis process produced an increase in oxygen, which in turn shielded the surface from the harmful radiation and permitted an expansion of life onto the land. This simplified chain of events leading to today's atmosphere, while plausible in light of current geologic and biologic evidence, will undoubtedly undergo considerable refinement and modification in the future.

One of the barriers of communication between scientists and nonscientists is the scientist's use of units unfamiliar to the nonscientist. This difficulty is especially pronounced in the United States, where the scientific community generally uses metric units (e.g., meters, kilometers, degrees Celsius), while the general public utilizes the more unwieldy English units (e.g., feet, miles, degrees Fahrenheit.)*

In writing a book for people who are accustomed to the English system, compromises must be made between a rigid adherence to the more rational metric system (which would lead to many unfamiliar units and hamper communication) and the complete use of the familiar English system (which would perpetuate the unwieldy system). Added to the dilemma is the fact that some meteorological variables are commonly expressed in both units; for example, surface temperatures on weather maps are plotted in degrees Fahrenheit (°F), but temperatures aloft are expressed in degrees Celsius (°C). In this book, the metric system is emphasized, but, for convenience, English units are also frequently given. An attempt is made to introduce the metric system without requiring memorization of a set of conversion factors.

Besides conversion difficulties, many people frequently have only vague ideas of the representative distance scales associated with meteorological phenomena or of the typical and extreme values of many meteorological variables. Unless you know the characteristic order of magnitude of a given variable, you may be unable to appreciate the meaning or significance of the numerical values given in the text. For example, you may not be impressed by the surface pressure value of 908 millibars (mb) recorded in Hurricane Camille unless you know that the average sea-level pressure on the earth is about 1013 millibars, and that pressures lower than 910 millibars have only been recorded a few times in history. Therefore, in addition to discussing some of the commonly used units in meteorology, this section presents typical values and extremes of several meteorological phenomena to give you a feeling for what is normal and abnormal in the weather.

2.2.1 TEMPERATURE

Temperatures on surface weather maps are plotted in degrees Fahrenheit (°F); on upper-level maps, degrees Celsius (°C) are used. The conversion from °F to °C is given by

$$T(°C) = 5/9[T(°F) - 32] \tag{2.3}$$

---

*It is ironic that the English have recently joined the majority of nations in adopting the metric system, so that the United States now stands alone among major nations in its use of the archaic and inconvenient English system. However, some progress was made on December 2, 1975, when President Ford signed into law a bill creating the U.S. Metric Board to coordinate voluntary plans to convert the United States to the metric system. However, the board has no power to compel use of the metric system, and no target date was set for the completion of the conversion process. Perhaps it should be renamed the "U.S. system" until the United States finally abandons it in favor of the much easier metric system.

One Celsius degree (C°) is 1.8 times bigger than a Fahrenheit degree (F°); for example, an increase in temperature of 1C° equals an increase of 1.8F°. Therefore, if the temperatures on weather broadcasts are reported in °C rather than °F, some precision is lost. For example, a temperature of 32°F represents a range from 31.50 to 32.49°F. The corresponding temperature of 0°C, however, represents a range from 31.11 to 32.88°F. When the temperature is close to freezing, this loss of information can be important. Figure 2.1 compares the Celsius and Fahrenheit scales and gives several benchmark temperatures for reference. In theoretical work, degrees absolute (°A) or kelvins (K) are used. The temperatures in degrees absolute or kelvins are the same and are related to the temperature in degrees Celsius by the expression

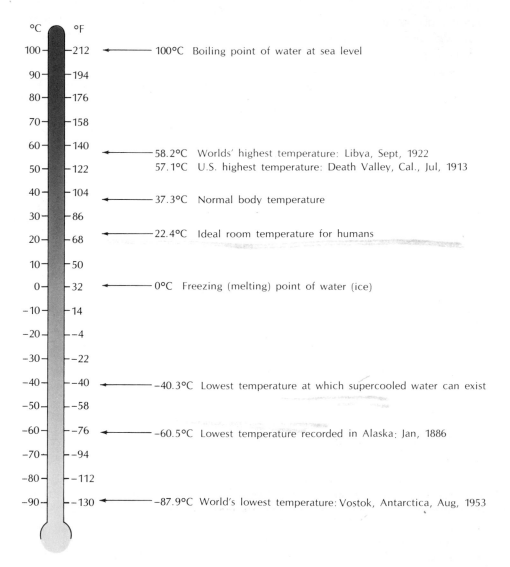

FIGURE 2.1 *Comparison of Fahrenheit and Celsius temperature scales.*

$$T(\text{K or }°A) = T(°C) + 273 \qquad \textbf{(2.4)}$$

2.2.2 Length

The units of length commonly used to express horizontal distances are kilometers (km), statute miles (stat. mi), nautical miles (naut. mi), and degrees of latitude (° lat.). Relationships among these are the following:

$$1 \text{ naut. mi} = 1.15 \text{ stat. mi}$$
$$1° \text{ lat.} = 60 \text{ naut. mi}$$
$$1 \text{ km} = 5/8 \text{ stat. mi}$$
$$1° \text{ lat.} = 111 \text{ km}$$

Degrees of latitude are useful when working with weather maps. Thus, the great circle distance between New York and San Francisco is 2571 stat. mi, 2233 naut. mi, 4138 km, or 37.2° lat. (figure 2.2). Vertical distances are usually expressed in thousands of feet, kilometers, or miles

$$1 \text{ km} = 3280.8 \text{ ft}$$
$$1 \text{ km} = 0.621 \text{ stat. mi}$$

The heights of typical atmospheric phenomena are given in figure 2.3.

2.23 Horizontal Wind Velocity

Wind speeds are plotted on weather maps in knots (kt) and reported over the radio in miles per hour (mi/h). A knot is a nautical mile per hour and is slightly faster than a statute mile per hour (mi/h). One knot equals 1.15 mi/h. For most order-of-magnitude purposes, they can be considered the same. The effect of wind velocities that may be observed on land is given by the Beaufort scale (figure 2.4), devised in 1806 by Sir Francis Beaufort. He originally described the effects of the wind on a full-rigged man-of-war; the effects of the same wind speeds on land-based phenomena were added later.

2.2.4 Vertical Velocity

The up-or-down component of air motion is the vertical velocity. The magnitudes of the vertical velocity depend on the scale of the weather system. For large, synoptic-scale systems, they are much smaller than the horizontal velocities, averaging a few centimeters per second (0.1 mi/h), truly a snail's pace. These velocities are too small to measure directly; instead, they must be inferred from the other meteorological variables. However, in small-scale systems such as in thunderstorms, in individual cumulus clouds, or in the vicinity of steep terrain, the vertical velocity may exceed several tens of meters per second (20–60 km/h). These are strong enough to be measured directly by instrumented aircraft. Furthermore, vertical motions associated with turbulent gusts near the ground may exceed several meters per second over short time periods, as shown in figure 2.9.

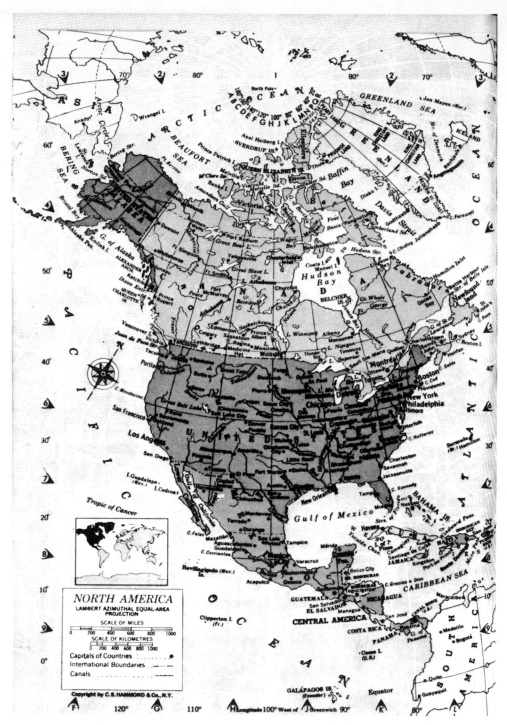

FIGURE 2.2 *Map of North America.*

2.2.5 Pressure

Pressure is the force exerted by the atmosphere per unit area of surface. In meteorology, the most commonly used units of pressure are millibars (mb), but inches of mercury (in. Hg) are also reported. Note that "inches of mercury" are not really units of pressure, because inches are not the correct dimensions for force per unit area. When we say that the pressure is 30 inches of mercury, we mean that the atmospheric pressure is sufficient to support a column of mercury 30 inches high. The conversion from inches of mercury to millibars is

$$1 \text{ in. Hg} = 33.86 \text{ mb}$$

Some typical and extreme values of sea level pressure are given in figure 2.5.

Since 1975 or so, an attempt has been made to introduce a set of standard metric units for this variable. In particular, the standard unit for pressure is the kilopascal, which equals 10 millibars. Thus, a typical sea-level pressure of 1015 millibars is now sometimes given as 101.5 kilopascals.

2.2.6 Moisture

As discussed in chapter 6, many variables are used to describe the amount of water vapor in a parcel of air. Here we introduce three of the most commonly used variables, the relative humidity, dew point temperature, and wet-bulb temperature. At a given temperature there is a maximum amount of water vapor that can exist in a parcel of air; if the air actually contains this maximum amount it is said to be saturated. The relative humidity is the ratio of the actual amount of vapor to the maximum possible amount, so it represents the percent of saturation of the air.

Because the maximum amount of water that can be contained in a parcel of air decreases with decreasing temperature, if a given parcel of air is cooled without change of actual moisture content, the relative humidity will increase. If the temperature is reduced sufficiently, the relative humidity will reach 100 percent and the air will be saturated. This process often occurs near the ground on calm, clear, summer nights, and the result is dew. Thus, the temperature to which air must be cooled in order to reach saturation is called the dew point.

Everyone has experienced the cooling effect of evaporation upon stepping out of a lake or shower. Body heat energy is used to evaporate the water, and the loss of heat is felt as a cooling. Similarly, if rain falls into unsaturated (relative humidity less than 100 percent) air, energy from the air is used to evaporate the water, and the air cools. Cooling will continue until evaporation ceases, which occurs when the air becomes saturated. The temperature of the air at this point is called the wet-bulb temperature. The wet-bulb temperature may be measured by wrapping a wet rag around the bulb of a thermometer. If the thermometer is well ventilated, the lowest temperature reached by the thermometer during the evaporation of the water is the wet-bulb temperature.

29

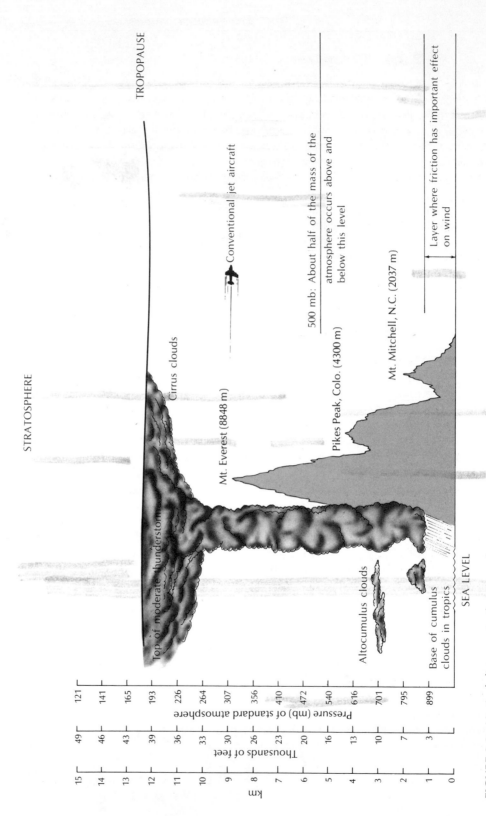

FIGURE 2.3 *Vertical distances in the atmosphere.*

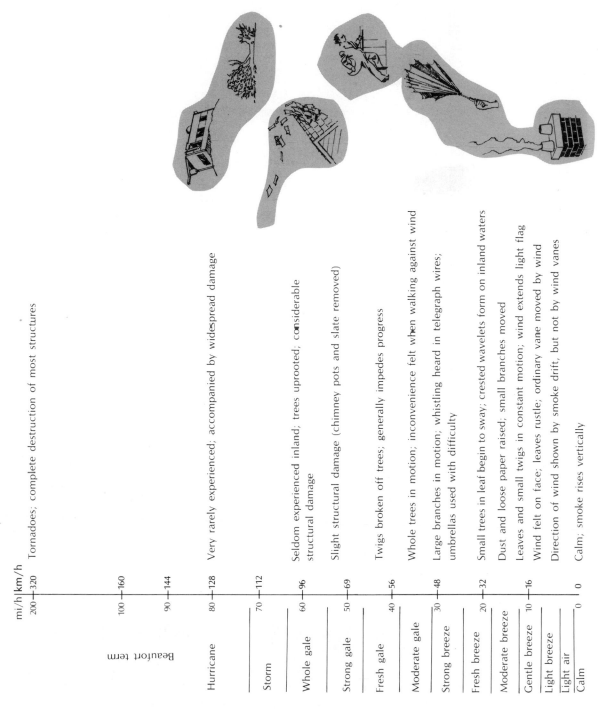

mi/h	km/h	Beaufort term	
200	320		Tornadoes; complete destruction of most structures
		Hurricane	
100	160		
90	144		
80	128		Very rarely experienced; accompanied by widespread damage
70	112	Storm	
60	96	Whole gale	Seldom experienced inland; trees uprooted; considerable structural damage
50	69	Strong gale	Slight structural damage (chimney pots and slate removed)
40	56	Fresh gale	Twigs broken off trees; generally impedes progress
30	48	Moderate gale	Whole trees in motion; inconvenience felt when walking against wind
		Strong breeze	Large branches in motion; whistling heard in telegraph wires; umbrellas used with difficulty
20	32	Fresh breeze	Small trees in leaf begin to sway; crested wavelets form on inland waters
		Moderate breeze	Dust and loose paper raised; small branches moved
10	16	Gentle breeze	Leaves and small twigs in constant motion; wind extends light flag
		Light breeze	Wind felt on face; leaves rustle; ordinary vane moved by wind
		Light air	Direction of wind shown by smoke drift, but not by wind vanes
0	0	Calm	Calm; smoke rises vertically

FIGURE 2.4 Beaufort's description of wind effects.

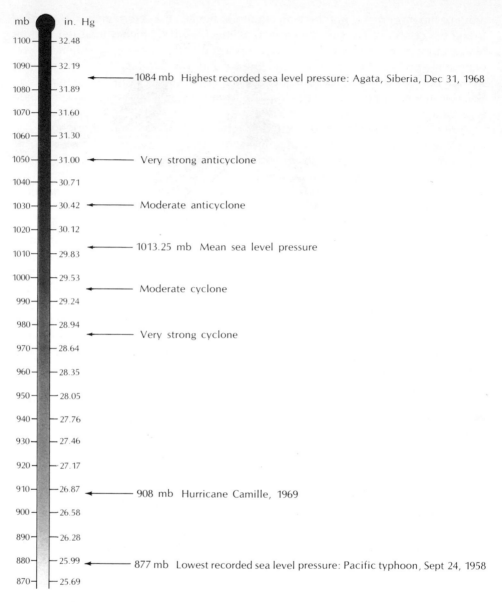

FIGURE 2.5 *Representative barometric pressures in the atmosphere expressed in millibars and inches of mercury.*

2.2.7 ATMOSPHERIC MEASUREMENTS NEAR THE GROUND

Almost everyone has used a liquid-in-glass thermometer to estimate the air temperature. These thermometers usually contain either mercury or alcohol. It is important to note that the thermometer measures the temperature of the instrument, not the temperature of the air. If a thermometer is exposed to the direct rays of the sun, it can read as much as 10C° higher than the true air temperature. Conversely, a thermometer exposed to the clear sky at night will radiate and cool faster than the surrounding air and read too low. Thus, thermometers are put into well-ventilated shelters for the best estimate of the air temperature. Because temperature often varies rapidly in the vertical, a standard observation level of about 2 meters is recommended.

Temperature may also vary rather greatly over short horizontal distances; therefore, the shelter is usually located in a representative, well-exposed location. Even so, it is futile for most purposes to read the temperature to an accuracy better than 0.5C°, because small-scale variations are at least that large.

The true air temperature varies on time scales of seconds, minutes, and hours. Figure 2.6 shows the rapid variations that can occur over 1 minute. In general, these variations are not representative of large horizontal distances, and so for synoptic purposes, the representative temperature should be an average over 15 minutes or so. This average is accomplished in part by using ordinary liquid-in-glass thermometers, which respond sluggishly to temperature changes and thus naturally smooth out temperature variations over time periods shorter than a minute.

Specialists in turbulence and small-scale meteorology are often interested in the rapid temperature fluctuations. One way to measure the high frequency variations is with a sonic thermometer. This instrument measures the speed of sound, which depends on temperature, between two points and thus gives an estimate of the mean temperature along the path of the sound.

The oldest and still most widely used weather map is the sea-level pressure chart. At weather stations, the station pressure is measured with a mercury barometer, which is designed so that the weight of a column of mercury just bal-

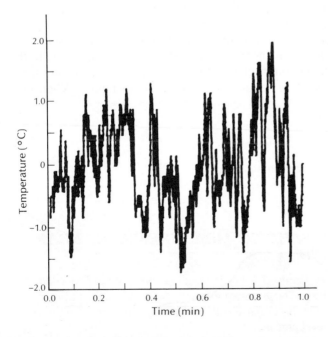

FIGURE 2.6 *Departures of temperatures from the mean in degrees Celsius over a 1-minute period at an elevation of 2 meters above the ground. The data were taken in March, 1977 at midday under clear skies near State College, Pennsylvania.*

ances the total weight of the atmosphere at the elevation of the station. Because pressure varies much more rapidly in the vertical than in the horizontal, if the station pressure were plotted on weather maps, it would simply reflect variations of terrain height rather than the important horizontal variations of pressure on a constant-level surface. Therefore, all station pressures are converted to what the pressure would be if the station were located at sea level—a process known as the *reduction* of the station pressure to sea-level pressure. To reduce the pressure to sea level, an amount of pressure is added which corresponds to the weight of the hypothetical column of air between the station and sea level (figure 2.7). This correction depends on the assumed temperature of the fictitious layer, which is taken to be the average of the present temperature at the station and the temperature twelve hours earlier. If this value is unrepresentative, the reduced sea-level pressure will also be unrepresentative.

Figure 2.7 shows that, although the pressure at the mountain station *B* is much lower than the sea-level station at *A*, the lower value at *B* is caused solely by differences in elevation. When the pressure is reduced to sea level, the pressure is actually higher at *B*. It is the small variation of pressure on a horizontal surface—in this case, sea level—that is important in causing the winds to blow.

Pressure is the only variable measured indoors. One problem with the mercury barometer is its tendency to change with temperature as well as with pressure. To control this effect, the pressure measurements are made near temperatures of 26°C(78°F), and the measured pressure is corrected to a temperature of exactly 26°C.

The most common type of barometer in the home is the *aneroid barometer*. In the aneroid barometer, a coil of hollow tubing is evacuated and held apart by a spring. Increasing pressure compresses the tubing which moves a needle through a mechanical linkage. The aneroid barometer can also drive a pen which records the changing pressure on a chart. This latter device, which gives a continuous time history of the pressure, is called a barograph.

Humidity is measured in a number of ways. At most weather stations, a wet-bulb thermometer is read simultaneously with a dry-bulb thermometer. Both are well ventilated. If the air has high relative humidity, little evaporation takes place, and the wet-bulb temperature remains near the air temperature. When the

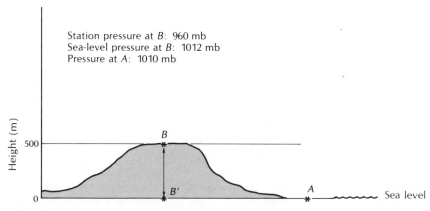

FIGURE 2.7   *Reduction of station pressure to sea level pressure.*

air is dry, the wet-bulb evaporates rapidly and its temperature becomes colder than the air temperature. The difference between the temperatures of the two thermometers—the wet-bulb depression—is an indication of relative humidity. Table 2.3 gives the relationship between relative humidity, temperature, and depression of the wet bulb. Thus, for an air temperature of 60°F and a wet-bulb depression of 10°F, the relative humidity is 48 percent. Given the wet-bulb and dry-bulb temperatures and the station pressure, other moisture variables such as the dew point can also be computed.

TABLE 2.3 *Relative humidity (percent)*
(pressure = 30 in. Hg or 1016 mb)

Air temperature (°F)	Depression of wet-bulb thermometer (°F)														
	1	2	3	4	5	6	7	8	9	10	15	20	25	30	35
20	85	70	55	40	26	12									
25	87	74	62	49	37	25	13	1							
30	89	78	67	56	46	36	26	16	6						
35	91	81	72	63	54	45	36	29	19	10					
40	92	83	75	68	60	52	45	37	29	22					
45	93	86	78	71	64	57	51	44	38	31					
50	93	87	80	74	67	61	55	49	43	38	10				
55	94	88	82	76	70	65	59	54	49	43	19				
60	94	89	84	78	73	68	63	58	53	48	26	5			
65	95	90	85	80	75	70	66	61	56	52	31	12			
70	95	90	86	81	77	72	68	64	59	55	36	19	3		
75	96	91	86	82	78	74	70	66	62	58	40	24	9		
80	96	91	87	83	79	75	72	68	64	61	44	29	15	3	
85	96	92	88	84	80	76	73	69	66	62	46	32	20	8	
90	96	92	89	85	81	78	74	71	68	65	49	36	24	13	3
95	96	93	89	85	82	79	75	72	69	66	51	38	27	17	7
100	96	93	89	86	83	80	77	73	70	68	54	41	30	21	12
105	97	93	90	87	83	80	77	74	71	69	55	43	33	23	15
110	97	93	90	87	84	81	78	75	73	70	57	46	36	26	18
115	97	94	91	88	85	82	79	76	74	71	58	47	37	28	21

As almost everyone knows, hair responds strongly to changes in humidity. In fact, humidity can be recorded by a hair hygrograph. The hair hygrograph is based on the fact that human hair (especially blond hair) expands with increasing relative humidity. Through a mechanical linkage, several strands of hair are made to drive a pen, which produces a record of relative humidity.

Wind speeds are measured by anemometers or recorded by anemographs. For estimating large-scale average winds, the sensor is usually either a windmill-type rotor or a *cup anemometer*. Anemometers react to wind changes much faster than thermometers react to temperature changes. If you inspect the record of an anemograph (figure 2.8), not only the important one-hour average wind appears, but also gusts and lulls, which typically appear every few minutes. Thus, anemograph records can be used not only to estimate important average winds but also to study the characteristics of turbulence.

FIGURE 2.8 (a) Wind direction variation at 2 meters. The number 270 refers to a west wind, 300 to a west-northwest wind, and 330 to a north-northwest wind. See caption for figure 2.6 for description of site. (b) Wind-speed variation at 2 meters for the same period as (a).

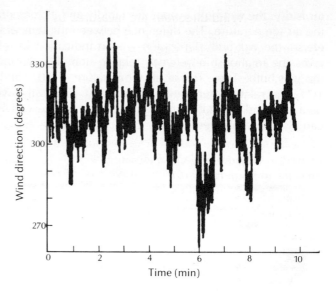

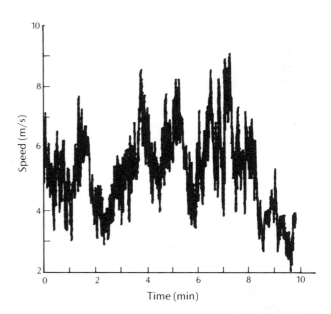

If the appropriate time-averaged wind is not obtained, the reported wind speed and direction may not be representative of the general wind pattern. This is one reason why surface winds often appear so variable on weather maps. In addition to fluctuating with time, the wind speed also varies rapidly in the vertical. Therefore, wind speeds should be measured at a constant height over uniform terrain, say 10 meters above the surface. In practice, however, anemometers are often on top of buildings of various heights over nonuniform terrain. Obstacles may also affect the wind speed and direction. It is no wonder that wind speeds (and directions) are often erratic.

Wind directions are measured by wind vanes. Again, the often used one-minute averages are not always representative of the synoptically important one-hour average wind direction. Unlike wind speed, however, wind direction does not change systematically with height near the ground, at least over uniform terrain.

Vertical motions can be measured from bidirectional vanes or from fluctuations of the speed of sound traveling vertically. At most weather stations, vertical motions are not measured because averages over the synoptic time scale of an hour or so are too small to measure. Variations of the vertical motion from second to second are large, however, as shown in figure 2.9. These vertical gusts play an important role in the dispersion of contaminants, and therefore vertical motion fluctuations (rather than averages) are measured and recorded at locations where air pollution meteorology is studied.

2.2.8 METEOROLOGICAL MEASUREMENTS IN THE FREE ATMOSPHERE

The basic instrument for measuring atmospheric variables up to heights of about 30 kilometers is the *radiosonde*. The radiosonde is an instrument package which contains sensors of temperature, pressure, and humidity, as well as a radio transmitter, and is carried aloft by a balloon filled with helium (figure 2.10). It is tracked by a receiver at the ground which can determine the angle between the ground and the balloon (balloon elevation angle) and the compass direction of the balloon (balloon azimuth). The balloon's elevation h above the ground (figure 2.10) can be inferred from the recorded pressure. Because the balloon's position is known at all times, the motion of the radiosonde, which moves with the wind, can be used to determine the upper-level winds.

Weather information is obtained from radiosondes only on the way up. Many radiosondes are never recovered, and if they are, they tend to be well

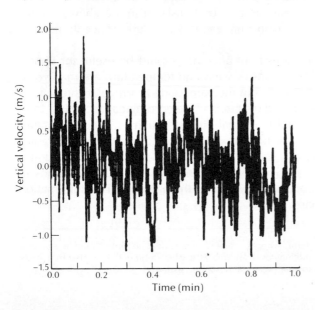

FIGURE 2.9 *Vertical velocity over a 1-minute period at an elevation of 2 meters above the surface. See caption for figure 2.6 for details of site.*

FIGURE 2.10 *Elevation angle and azimuth angle used in tracking a radiosonde.*

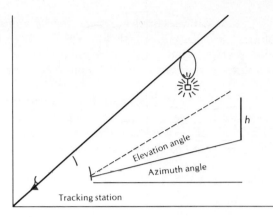

battered. Therefore, radiosonde observations are expensive, and only relatively wealthy nations can afford dense networks. Thus, radiosonde coverage in the world is spotty. Figure 2.11 shows the regions of the world that are within 280 kilometers of a radiosonde station. This distance is assumed to be small enough to adequately resolve the large-scale variation of the winds, temperature, and moisture fields. As shown by figure 2.11, the major gaps in the radiosonde network are over the oceans and much of South America, Africa, and Antarctica.

An advantage of radiosonde observations is that they are made *synoptically,** that is, at the same time all over the world. By international agreement, observations at all radiosonde stations are made at Greenwich (England) midnight, and at most stations also at Greenwich noon.

Pressure is sensed on radiosondes by a small aneroid barometer. As the pressure decreases, the barometer determines the transmission sequence of temperature and humidity. The temperature is determined from _thermistors_, small ceramic beads in which electrical resistance changes with temperature. The changes in resistance then change the frequency of the radio signal, which can be decoded back at the receiving station. Humidity is sensed by a carbon strip which expands in humid air. This expansion can also be translated into a change of frequency. Special radiosondes are also instrumented to measure concentrations of ozone.

In principle, observations from airplanes should be useful for upper-air synoptic analysis. In fact, observations of wind and temperature are made routinely from commercial military aircraft, but they are made at irregular intervals in space and time and have not been used in standard meteorological analysis. However, airplanes have been very useful in special projects, such as in detailed analysis of the jet stream, of atmospheric turbulence, and of concentrations of radioactive particles and trace gases. In addition, airplanes are still used for reconnaisance of severe storms, particularly hurricanes. The clouds and winds in hurricanes can be monitored from planes flying into these storms. This information is often included on weather maps.

---

*In meteorology, the term *synoptic* means, "coincident in time," and a synoptic weather map shows meteorological conditions over an area at a given moment. It is also frequently used to denote large-scale, as opposed to small-scale, atmospheric patterns.

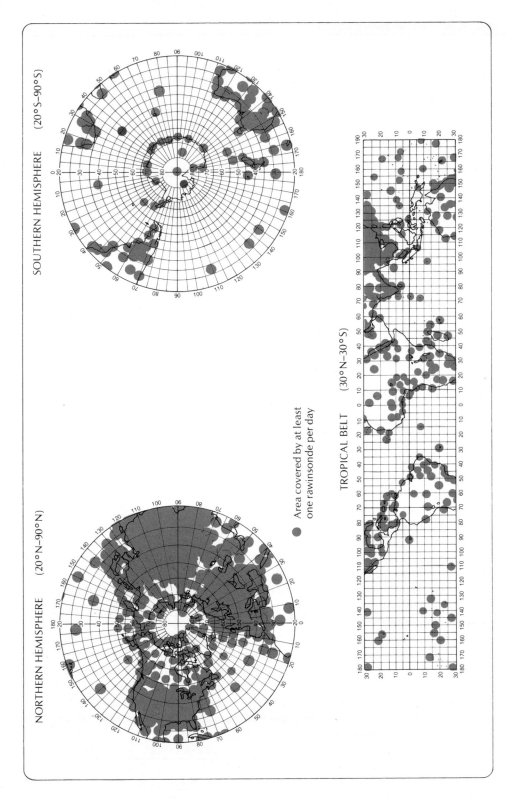

FIGURE 2.11 *Regions of the world within 280 kilometers of a radiosonde station.*

SOUTHERN HEMISPHERE (20°S–90°S)

NORTHERN HEMISPHERE (20°N–90°N)

TROPICAL BELT (30°N–30°S)

Area covered by at least
one rawinsonde per day

Tetroons, balloons that maintain a constant level in the atmosphere, have often been proposed for meteorological measurements, particularly of winds. If successive positions of such balloons are monitored by a satellite, the winds can be found. So far, tetroon systems have mostly been experimental. Some have been very successful, circling the globe for over a year. One of the difficulties is that the tetroons collect condensation or ice in humid areas and crash. Only at levels above clouds, say 8 or 10 kilometers, do they survive for a long time.

Observations above 30 kilometers come from a rather sparse rocket network. Small rockets are fired on alternate days to heights of typically 60 kilometers. At the top of the trajectory, parachutes are released which carry a thermistor to record temperature. Winds are determined by tracking the falling parachutes by radar. Except for these rockets, most observations in the upper atmosphere and in the mesosphere are made by remote sensing from the ground or from satellites.

2.2.9   REMOTE SENSING FROM THE GROUND

There are many possibilities for exploring atmospheric properties aloft by using instruments at the ground, but for the most part, measurements from such instruments are made for research rather than for operational purposes. They are not routinely included on conventional weather maps. An exception is the detection and analysis of precipitation by radar.

In principle, weather variables can be sensed by the propagation of two types of waves, acoustic and electromagnetic. Among the electromagnetic waves, there are many options, including ultraviolet, visible, infrared, microwaves, and conventional radio waves. The radio waves give information only about conditions in the atmosphere above 50 kilometers or so and will not be discussed here. The others are important for the exploration of the lower atmosphere.

Remote sensing can be made in two ways, active or passive. In the active mode, a wave is emitted at the ground. Its direction of propagation is altered by the atmosphere. It may either be scattered back toward the source (back scatter) or scattered ahead (forward scatter). In either case, the characteristics of the scattered signal provide information about the part of the atmosphere through which the signal is passing. This information may include the intensity of turbulence, the lapse rate, and the mean wind velocity.

Active remote-sensing systems include radar, lidar, and sodar (also called *acdar*). Of these, *radar* is the system most widely used in synoptic meteorology. Radar emits high-frequency radio waves (microwaves) with typical wavelengths of 0.1−10 centimeters. These waves can penetrate clouds, but they are reflected by precipitation. The time it takes the wave to reach the precipitation and return gives the distance of the precipitation. Furthermore, if the raindrops or snowflakes are moving toward or away from the radar transmitter, the wavelength of the returning signal will be altered slightly (the Doppler shift). The amount of the change is therefore a measure of the velocity of the precipitation particles.

The great advantage of radar is that it can detect precipitation between stations. The evolution, motion, and structure of the precipitation systems can be monitored and used to make short-range predictions. Besides monitoring precipitation, radar is also used to track meteorological balloons. The motion of the balloon gives wind information. The trajectory of the balloon is often used in estimating the trajectories of pollution. Another artificial target for Doppler radar is chaff made from thin metal strips which follow the air motion. Even swarms of insects have been useful radar targets.

Occasionally, a radar may show an echo when there are no visible targets present. These echoes are called *angels*. Angels are caused by inversion layers, discontinuities in moisture, or turbulent regions.

Lidar is similar to radar except that in lidar, the radiation is in the form of visible or infrared laser beams rather than microwaves. These beams of light are extremely narrow. They are applied primarily to detect particles in the atmosphere—in smoke plumes, in the stratosphere, or even in the lower thermosphere. Doppler lasers can measure the velocities of the particles.

As with radars, lidars can get returns from turbulent patches in which there are temperature fluctuations. However, the useful range of these measurements is only a few kilometers.

For many years, acoustic methods have been used to explore the upper stratosphere. Sound waves produced by loud noises such as explosions are scattered forward toward the ground.

Sodars (or *acdars*) are acoustic sounding devices that emit periodic beeps. The sound impulses are then reflected from regions of turbulent temperature fluctuations. The principal use of sodar is to measure the thickness of the turbulent mixed layer or friction layer. Above this turbulent layer, no sodar returns occur. The thickness of the mixed layer plays an important role in the diffusion of pollutants (see chapter 13). Figure 2.12 shows the growth of a mixed layer during the day as revealed by an acoustic sounder. The mixed layer grows from around 170 meters at 8 A.M. local time to 870 meters by noon.

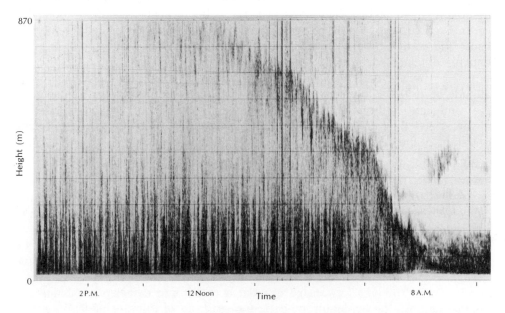

FIGURE 2.12 *Growth of the depth of the mixed layer as revealed by an acoustic sounder.*

41

In contrast to active remote-sensing systems (radiometers), passive systems monitor radiation emitted by the atmosphere in the infrared, microwave, or visible portion of the spectrum. The radiation emitted by a given layer of the atmosphere may be used to estimate the temperature of this layer if the emissivity is known. The absorption of solar short-wave radiation is a useful indication of the distribution and amounts of trace gases such as ozone or oxides of nitrogen.

Passive remote sensing from the ground is not as common as the ground-based active systems. Passive systems are usually flown on balloons or on satellites. An exception is the Dobson Spectrophotometer, which measures the total mass of ozone above the ground. It measures the intensity of two ultraviolet wavelengths arriving from the sun. One wavelength is strongly absorbed while the other is not. The ratio of the intensities of the two wavelengths is a measure of the total amount of ozone present.

2.2.10 OBSERVATIONS FROM WEATHER SATELLITES

Weather satellites have a tremendous advantage over other observing platforms in that they can monitor atmosphere conditions over the whole world in a spatially continuous manner. Satellites can observe the temperature, moisture content, and cloud cover over uninhabited land regions and over the oceans, as well as over densely populated areas. The disadvantage is that the data received by the satellite represent an average over a given horizontal distance and vertical depth, and so some smoothing of the data is inevitable.

Weather satellites may either orbit the earth around the poles (polar-orbiting satellites) or remain over a fixed spot on the equator (geostationary satellites). The polar-orbiting satellites have orbits that are nearly parallel to the earth's meridians (figure 2.13). They cross over the poles in each revolution about the earth. Each successive orbit passes farther to the west than the preceding one because the earth turns underneath the satellite's orbit. Thus, north-south strips of data are obtained that pertain to nearly the same local time on earth.

A geostationary weather satellite remains fixed over the equator at an altitude of 36,000 kilometers (figure 2.14). It revolves around the earth's center at the same rate as the earth rotates about its axis; therefore, a stationary satellite always views the same portion of the earth. A stationary satellite can view about one-third of the earth's surface area. A great advantage of the geostationary satel-

FIGURE 2.13 *Polar-orbiting satellite on successive orbits. The orbit of the satellite is fixed with respect to the stars, but the earth rotates under the satellite. Thus, each successive orbit is farther west than the previous one.*

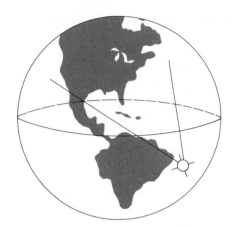

FIGURE 2.14 *Geostationary satellite over the equator which rotates with the earth and so always stays over the same spot on the earth.*

lites is that the variation of weather systems on earth with time can be easily seen by comparing pictures of the same area at short time intervals (in approximately half-hour increments). Time-lapse movies can be made from the photographs to study the movement of weather systems. Because clouds tend to move with the wind, cloud motions can be used to estimate wind direction and speeds.

Most sensors deployed on satellites are passive. The most familiar satellite observations are photographs taken with ordinary visible light. These show cloud patterns, water areas, mountains, variations in vegetation, and snow cover.

Infrared pictures are taken with radiation of wavelength about 11 micrometers, which corresponds to infrared wavelengths (see figure 4.2). These pictures show the intensity of infrared radiation emitted by the earth's surface or cloud tops because the clear atmosphere is transparent to this wavelength.

An advantage of infrared pictures over visible pictures is that infrared radiation is received both day and night. Also, the amount of radiation received is a measure of the temperature of the radiating surface—either the ground or clouds. If the infrared radiation intensity is low, the radiating surface is cold, and the radiation probably comes from a thick cloud with a high top, or from a thin, high cloud. These two possibilities can be distinguished by visible pictures which would show the thick cloud to be brighter. High infrared radiation intensity imples warm temperatures and therefore low clouds or no clouds at all. These two possibilities again can be distinguished on the visible pictures. Thus, the infrared pictures allow the determination not only of horizontal distribution of clouds but also of their heights.

A visible and infrared satellite photograph of a storm over the Great Lakes region may be compared in figures 3.6 and 3.10, respectively. These figures are discussed in detail in section 3.2; for now we simply note the difference in appearance of the clouds over Illinois, Indiana, and Ohio. In the visible picture, these clouds appear nearly as bright as the other clouds in the photograph. In the infrared picture, however, they appear only slightly whiter than the dark ground to the south. Thus, we can infer that these are low clouds.

In clear areas, the infrared pictures allow the determination of surface temperatures, in particular the locations of warm and cold regions in the oceans. These permit the study of the ocean currents.

In addition to cameras, satellites carry instruments which measure radiation intensity in narrow wavelength bands. Of particular importance are radiation measurements in several narrow bands in the neighborhood of 15 micrometers. In that spectral region, carbon dioxide (CO_2) absorbs and emits strongly. Fortunately, the distribution of CO_2 is remarkably constant, except very close to the ground. This means that in the 15-micrometer region, the radiation emitted only depends on the temperature of the CO_2. But the CO_2 at what level? The answer to this depends on the exact wavelengths. At some wavelengths, CO_2 absorbs very strongly. At these wavelengths, the radiation reaching the satellite comes from the high atmosphere—the stratosphere (figure 2.15). At other wavelengths, CO_2 does not absorb so strongly, and the radiation reaches the satellite from low in the atmosphere. Thus, the intensity of the radiation is a measure of temperature, and the wavelength is a measure of the height.

By making possible the estimation of the temperature distribution, 15-micrometer radiation measurements also give information about important properties of the wind field. As we shall see, the vertical variation of the wind is related to the horizontal temperature differences. Satellite data are especially well suited for the estimation of horizontal temperature gradients in clear areas and above clouds, which then give good indications of vertical wind shear.

Temperatures measured from satellites are now being introduced into ordinary upper-air charts, but it is not clear whether such charts are significantly improved by these additional data. The main difficulty is that infrared radiation does not penetrate clouds, so the temperature below the clouds cannot be determined by this method. This difficulty can be overcome by using short microwaves, which do penetrate clouds. In certain wavelengths, oxygen emits microwave radiation, the amount depending on the temperature of the oxygen. Radiation in these wavelengths can be analyzed in the same manner as 15-micrometer radiation.

Water vapor also radiates in certain regions of the infrared. Radiation in these wavelengths depends both on the distribution of water vapor and on the temperature. If the temperature is known from radiation emitted by CO_2 or oxygen,

FIGURE 2.15 *Radiation received in different infrared wavelengths from carbon dioxide (CO_2).*

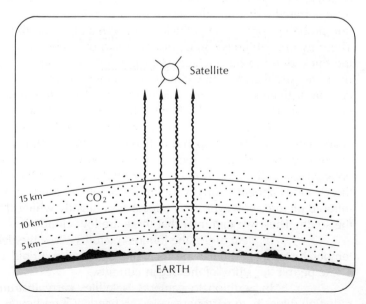

radiation in the wavelengths emitted most strongly by water vapor permits the computation of the water-vapor distributions.

Finally, it is possible to measure nonstandard variables from satellites. For example, the total amount of ozone can be found from satellites by measuring the ultraviolet radiation. Concentrations of other trace gases in the stratosphere can also be estimated by viewing radiation emitted by these gases from a satellite pointed not straight down, but instead almost horizontally, through the upper regions of the atmosphere.

Questions

1   How do the Blind River uranium deposits indicate that there was very little oxygen during the first quarter of the earth's life?

2   How can you distinguish high and low clouds if you have both an infrared and a visible satellite photograph?

3   What is the major constituent of volcanic emissions?

4   Why was the first plant life limited to the upper layers of water bodies?

5   Convert 75°F to degrees Celsius and Kelvins.

6   If the average sea-level pressure is 1013 mb, how high is the column of mercury in a barometer, on the average? Give answer in inches and in centimeters.

7   What is the approximate pressure at the top of Pikes Peak, Colorado?

8   Suppose you hang a wet shirt out to dry on a windy day and measure the temperature of the wet shirt to be 40°F. If the air temperature is 60°F, what is the relative humidity?

9   Give two examples of an active remote sensing system and one example of a passive system.

Problems

10   Suppose that the elevation angle of a weather balloon is 10° and the pressure reported by the balloon corresponds to a height of 5 km. How far away is the point directly under the balloon on the surface of the earth?

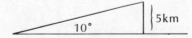

11   Suppose the temperature of an air parcel near the ground falls during the night, as indicated in the table below. Enter ↑ for increase, ↓ for decrease, or 0 for no change in the table to indicate the change in each variable through the night. At what time does fog form?

	Temperature	Dewpoint	Wet bulb	Relative humidity
Sunset	60	40	50	48
10 P.M.	55	?	?	?
12 midnight	50	?	?	?
3 A.M.	40	?	?	?
6 A.M.	35	?	?	?

THREE

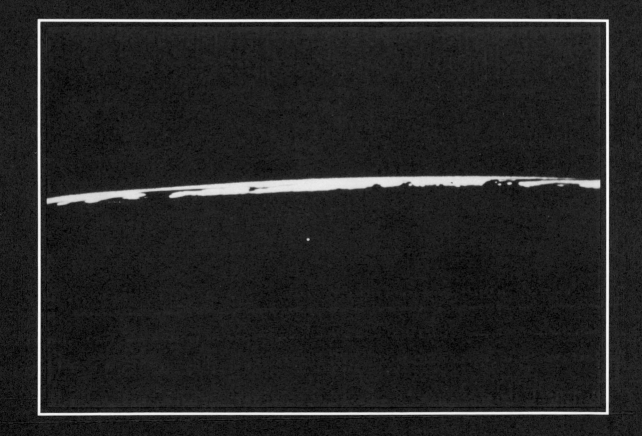

An Overview of Atmospheric Structure

sounding • ozone • troposphere • stratosophere • thermosphere • constant pressure chart • horizontal scale • synoptic • trough • ridge • isotherm • inversion • noctilucent cloud • ion • aurora borealis • aurora australis • front • cyclone • anticyclone • convergence • air mass • mesoscale • turbulence • dewpoint depression

In the following chapters we will explore the structure and life cycle of many atmospheric systems such as thunderstorms, jet streams, extratropical cyclones, and hurricanes. We will also look at the climates of the world, especially that of North America. Before getting into these details, however, we present an overall view of the vertical and horizontal structure of the atmosphere in this chapter and introduce the graphs and weather maps that will later illustrate the structure of the atmosphere at various levels and over different time and space scales.

3.1   VERTICAL STRUCTURE OF THE ATMOSPHERE

As the radiosonde rises through the atmosphere, it transmits information on the temperature, dewpoint, and wind which may be plotted on a graph to depict the vertical structure of the lower atmosphere. The resulting graph, called a *sounding,* summarizes the vertical variation of temperature, moisture, and wind. The variation of these quantities is very important in understanding and forecasting severe weather such as thunderstorms, or air pollution episodes in which pollutants are trapped close to the surface by a layer of light, warm air overlying dense, cool air.

Figure 3.1a shows a vertical sounding of temperature and dewpoint for Dayton, Ohio, at 7 P.M. EST on October 23, 1979. This sounding was taken about 3 hours later than the satellite photographs in figures 3.6 and 3.10. The horizontal lines, labelled in pressure on the left and approximate height above sealevel on the right, denote the level in the atmosphere. The solid lines sloping upward toward the right are lines of equal temperature, *isotherms*. The temperature at each level is obtained by interpolation between the isotherms, if necessary; thus, the temperature at a pressure of 600 mb (approximate height 4.2 km above sea level) is about −11°C. The dewpoint is obtained from the dewpoint sounding (dashed line) in the same way; thus, the dewpoint at 600 mb is about −20°C.

The sounding Figure 3.1a shows several interesting characteristics. From the surface to a pressure of about 720 mb (height 2.8 km), the temperature decreases from +4°C to −10°C. The air in this layer is very moist, as shown by the near equality of the temperature and dewpoint. In the layer between 720 and 680 mb (height 2.8 to 3.2 km), the temperature increases with elevation, reaching a maximum temperature of about −6°C at 680 mb. This layer in which the tem-

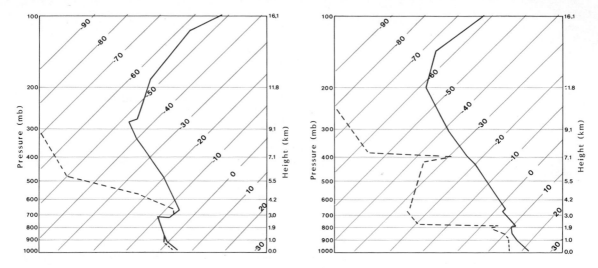

FIGURE 3.1 *(A) Temperature (solid line) and dewpoint (dashed line) sounding for Dayton, Ohio, at 7 P.M. EST on October 23, 1979. (B) Temperature (solid line) and dewpoint (dashed line) sounding for Cape Hatteras, North Carolina, at 7 P.M. EST on October 23, 1979.*

perature increases instead of decreases with elevation is called an *inversion*. Above the inversion, the temperature decreases again with elevation until a pressure of 290 mb is reached (height 9.4 km). This layer is much drier than the lowest layer, as indicated by the large difference between the temperature and dewpoint. From the level of 290 mb to the top of the sounding (100 mb or 16.1 km), the temperature is nearly constant with elevation and the air is extremely dry.

Figure 3.1b shows a sounding for Cape Hatteras, on the coast of North Carolina, at the same time as the Dayton sounding shown in figure 3.1a. A comparison of the two soundings shows a considerable difference. Cape Hatteras is about 20°C warmer in the low levels than Dayton, and although the layer from the surface to about 800 mb (height 1.9 km) is moist at Cape Hatteras, it is saturated over Dayton. Furthermore, the prominent inversion between 720 and 680 mb levels is not present over Cape Hatteras; instead, two weak inversions are present near the 790 and 680 mb levels. At higher levels, Cape Hatteras shows a thin layer of moisture at about 400 mb (height 7.1 km) which is not present over Dayton. Finally, the level at which the temperature becomes nearly constant with elevation is higher over Cape Hatteras than over Dayton.

These two soundings illustrate the considerable variation in vertical distributions of temperature and moisture at different locations over the Earth and at different times of the day and year. It is useful to consider the mean vertical structure of the atmosphere which is obtained by averaging many such soundings over the Earth at various times of the year. Figure 3.2 shows a graph of the mean temperature up to a height of 120 km and, although it is smoother than most individual soundings (compare with figure 3.1), it still shows important variations. On the average, temperature decreases with height from the surface to about 8 km. It then begins to increase with elevation, and reaches a maximum at about 40 km;

49

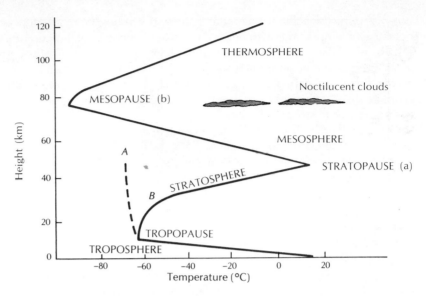

FIGURE 3.2 *Vertical distribution of temperature to a height of 120 kilometers. Heavy line, observed; dashed line, without high-level heating.*

the temperature at this level is almost as high as the average surface temperature. Above 40 km, the temperature decreases rapidly, and reaches a minimum of −85°C at an elevation of about 78 km. Above this level, the temperature increases again until the outer fringes of the atmosphere are reached. The explanation for these alternate layers of temperature decrease and increase with height is found in the vertical distribution of heat sources in the atmosphere. In the next section we describe the processes that determine the temperature structure of the various atmospheric layers.

3.1.1 THE TROPOSPHERE

The cloudless atmosphere is quite transparent to visible sunlight, so the ground is warmed more directly than the atmsophere. Only a small portion of the sun's radiation is trapped by the atmosphere; instead, the air is warmed indirectly by the upward transport of heat from below. As the ground is heated, convection currents are initiated which transport heat up into the atmosphere.

Another heat source for the atmosphere is the release of latent heat when water vapor condenses. Some of the energy of the sun is used for evaporation of water at the ground. The invisible water vapor is then carried into the atmosphere, where it condenses into water droplets we see as clouds. This process releases heat to the air.

In general, the indirect heating processes of the atmosphere decrease with height up to 10 kilometers or so. The reduction of the heating processes with height in the troposphere, and the cooling of the air by long-wave radiation (see sec-

50

tion 4.2), which reaches a maximum in the upper troposphere, are responsible for the decrease of temperature with height. The rate of decrease of temperature with height is called the *lapse rate*. Its average value is 6.5C°/km (about 3.5F°/1000 ft).

The general decrease of temperature with height has been known for a long time, and early scientists believed the temperature decreased uniformly to absolute zero at great heights. It was a great surprise when, around 1900, balloonists found that this decrease stops abruptly just above 9 kilometers (30,000 feet) or so in middle latitudes. Above this height, the temperature remains constant. At first, meteorologists thought that this observation was in error, for the constant temperatures would require a heat source at this altitude and no such heating mechanisms was known. However, as discussed in the following section, the presence of ozone above this height causes heating by the direct absorption of sunlight.

The region of the atmosphere nearest to the earth, in which temperature generally falls with height, is called the *troposphere* ("weather sphere"). The next-higher region, where the temperature no longer decreases, is the *stratosphere*, and the surface separating the two "spheres" is called *tropopause* ("end of troposphere"). As the name implies, almost all the "weather"—rain, clouds, and snow—occurs in the troposphere. Only occasionally will a violent thunderstorm break through the tropopause into the stratosphere. Usually the stratosphere is relatively quiet and dry.

Figure 3.3 shows vertical cross sections of the mean temperatures averaged around latitude circles for the months of January and July. The heavy lines denoting the tropopause separates the troposphere and the stratosphere. This figure shows that the tropopause and the stratosphere have quite complicated properties. For example, the tropopause is higher and colder at the equator than at middle and high latitudes. Further, it usually has breaks in it. The cold equatorial tropopause has a surprising consequence. At 15 kilmeters (50,000 feet) and just above, the equator is actually much colder than the regions farther north or south, just opposite the situation close to the ground.

3.1.2 THE STRATOSPHERE AND OZONE

The layer above the troposphere in which the temperature increases with height is called the *stratosphere*. The stratosphere is relatively stable, with warm air overlying cool air, so the vertical motions tend to be weak. In addition to the lack of vertical motion, an extremely low moisture content (typically three parts per million) prevails. Therefore, clouds occur only in very restricted regions, e.g., in the antarctic during winter, and there is never any precipitation. However, the stratosphere is not motionless. There are strong horizontal winds in the stratosphere, particularly near the bottom, and at higher levels in the polar regions during the winter. Winds of 300 km/h (186 mi/h) are not unusual.

The temperature decreases upward in the stratopshere because of the presence of ozone, which is most dense between 20 and 32 kilometers. The total amount of ozone is actually very small. If all the ozone were taken down to the ground from the stratosphere, where it would be compressed by the weight of the overlying atmosphere, it would form a layer less than a centimeter thick.

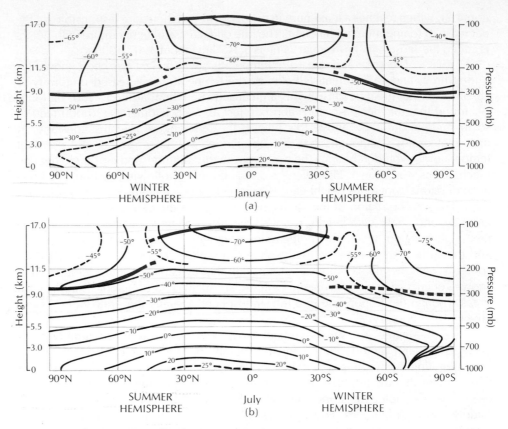

FIGURE 3.3 *Average vertical temperature distribution in (a) January and (b) July.*

Ozone is so active, however, that a concentration of as little as one part in a million may have important effects. Unlike other gases in the atmosphere, ozone has the ability to absorb ultraviolet radiation. This absorption of energy heats the stratosphere, so that it is much warmer than it would be without ozone. At the same time, the ozone shields animals, plants, and people from dangerous radiation which otherwise would make the world uninhabitable for the familiar forms of life. In fact, the small amounts of ultraviolet radiation that do penetrate the ozone blanket cause sunburning and appear to be an important cause of skin cancer in humans.

If no ozone were present, the average vertical temperature distribution would resemble curve *A* in figure 3.2. The actual average distribution looks like curve *B*, with the temperature peak at point *a* because of ozone heating. This point is called the *stratopause*, the end of the stratosphere. Here the temperature is almost as high as at the ground, attesting to the efficiency of ozone in absorbing solar energy.

The warm stratopause has the peculiar property of reflecting sound waves, to such an extent that powerful explosions can often be heard hundreds of

Plate la (above)   A lightning stroke darts from the side of an evening thunderstorn over Miami, Florida (courtesy of Peter Black).

Plate 1b (below)   Ice crystals assume the shape of a bird in an unusual cirrus cloud formation. The "beak" is a small funnel cloud (courtesy of Richard A. Anthes).

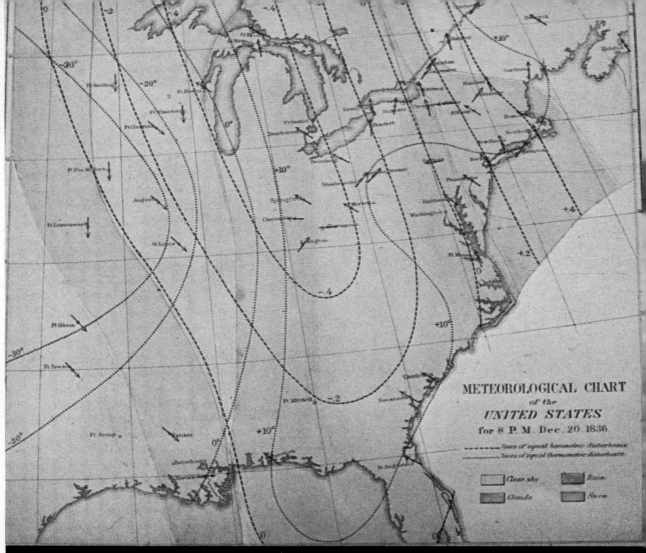

METEOROLOGICAL CHART
of the
UNITED STATES
for 8 P.M. Dec. 20, 1836.

- - - - - - lines of equal barometric disturbance
·········· lines of equal thermometric disturbance

Clear sky   Rain
Clouds   Snow

Plate 2a (above) Loomis' analysis of a winter storm of December, 1836 [from Elias Loomis, "On Certain Storms of Europe and America, December, 1836," *Smithsonian Contributions to Knowledge*, Vol. 11 (Washington, D.C.: Smithsonian Institution, 1860) pp. 27–39].

Plate 2b (below) Ley's model of the extratropical cyclone [from Robert H. Scott, *Weather Charts and Storm Warnings* (London: Longmans, Green and Co., 1877)].

kilometers away. In fact, this warm region was discovered because the guns at Queen Victoria's funeral in 1901 were heard in Germany, but not between England and Germany.

Since ozone is so important to our well-being, we will try to explain why it is mostly formed at some distance above the ground. Ozone is a form of oxygen in which three *atoms* of oxygen combine to form one *molecule* of ozone. An atom of oxygen is denoted by the symbol O, and a molecule of ozone is represented by O_3. In the lower atmosphere, practically all the oxygen occurs in molecules consisting of two atoms and is denoted by O_2. Very high up, say above 110 kilometers, O_2 is divorced by intense ultraviolet radiation and occurs mostly as single oxygen atoms, O.

To produce ozone, we have to combine atomic oxygen, O, and ordinary oxygen, O_2. If O and O_2 merely collide, however, they just bounce off each other and do not form O_3. To make ozone and satisfy all physical requirements, we need a third molecule, M, involved in the interaction, which is represented by

$$O + O_2 + M = O_3 + M \tag{3.1}$$

This molecule M acts like a minister at a wedding; O and O_2 get married, and M walks away after the wedding is completed. Chemists call the role of M that of *catalyst*. In other terms, the formation of ozone requires a triple collision, which is rare at very high levels (above 60 kilometers), because there are so few molecules that it is extremely unlikely that three of them will arrive at the same place at the same time. Thus, there is almost no ozone above 60 kilometers. On the other hand, there is almost no atomic oxygen, O, below approximately 16 kilometers, so the formation of ozone is unlikely there, too. These are the reasons most of the ozone is found in the stratosphere.

3.1.3 THE ATMOSPHERE ABOVE THE STRATOSPHERE

Figure 3.2 shows that the temperature drops again above the stratopause (above 50 kilometers). This region of decreasing temperature is called the *mesosphere*. If there were no special heating mechanism above the stratopause, the temperature would keep on dropping for great distances. Actually, however, the temperature reaches a minimum at an elevation of about 80 kilometers, a region called the *mesopause*, the coldest region in the atmosphere. The temperature at the mesopause may fall to $-100°C$ ($-150°F$) at the North Pole in summer. In fact, the mesopause can be so cold that the tiny amount of water vapor in this region forms ice clouds, called *noctilucent* clouds, which can be seen when the sun hits them after sunset.

The region above the mesopause is called the *thermosphere* ("hot sphere"), where the temperature may actually go up to several thousand degrees. One reason for these hot temperatures is that ultraviolet sunlight can be absorbed by oxygen in the thermosphere. Also, there are so few molecules to be heated (the density is quite low) that a little energy absorbed can produce a large temperature

53

increase. Because there are quite a few loose electrons and positive ions in the thermosphere, this highest part of the atmosphere is also called the *ionosphere*.*

By this time, you may wonder why the distant thermosphere is discussed at all in a survey book on meteorology, and how, if at all, it affects you. As far as we know, the thermosphere has very little effect on our daily weather. The only contact most of us are likely to have with the thermosphere is through its effect on ordinary AM radio reception. Loose electrons have the ability to reflect radio waves, which explain why stations far away can be heard at night. During daytime, long-distance radio transmission is not as good because there is too much ionization (and therefore, too many electrons) in the lower thermosphere. These electrons absorb the radio waves on the way up to the reflecting layer.

Because radio waves can be reflected and absorbed in the thermosphere, scientists can devise experiments with radio reception to determine the nature of the thermosphere. From such experiments, we know that the thermosphere has huge diurnal variations, both in temperature and number of electrons. Also, the thermosphere becomes most active every eleven years, when the sun is disturbed by large solar storms, called *sunspots*. These storms, which consist of relatively cool gases in the solar atmosphere, appear as dark spots on the face of the sun. Around the sunspot areas, bright flares of short duration appear, and the outer portions of the solar atmosphere become extraordinarily hot. These flares also disturb radio propagation on earth. A day or so after the appearance of solar flares, particles arrive from the sun which collide with particles of the upper atmosphere. These collisions excite the atoms, which then emit light. Thus, the *northern* and *southern lights (the aurora borealis* and *aurora australis,* respectively) usually appear a day or so after a major solar disturbance. A beautiful example is shown in Plate 16.

Auroras occur at high latitudes because the earth's magnetic field deflects the incoming beams of particles toward the poles. During the bombardment of the earth's upper atmosphere by the solar particles, the magnetic field of the earth is greatly disturbed. Even at the ground, a compass needle undergoes irregular, and apparently mysterious, movements, which are associated with upper-atmosphere electric disturbances. These disturbances are called *magnetic storms*. The main practical implication of solar disturbances and associated upper-atmosphere phenomena is that communications over long distances become difficult. Not only is radio propagation impeded, but also irregular currents are induced into cables under the oceans, making difficult communication by wire as well.

3.1.4 VERTICAL DISTRIBUTION OF DENSITY AND PRESSURE

So far, we have mainly discussed the rather complex vertical distribution of temperature. In a general way, the vertical distributions of density and pressure are much simpler: both decrease rapidly with height. For example, at the cruising altitude of conventional jet aircraft (10 km), the weight of

---

*An *ion* is an electrically charged atom or particle.

the overlying atmosphere is only about one-fourth of what it is at the ground; the pressure there is only about one-fourth of sea level pressure.

There is a convenient rule about variation of pressure with height, accurate to an altitude of about 60 miles: for every 10 miles of elevation, the pressure decreases by a factor of ten. Because the pressure at sea level is about 1000 millibars, at 10 miles up, it is about 100 millibars; 20 miles up, 10 millibars; 30 miles up, 1 millibar; and so forth. The same approximation holds fairly well for air density. Thus, in the lower thermosphere (figure 3.2), both density and pressure are only about 1/1,000,000 of their surface values.

The very small pressure and densities in the thermosphere have an important implication. There have always been scientists who have thought that solar disturbances should affect weather because they profoundly affect the thermosphere. But because the thermosphere contains only about 1/100,000 of the total mass in the atmosphere, it is easy to see that it can be violently disturbed and still have little effect on the weather near the ground.

HORIZONTAL STRUCTURE OF THE ATMOSPHERE 3.2

While a vertical sounding gives a detailed profile of the temperature, moisture, and winds in the atmosphere above a point on the earth's surface, it does not provide information about horizontal variation in a way that is easy to interpret. Since horizontal variations are extremely important in understanding and forecasting the weather, analyses of pressure, temperature, moisture, and winds on horizontal surfaces are essential. The most common surface for depicting weather variables is the surface of the earth itself, and almost every television and radio station and newspaper presents a version of the surface weather map.

3.2.1 SURFACE WEATHER MAPS

There are approximately 600 weather stations over North America that report some surface weather data at least once every hour. In addition to the meteorological requirements for these surface data, one of the major reasons for this large number of stations is the demand of the aviation industry for timely weather reports.

Most surface weather stations report temperature, dewpoint, sea-level pressure (the pressure that would exist if the station were located at sea-level), and wind direction and speed. They also indicate the horizontal visibility (if under 10 miles), cloud amounts, and precipitation (if any). These data are plotted on a horizontal map at the location of each observation according to the plotting model shown in figure 3.4. In the plotting model the winds are represented by an "arrow," with the direction *from which the wind is blowing* indicated by the direction of the arrow and the speed (in knots) indicated by the number and length of the "feathers". One may imagine the circle, which represents the station location, as an apple, with the arrow sticking into the apple; the direction from which the arrow came is the wind direction. The wind speed is depicted to the nearest 5 knots: each full-length feather represents 10 knots, and a half-length feather represents 5 knots. The examples shown in figure 3.4 will be helpful in learning to decode plotted weather data. In *a* the wind is *from* the northwest at 20 knots, whereas in *b* it is *from* the south at 10 knots. In *c* and *d* the winds are from the southeast at 15 knots and from the northeast at 10 knots, respectively. If the wind is 5 knots, the half-feather is drawn at an angle to the shaft of the arrow, so that it will not be mistaken for a full-length feather. In *e*, therefore, the wind is from the northeast at 5 knots.

The temperature and dewpoint (in °F in the U.S., and °C elsewhere) are plotted to the left of the station circle, with temperature on top. Thus in figure 3.4a the temperature is 40°F and the dewpoint is 20°F. The closer the dewpoint is to the temperature, the greater the relative humidity. When they are equal, as in figure 3.4d, the relative humidity is 100%.

The amount of cloud cover is represented by the shaded fraction of the station circle. In *a* the unshaded circle denotes clear skies. In *b* half of the sky is covered by clouds, while in *c* the sky is totally covered. If precipitation or fog obscures the sky so that it is impossible to tell the amount and type of clouds, a cross mark is placed in the circle, as in *d*.

A symbol indicated the present "weather" (e.g., precipitation, haze, fog) and the visibility (in miles) are plotted on the left between the temperature and dewpoint. Some of the common symbols are given in figure 3.5. Figure 3.4c depicts a light rainshower with a visibility of 2 miles. In figure 3.4d light snow has reduced the visibility to 1/2 mile.

Figure 3.6 shows a number of surface observations plotted on a visible satellite photograph of the eastern United States and southeastern Canada at 3:30 P.M. EST on October 23, 1979. Clouds show up as bright areas, since clouds reflect most of the sunlight (as much as 90 percent for thick clouds) striking their tops. In contrast, the ground reflects only about 5 to 10 percent of the incoming sunlight and hence appears dark.

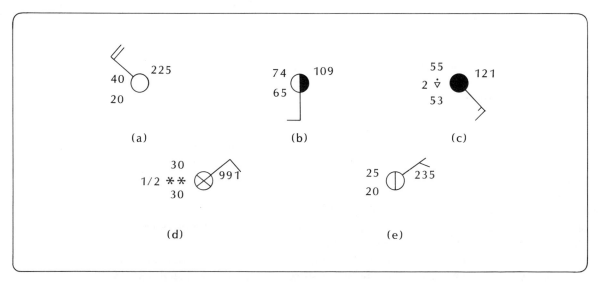

FIGURE 3.4 *Examples of plotted surface weather data.*

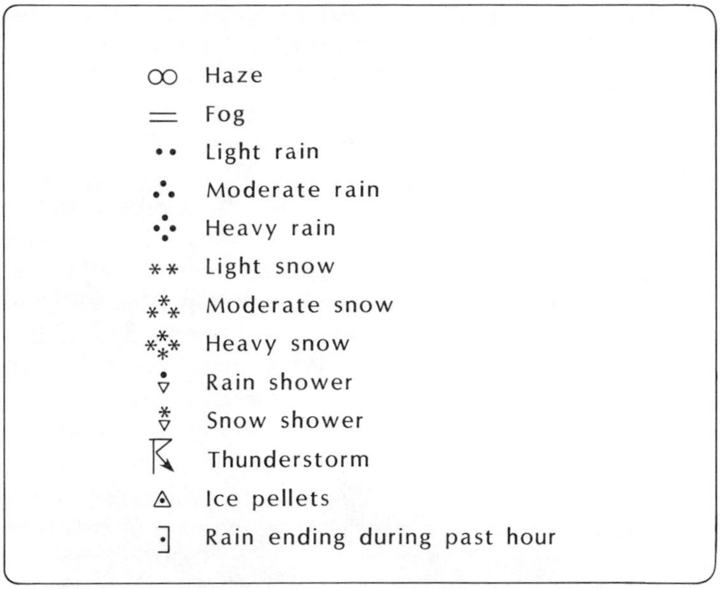
Haze
Fog
Light rain
Moderate rain
Heavy rain
Light snow
Moderate snow
Heavy snow
Rain shower
Snow shower
Thunderstorm
Ice pellets
Rain ending during past hour

FIGURE 3.5 *Common weather symbols.*

The most striking feature on the satellite picture is the nearly circular mass of clouds centered over Michigan, with a "tail" of clouds extending around the northeastern side from Ontario southward across New York and Pennsylvania, and then forming a graceful arc which sweeps across Virginia and the Carolinas, Georgia, northern Florida, and finally out into the Gulf of Mexico. We will see that this arc of clouds is associated with a *cold front* which marks the leading edge of a cool, dry flow of air moving eastward across the eastern United States.

The present weather plotted on the satellite photographs indicates that some of the clouds in the circular area and along the cold front are producing rain and even snow. Temperatures in the 20s and 30s (°F) and northerly winds of 10 to 20 knots over the upper Midwest and southern Canada are presenting people in these regions with a early taste of winter. Evidently a major autumn storm, or *cyclone,* is affecting the Great Lakes region this day.

Surface data add considerable information to the photograph. We note that the lowest pressures are located near the center of the swirling cloud mass in upper Michigan, while the highest pressures are located where the skies are predominantly clear, along a line from central Canada to Texas. Although it is possible to obtain an idea of the pressure variation simply from the plotted data alone, a much more useful picture emerges if we connect points on the map having equal sea-level pressures in order to obtain an *analysis,* as shown in figure 3.7. Lines of equal pressure are called *isobars,* and those in figure 3.7 reveal a large region of low pressure associated with the circular cloud mass. On the other hand, where the pressure is relatively high, there are generally fewer clouds. Such high pressure areas are called *anticyclones.*

Although the lowest pressure on the sea-level pressure map is located over northern Michigan, a zone of relatively low pressures called a *trough* extends southward from the low center. This trough coincides rather well with the arc of

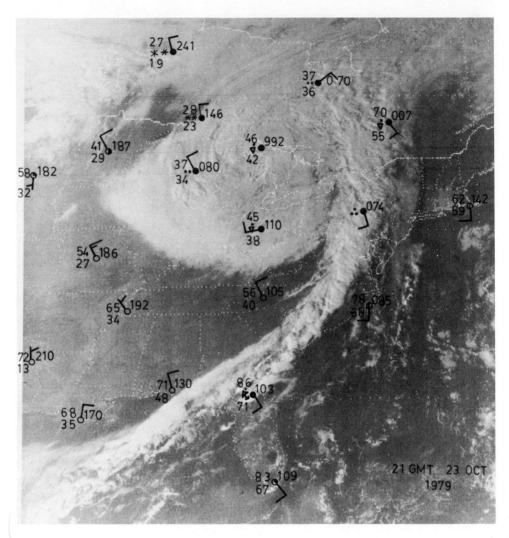

FIGURE 3.6 *Visible satellite photograph at 3:30 P.M. EST on October 23, 1979, and surface data for 4:00 P.M. EST on October 23.*

clouds in the satellite photograph. In contrast, along the axis of relatively high pressure called a *ridge,* which connects the two centers of high pressure in Canada and Texas, the clouds are absent.

The surface wind observations reveal a systematic flow or circulation of air around the center of low pressure. Over the Northern Hemisphere, the sense of the circulation around such large-scale low-pressure centers is always counter-clockwise, or *cyclonic.* In contrast, the circulation around centers of high pressure is clockwise in the Northern Hemisphere, or *anticyclonic.*

A careful inspection of the surface winds reveals a general flow of air toward lower pressure, in addition to the cyclonic circulation around the low. As

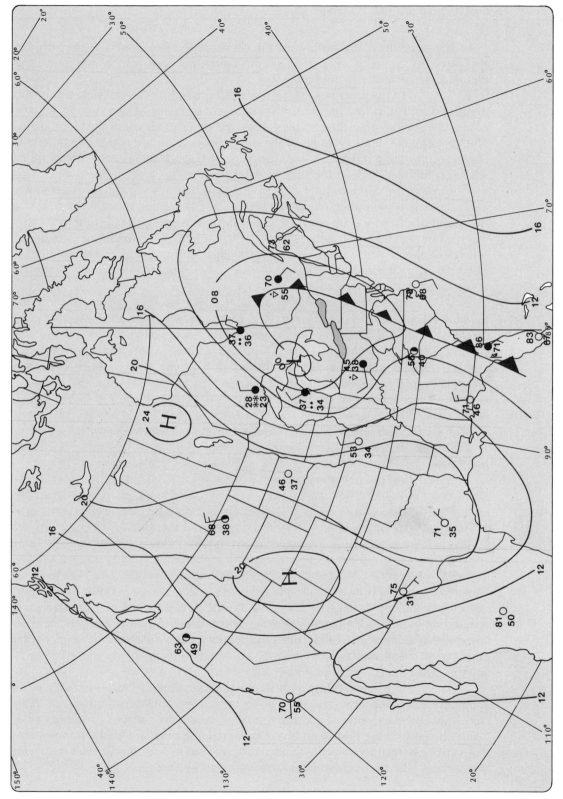

FIGURE 3.7 *Sea level pressure analysis at 4 P.M. on October 23, 1979.*

59

this surface air flows inward toward the center of the low from all sides, it must rise. As we shall see in later chapters, this rising motion is responsible for all the clouds and precipitation around the low.

The general flow of air toward the band of clouds lying within the elongated trough of low pressure indicates a convergence of two distinct streams of air, one stream originating northwest and the other southeast of the cloud band. The temperatures and dewpoints of these two flows of air are noticeably different. For example, the temperature and dewpoint at Cape Hatteras are 78 and 68°F, while on the other side of the front in Asheville, North Carolina, the temperature and dewpoint are 56 and 40°F. Indeed, the flow from the northwest is generally cool and dry, while the flow from the southeast is warm and moist. When we say dry and moist here, we are referring to the absolute amount of moisture in the air, as given by the dewpoint, rather than the relative humidity, which is a measure of how close the air is to saturation. Thus air with a temperature and dewpoint of 68 and 65°F is unsaturated, but it contains more water vapor than saturated air with a temperature and dewpoint of 40°F.

Large-scale volumes of air with similar moisture contents and temperature throughout are called air masses. There are two major air masses in figure 3.6. Over the past few days the air east of the cloud band has been over the Atlantic Ocean and Gulf of Mexico and has become warm and moist through contact with the water; this air mass is termed *maritime tropical*. In contrast, the air on the northwest side of the band is cooler and drier, having come from the central United States and southern Canada; this second air mass is called *continental polar*. To be sure, the air over Texas has spent the past few days at more southerly latitudes than the air over Minnesota, and is therefore warmer. Nevertheless, the horizontal contrasts of temperature and moisture across the cloud band are greater than the horizontal contrasts within each air mass.

It should be apparent by now that the cloud band and sea-level pressure trough represent a boundary between the two air masses. Such boundaries are called *fronts*, named after the boundaries between opposing armies in World War I. The type of front is determined by the direction of movement. If cooler, dry air is advancing and replacing the warm, moist air, the front is a *cold front*. In contrast, if the cold air is retreating and is being replaced by warmer air, the front is called a *warm front*. The boundary in figure 3.6 is advancing toward the east and, therefore, is a cold front. Surface characteristics which usually occur in association with fronts include a trough of low pressure, the converging of two air streams, a marked difference in temperature and dewpoint on either side of the front, and an increased likelihood of clouds and precipitation. Later chapters will explore the structure of fronts and their relationship to clouds, precipitation, and jet streams aloft in greater detail.

In spite of its importance in meteorology, the surface map only gives a partial view of the atmosphere at a given time. The formation and decay of highs and lows, their movement, and the formation of clouds and precipitation all depend crucially on the flow of air above the surface, particularly in the troposphere. Thus it is essential to map the temperature, moisture, and winds at levels above the surface. The next section introduces upper-air analyses.

60

3.2.2 UPPER AIR MAPS

After seeing the isobar pattern at sea level, we might wonder what the isobars look like at other elevations in the atmosphere. Do highs and lows exist aloft, and if they do, have they the same appearance as the surface-pressure systems? We could answer these questions by plotting the pressures at another fixed elevation in the atmosphere, for example 5 kilometers. In practice, however, most upper-air weather data are analyzed at a constant-pressure level rather than at a constant-height level. A constant-pressure level is a two-dimensional surface (something like a sheet or a rug) on which the pressure is identical everywhere (see figure 3.8). Constant-pressure surfaces are very nearly parallel to constant-height surfaces, but they can bend and buckle a little. The typical variation in height of the 500-millibar surface is from 4.8 to 5.8 kilometers (16,000 to 19,000 feet).

Although constant-height charts may seem more straightforward, constant-pressure charts offer several benefits. First, aircraft tend to fly on constant-pressure surfaces rather than on constant-height surfaces, since the aircraft altimeter

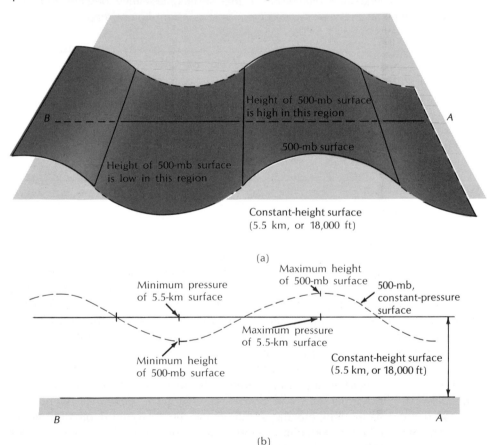

(a)

(b)

FIGURE 3.8 *Relationship of constant-pressure and constant-height surfaces.*

61

is really a barometer. Second, and more important, the equations used in meteorology are simpler if the variables are all defined on constant-pressure surfaces. Neither of these reasons directly concerns us in this book; for our purposes, it would be just as easy to work with analyses on constant-height surfaces. However, all the upper-air weather maps we are likely to encounter will be on constant-pressure surfaces, so we need to understand how these analyses relate to their counterparts on constant-height surfaces.

Actually, the interpretation of constant-pressure-surface analyses is not really difficult. The main thing to remember is that pressure variations on a constant-height surface correspond to height variations on a constant-pressure surface. In figure 3.8, for example, where the height of the constant-pressure surface is high (above 5.5 km), the pressure of the constant-height surface is also high (greater than 500 mb). Likewise, where the height of the 500-millibar surface is low (below 5.5 km), the pressure of the constant-height surface is also low (less than 500 mb). Thus, instead of drawing isobars on a constant-pressure surface, we draw contours of equal elevation. Thus, highs and lows on a 500-millibar map are really regions of high and elevations of that surface, but they behave and look exactly like high- and low-pressure centers on a constant-height map.

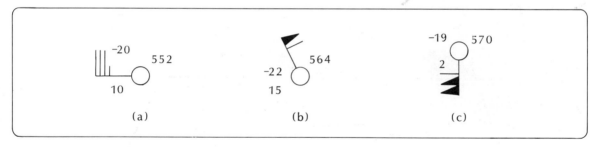

FIGURE 3.9 *Examples of temperature, dewpoint depression, wind, and height as plotted on a 500-millibar constant-pressure surface.*

Data on upper-level, constant pressure charts are plotted according to a convention very similar to that of the surface map, although information on clouds, present weather, and visibility are not plotted. As illustrated in figure 3.9, which represents typical data on the 500-millibar surface, the winds are depicted as they are on the surface map. Note that winds of 50 knots are indicated by the solid triangular flag; thus the winds in figure 3.9b are from the northwest at 60 knots, but at c they are from the south at 110 knots.

The temperature on constant pressure surfaces is plotted and analyzed in degrees Celsius. The moisture content of the air is indicated by the difference between the temperature and dewpoint, called the *dewpoint depression*. Thus, in figure 3.9a the temperature is −20°C and the dewpoint depression is 10° C, which means that the dewpoint is −30°C. If the dewpoint depression exceeds 30°C, an x is plotted.

The *height* of a constant pressure surface, which corresponds to the *pressure* of a constant height surface, is plotted on the upper right side of the

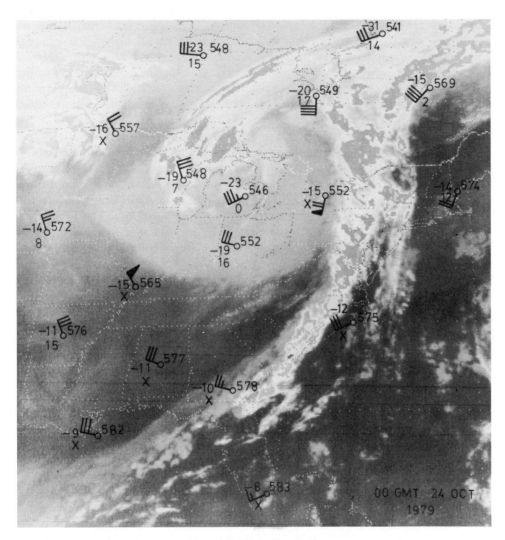

FIGURE 3.10 *Infrared photograph at 4 P.M. EST on October 23, 1979, with 500-millibar data for 7 P.M. EST on October 23.*

station circle in decameters. The height in meters is obtained by adding a zero to the plotted number, so that the height of the 500-millibar surface over station *a* is 5520 m.

   Figure 3.10 shows some 500-millibar data plotted on an infrared photograph of the same storm depicted in figure 3.6. This picture was taken at 4 P.M. EST, or one half hour later than the visible photograph of figure 3.6. As discussed earlier, infrared pictures show the temperature of the clouds or surface of the earth in the "view" of the satellite. Radiation seen by the satellite is converted to a temperature, which is then converted to a shade of gray ranging from black to white. Horizontal variations of this gray scale indicate horizontal variations of temperature. Thus, the warm Atlantic ocean in the southeast appears uniformly dark. Strong contrasts occur along the edges of cloud patterns such as the frontal band along the East Coast. Note that much of the circular cloud mass over the Midwest appears as only a slightly lighter shade of gray than the nearby ground (compare northern and southern Illinois, for example). The visible picture shows that these clouds are bright and nearly unbroken; because the infrared photograph indicates that they are nearly the same temperature as the ground, however, we can infer that they are low clouds. This inference is correct, as shown by the sounding at Dayton, Ohio, in figure 3.1a. The saturated layer from about 2 to 3 km indicates the cloud layers. Above 3 km, the atmosphere dries out rapidly.

The 500-millibar data at 7 P.M. EST, plotted in figure 3.10, indicate quite a difference in air flow at this level compared to the surface flow. Winds are much stronger, averaging near 40 knots with some at nearly 100 knots. In general, westerly winds prevail, although a strong perturbation exists over the Midwest. The heights of the 500-millibar surface, contoured in figure 3.11, are lowest over Michigan, and it is apparent that this upper-level low is part of the same low-pressure system we saw on the sea-level pressure analysis (figure 3.7). We also note that the winds bear a close relationship to the height contours, with air circulating around the low in a counterclockwise (cyclonic) manner. Finally, we note that temperatures are generally lower to the north and wherever the heights are lower. For example, the temperature over Flint, Michigan, is $-23°C$ and the height is 5460 m, whereas over Miami, Florida, the temperature and height are $-8°C$ and 5830 m, respectively. We will see in later chapters that the above relationships between temperatures, heights, and winds are not coincidental, but part of a general relationship which is very helpful in understanding the origin and structure of atmospheric circulation features like the storm in this example.

3.3   SCALES OF METEOROLOGICAL PHENOMENA

Even a casual look at the satellite photographs of figures 3.6 and 3.10 reveals a tremendous variation in the size, or scale, of clouds and cloud patterns. Over south Florida tiny cumulus clouds appear as mere dots, whereas the circular cloud pattern over the upper Midwest covers a region several thousands of kilometers in diameter. Although not revealed by the relatively few data plotted on these photographs, similar variations in size occur with temperature, moisture, and wind patterns. In fact, important variations in meteorological variables exist over distances ranging from a few centimeters to many thousands of kilometers. Obviously, to avoid confusion, we do not want to study simultaneously all three sizes of motion. In practice, we select our principal time and space scales of motion. The time scale is determined by the lifetime, or period, of the phenomenon; the space scale is determined by the typical size or wavelength.

Figure 3.12 gives the time and space scales for several atmospheric features. The concept of scale is somewhat imprecise; therefore the actual sizes or time periods of phenomena which belong to the same scale may vary considerably, perhaps by as much as one order of magnitude (factor of ten). Thus, in a sample of one hundred thunderstorms, individual storms may vary in diameter between 1

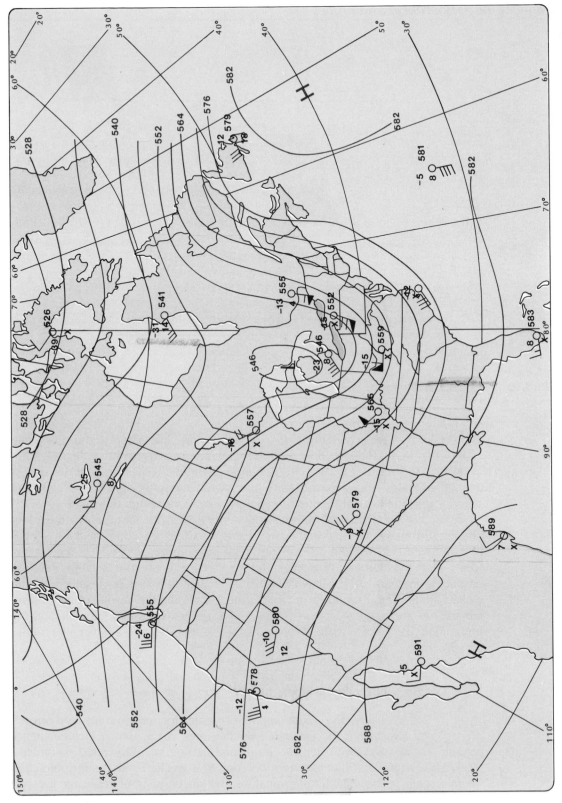

FIGURE 3.11 Height contours (decameters) of 500-millibar pressure surface for 7 P.M. EST on 23 October, 1979.

65

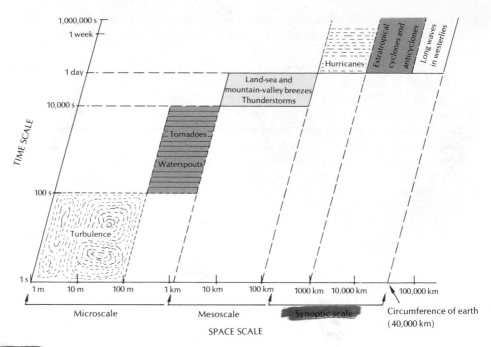

FIGURE 3.12 Horizontal scales of atmospheric motions.

and 8 kilometers and may last from thirty minutes to several hours. Yet, all of these thunderstorms have a horizontal space scale of a few kilometers and a time scale of about an hour. The important point is that thunderstorms never go through a life cycle in a second, nor does a single thunderstorm ever cover an area the size of the United States. Thunderstorms have their own niche in the spectrum of time and space, even if individual thunderstorms show variations within this niche.

As another example of a class of weather phenomena which is identifiable with particular time and space scales, consider the extratropical cyclones (low-pressure systems) and anticyclones (high-pressure systems) which affect many states over a period of a few days. To forecast the evolution of these weather-producing systems, we deal with horizontal scales that can be resolved on the weather map. These *synoptic,* or large, scales determine the density of the observing stations and range from several hundred to several thousand kilometers in the horizontal. The vertical scale of these weather systems is quite a bit smaller, about 10 kilometers.

3.3.1 THE SYNOPTIC SCALE

Probably the most important scale for determining tomorrow's weather is the synoptic, weather-map, scale, which includes atmospheric phenomena with typical horizontal scales of 800–8000 kilometers (500–5000 miles) and lifetimes from one day to a week. The basis for prediction on this scale is the weather-map analysis, which consists of reducing the vast amounts of meteorological data (pressures, temperatures, winds, humidities, clouds, precipitation) from hundreds of weather stations into meaningful patterns that can be interpreted by the meteorologist.

A familiar example of a weather analysis is the sea-level-pressure chart shown in figure 3.13 for December 25, 1973. By plotting the sea-level pressure for each station and drawing lines (isobars) through points of equal pressure (estimating values between stations, if necessary), a large number of data that would be virtually meaningless in tabular form are organized into patterns which reveal much about the weather. One purpose of drawing isobars is to summarize the enormous amount of information available on the map and to produce a visualization of large-scale pressure patterns; the second purpose is to isolate the large-scale features from features of a smaller scale, which are eliminated by drawing smooth lines. In figure 3.13 the isobars were required to fit exactly each observation; the lines are irregular, with many wiggles, especially in the Ohio Valley and the Rocky Mountains. One reason for the wiggles are inaccurate observations, but small-scale motions also contribute to the irregularity. Smoothing eliminates both errors and unwanted small scales. In figure 3.14 small-scale variations in the pressure pattern, unimportant for synoptic-scale analyses, have been removed.

In the smoothed sea-level-pressure chart, we see several examples of synoptic-scale features. A region of high pressure covers New England. The diameter of this "high" is roughly 30 degrees of latitude (3330 kilometers). A minimum of pressure occurs over Iowa, and the scale of this "low" is about 10 degrees of latitude (1110 kilometers). Another elongated low of about the same scale is centered near the Montana-Canada border.

If we were to look at the surface-pressure map for an hour later than the map shown in figure 3.14, we would see very small changes in the large-scale patterns, indicating that pressure systems with horizontal scales of 1000–3000 kilometers have time scales longer than an hour. On the other hand, if we looked at the surface map a month later, the highs and lows of December 25, 1973, would have disappeared, indicating that the time scale is less than a month. Finally, if we looked at hourly pressure maps and followed the movement of each particular high and low, we would find that they persist as identifiable features for two to five days. Thus, the time scale of highs and lows with horizontal scales of 1000 to 3000 kilometers is on the order of a few days.

Figure 3.15 shows the 500-millibar-height contour analysis for the same time as the surface-pressure analysis of figure 3.14 (December 25, 1973). The lines of equal elevation are labeled in feet. A comparison of the upper-air analysis with the surface analysis shows two important points. First, the height contour patterns have some similarity to the surface pressure patterns (note the relatively high heights—a *ridge*—over New England and the low heights—a *trough*—over the Midwest) but are generally smoother and more wavelike, with fewer closed centers of high or low heights. Second, the wave patterns in the height contours are about the same scale as the surface pressure patterns; the wavelength of the trough-ridge system over the United States is about 50 degrees of latitude, or 5550 kilometers. Again, if we were to watch these patterns from hour to hour, we would be able to follow individual troughs and ridges for several days before they lost their identity. So, the time and space scales of the upper-air

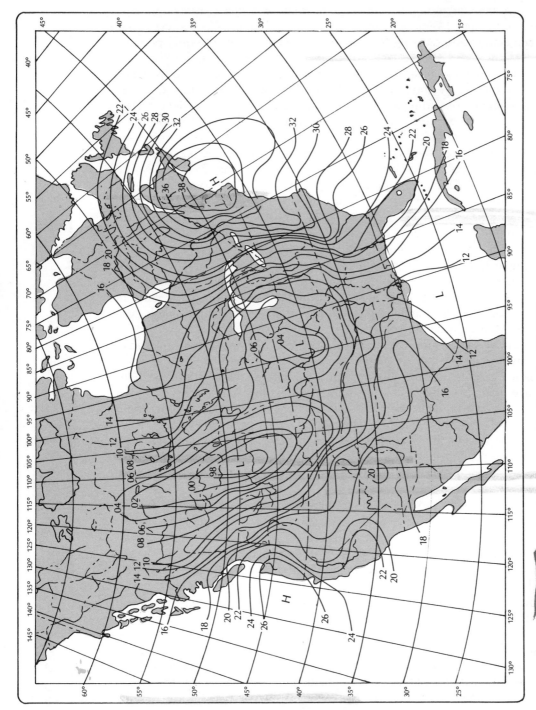

FIGURE 3.13 *Unsmoothed sea level pressure map for December 25, 1973.*

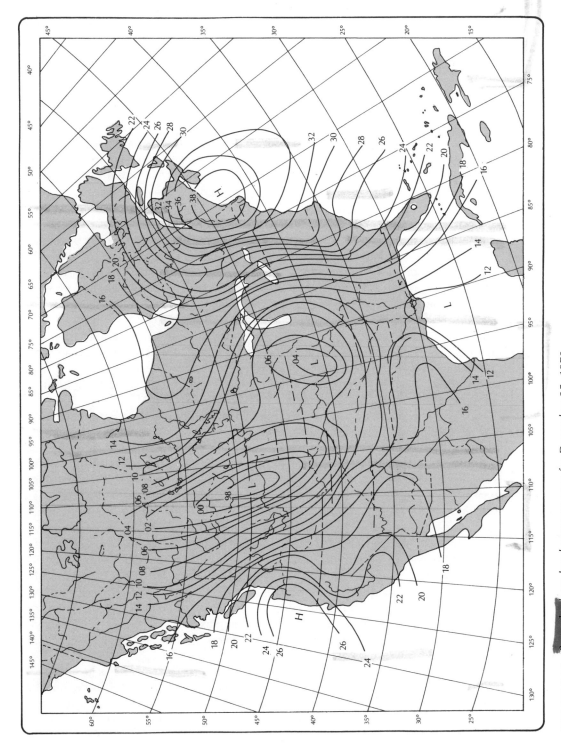

FIGURE 3.14 Smoothed sea level pressure map for December 25, 1973.

features are roughly the same as the smoothed highs and lows on the surface map. The evolution of weather systems of this scale is thus easily studied by having weather stations separated by about 250 kilometers and by taking observations every twelve hours; in fact, these are the space and time densities of upper-air observations over the United States.

3.3.2   THE MICROSCALE: TURBULENCE

Small-scale motions do not appear on large-scale weather maps, but they do exist. These *microscale* motions (*eddies*) are frequently chaotic, with dimensions from a centimeter to several hundred meters, and with time scales from a second to several minutes. This partcicular class of wildly fluctuating motions is called *turbulence*. Turbulence produced the rapid fluctuations in temperature and wind shown in figures 2.6, 2.8, and 2.9. One type of turbulence, *thermal turbulence*, is caused by heating from below, which leads to rising columns of air called *thermals*, or *convection currents (cells)*. When these convection cells contain enough moisture, condensation may occur, and we see cumulus clouds. Another form of turbulence is *mechanical turbulence* caused by air flowing over rough terrain. This turbulence is associated with the rapid change of wind with altitude which occurs just above the ground. The gusty nature of surface winds is caused by either mechanical or thermal turbulence.

The importance of small-scale turbulence lies in its ability to gradually change the large-scale conditions, especially close to the ground. Small-scale turbulence warms the air by bringing up heat from the surface, it adds moisture, and it slows down the average wind by mixing it with the slower-moving air from lower levels.

Even though the physical relationships discussed earlier theoretically can be used to describe the motions of each turbulent eddy, we are clearly incapable of observing, visualizing, and analyzing the millions of eddies forming and dying each minute over the earth. The representation of the effect of the rapidly varying small-scale motions on the slowly varying large scale is a problem which has not yet been solved satisfactorily. Because it is impossible to describe each individual eddy, we attempt to relate the total small-scale transport of heat, moisture, or velocity to some easily measured property of the large-scale weather system. For example, consider a bathtub with hot water on one side and cold on the other. If the water is stirred, heat is transferred by small-scale eddies from the warm to the cold side. It makes sense to assume that the amount of heat transported by these eddies is proportional to the temperature difference between the hot and cold water; the larger the temperature contrast, the larger the heat transport. This simplified assumption, which reduces a very complex physical problem to a single parameter (the difference in temperature between the hot and cold water), is an example of *parameterization*.

Meteorologists frequently use similar approximations, or parameterizations. For example, it is assumed that the evaporation rate at the surface of the ocean is proportional to the difference between the amount of water vapor just above the ocean and the amount of water vapor in the air higher up.

70

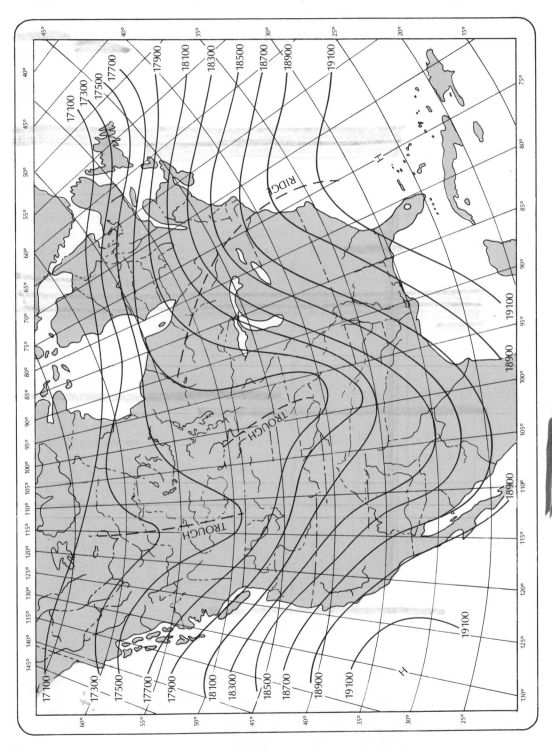

FIGURE 3.15 Height contours (in ft) of 500-millibar pressure surface for 7 A.M. EST on December 25, 1973. [m = ft (0.305)]

71

Unfortunately, not all small-scale motions can be treated as chaotic turbulence. For example, sea-breeze circulations, tornadoes, or local circulations caused by cities do not act as turbulence does but instead are quite organized. These systems belong to a third general class of scales, the *mesoscale*, or middle scale.

3.3.3 THE MESOSCALE

Between the large (weather-map, *synoptic*) scale and the smaller turbulent scale (*microscale*) lies the mesoscale. A typical example of motion on this scale is the *sea breeze*, which has horizontal dimensions of about 20 kilometers and a time scale of a day. Near the surface, air flows from the sea to the land, the sea breeze. At roughly 2 kilometers above the surface, a return flow of air is directed from land to sea. Another mesoscale circulation is the flow created by the differences in temperature between urban and rural areas. Still another example is the *squall line*, which is a system of large thunderstorm clouds in a line perhaps 200 kilometers long.

Mesoscale motions are especially difficult to study. For large-scale flow, we can ignore vertical accelerations, while in mesoscale flow, we cannot. For small-scale flow, we need not consider the earth's rotation; for mesoscale flow, we must. These problems in studying the mesoscale quantitatively, along with the difficulties in obtaining observations on this scale with conventional radiosonde balloons, have hampered research. Therefore, we know more about both the synoptic scale and the microscale than we do about the mesoscale. However, important weather phenomena, such as air-pollution episodes, thunderstorms, and tornadoes, are associated with mesoscale circulations. With advanced remote-sensing equipment to provide the detailed data necessary to resolve these mesoscale features, and with bigger and faster computers to process the vast amounts of data, there is hope for understanding this important scale.

3.3.4 INTERACTION OF SCALES OF MOTION

At a given time, meteorological observations may be influenced by many scales of motion. For example, the wind at a station may be controlled by turbulence, mountain waves, sea breezes, or convection cells, as well as by the much larger map structures—cyclones, anticyclones, and fronts. Thus, although for simplicity we frequently consider only one scale at a time, in reality all of the scales of motion are interacting, the small scales affecting the large scale and vice versa. As an example of scale interaction, consider a large body of cold, dry air moving over much warmer water, as happens frequently off Cape Hatteras, North Carolina, in winter. When cold air flows over a warm surface, the air becomes unstable. Convection cells that are quite small, perhaps 500 meters in horizontal diameter, develop. These cells bring dry, cold air down and moist, warm air up, with the result that the initially dry and cold air is warmed and filled with water vapor. The small-scale motion has affected the large-scale distribution of temperature and humidity.

The interaction of different scales of motion is one of the most difficult problems of quantitative meteorology, for it is impossible to treat numerically all relevant scales, which range from a centimeter to thousands of kilometers in space, and from a second to months in time. Because only one particular scale can be considered in detail at a time, the combined effects of all other scales must be approximated in some manner. Sometimes these approximations work, and the forecast rain arrives on time. At other times, they do not work very well, and the forecast rain becomes 30 centimeters of snow.

Questions

1   What were the temperature and dewpoint at 500 mb over Dayton, Ohio, at 7 P.M. EST October 23, 1979 (use figure 3.1a)?

2   What causes the warm stratopause (see Figure 3.2)?

3   What is the average temperature in July at 60°N at 700 mb pressure (see figure 3.3)?

4   What is the average height of the tropopause over the Equator in July (see Figure 3.3)?

5   If the height of the 500-mb surface is higher than normal, the pressure on the 5.5-km-height surface is probably higher or lower than normal at this location?

6   Describe the chemical reaction that forms ozone.

7   If you were flying at a pressure level of 1 millibar, what percent of the total atmosphere would be beneath you?

8   Interpret (decode) the weather symbols below.

9   Name three mesoscale weather phenomena.

Problems

10   Using the stations plotted on figure 3.11, plot a graph of temperature *vs.* height of the 500-mb surface. Let the height range from 5200 to 6000 m. What does this graph tell you about the correlation between temperature and height of the 500-mb surface?

11   Streamlines are lines which are everywhere tangent to the winds, and therefore depict the direction of the flow. Using a light-colored pencil, draw some streamline on the surface weather map (figure 3.7). Compare your streamlines in the vicinity of the low over Michigan with the early models shown in figure 1.3.

FOUR

Radiation

radiation • waves • wavelength • frequency • absorptivity •
emissivity • Kirchoff's law • Planck's law • Stefan-Boltzman law •
Wien's law • absolute zero • infrared • visible • cosine law •
atmospheric effect • greenhouse effect • scattering • absorption •
reflection • conduction • seasons • perihelion • aphelion • obliquity
of the ecliptic • solstice • equinox

One of the most striking features of the atmosphere is the fact that the air moves—the wind blows and clouds race across the sky. The constant motion of the air involves vast amounts of energy. Can you imagine the size of the fan required to move a cloud the size of a city?

What is the source of the energy that keeps the giant machine of our atmosphere running? The answer is, obviously, the sun; but how the sun makes the wind blow becomes the story of radiation in the atmosphere.

The major links relating the sun to the winds in our atmosphere are easy to present. The imbalance in how radiative energy is lost and gained by various parts of our planet results in regions of relative warmth and coolness. There are latitudinal variations—the tropics are much warmer that the poles—and altitudinal variations—the lower layers of the atmosphere are much warmer than the higher layers. Like a pan of water being heated only on one side, these temperature imbalances cause convection. Latitudinal variation causes convection in the form of the large cyclonic storms through the midlatitudes, and altitudinal variation causes convection in the form of cumulus clouds and thunderstorms. Exactly how temperature variations cause winds and storms will be the subject of later chapters; this chapter is concerned with how radiation produces temperature variations.

4.1 WHAT IS RADIATION?

To most people, the word *radiation* is associated with atomic processes such as those in nuclear power plants or atomic bombs. However, radiation is not confined to intense nuclear reactions of this kind but instead represents the transfer of energy associated with a wide variety of familiar experiences, including radio, television, microwave ovens, and most importantly, sunlight. Radiation can be generally defined as the transfer of energy by the rapid oscillations of electromagnetic fields in space. These oscillations can be considered as traveling waves, with a characteristic wavelength (distance between successive crests or troughs), frequency (number of wave crests passing a point per time, or the number of oscillations per time), and speed (rate of travel of an identifiable part of the wave, such as the crest). The wavelength, frequency, and speed of a wave are illustrated in figure 4.1.

It may seem surprising that radio waves, television waves, microwaves in ovens, sunlight, and X rays are basically the same phenomenon—electromagnetic waves capable of transmitting energy through a vacuum. The basic

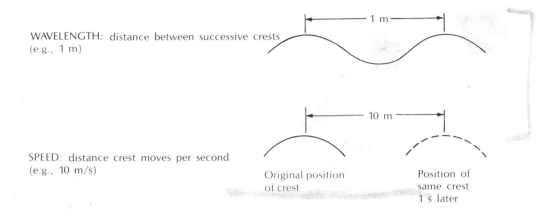

WAVELENGTH: distance between successive crests
(e.g., 1 m)

SPEED: distance crest moves per second
(e.g., 10 m/s)

Original position of crest

Position of same crest 1 s later

FREQUENCY: number of crests passing a given point per second; in above example, 10/s. In general, frequency = speed/wavelength

FIGURE 4.1 *Illustration of wavelength, frequency, and speed of a wave.*

difference between these forms of radiation is the wavelength, as indicated in figure 4.2. Also, because the speed of all these waves is the same [3×10^8 meters per second (m/s), equal to the speed of light], the frequency of each type of wave varies. For example, the frequency of AM radio waves of wavelength 500 meters is 6×10^5 oscillations per second. The frequency of visible radiation, whose wavelength is 0.5×10^6 meter, is 6×10^{14} oscillations per second.

4.1.1 *PROPAGATION OF ENERGY THROUGH A VACUUM*

Even though we are all familiar with the propagation of waves on lakes and oceans, it is admittedly difficult to visualize waves propagating through a vacuum. Indeed, it is tempting to envisage, as did some early scientists, a sort of mysterious plasma or ether filling the voids of space as water fills an ocean. However, if we accept the fact that gravitational forces act through a vacuum, or that a magnet can move a compass needle which is isolated in a vacuum, we can visualize more readily the concept of oscillating electromagnetic forces generated by vibrations of electrical charges at a point and act through empty space. The propagation of electromagnetic radiation through a vacuum and its subsequent effect on terrestrial matter are illustrated in figure 4.3. The oscillatory force field, which is the propagation of electromagnetic waves, is capable of causing charged electrons or other particles to move, thereby transferring energy to these particles. The rapid vibrations of the extremely hot [about 6000 kelvins (K)] molecules in the solar atmosphere create an oscillating electromagnetic force field, which propagates through empty space at the speed of light. When the electrical charges in the molecules of a gas, the ground, or our skin are subjected to this oscillating force, they begin to vibrate faster, increasing their *kinetic energy* (en-

77

ergy of motion). (This effect is somewhat analogous to waving a magnet back and forth near a compass needle that is isolated in a vacuum within a glass jar; the oscillating magnetic field causes the compass needle to swing, thereby transferring energy through the vacuum.) We feel this increase in kinetic energy as a rise in temperature. Thus, radiation acting through a vacuum can transmit heat energy from one group of molecules (the sun) to another group (the atmosphere, ground, or our bodies).

4.1.2 INTENSITY OF RADIATION

All objects are sources of radiation, but the amount and kind of radiation each object emits depend upon the temperature and emissivity of the object, which are measures of how efficiently the object emits radiation. The emissivity is defined as the ratio of the actual radiation emitted by a body to the maximum possible amount of radiation that could be emitted at the same temperature. Another way of describing the same concept is to use the absorptivity, which is a measure of the efficiency of a body in absorbing radiation. According to Kirchhoff's law, the absorptivity and emissivity of a body are equal; a good emitter is a good absorber. In the visible wavelengths, black coal dust is a good absorber (reflects very little visible sunlight). White snow, on the other hand, reflects most of the incident radiation in the visible range. A perfect absorber/emitter is called a black body and has an emissivity of 100 percent. A perfect reflector has an emissivity of 0.0 percent. In many applications, it is assumed that radiation is emitted by a black body; however, the emissivities of most objects and gases vary from about 10 to 95 percent.

An object or substance that is a good (poor) absorber of certain wavelengths is not necessarily a good (poor) absorber of all wavelengths. Thus, fresh snow is very nearly a white body (emissivity 5 percent) in visible wavelengths, whereas it is almost a black body (emissivity 98 percent) in infrared wavelengths.

If we wish to calculate the radiation emitted by a square meter of any body, we first calculate the radiation emitted by a perfect radiator (black body) of the same temperature and then multiply it by the emissivity. For the sun, and for most solids and liquids (even thick clouds), the emissivity is almost equal to one, so the black-body radiation equals the total radiation. However, the variation of the emissivity of air with the wavelength is quite complicated. At some wavelengths, particularly between 8 and 11 micrometers,* the emissivity is near zero. In this wavelength region, therefore, air emits (and absorbs) hardly any radiation; air is transparent to these wavelengths. At other wavelengths, for instance near 15 micrometers, the emissivity is almost one, and air radiates as a black body. In these wavelengths, air absorbs perfectly and is opaque (transmits no radiation).

The distribution of radiation emitted by a black body with wavelength, at a given temperature, is given by Planck's law. Figure 4.4 shows Planck's law in graphical form. On this graph, the temperature is given in absolute degrees,

---

*A micrometer (μm) is 10^{-6} meter, or 10^{-4} centimeter.

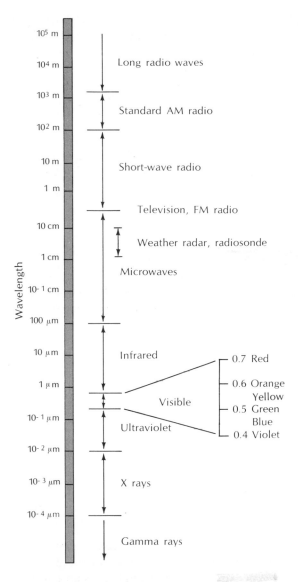

FIGURE 4.2 *Electromagnetic spectrum.*

or kelvins (K), found by adding 273 to the number of degrees Celsius. Temperatures in kelvins are never negative, and a temperature of zero on this scale is called *absolute zero*. If a body could be cooled to this temperature (an unattainable state), all molecules would stop moving.

Figure 4.4 shows that the total amount of radiation (the area under the curves) is much larger for hot bodies than for cold ones. The change of radiation intensity with temperature is even greater than casual inspection of figure 4.4 would indicate, because the graph for the temperature of the sun is not drawn on the same scale as the graphs for the temperatures on earth. Actually, the ratio of the radiation from a black body of a temperature of 6000 K (nearly the sun's temperature) to that from a black body of 300 K (the temperature of rather warm

79

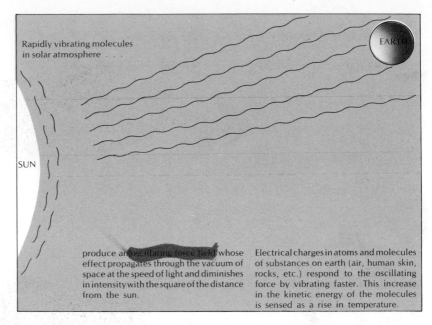

Rapidly vibrating molecules
in solar atmosphere . . .

EARTH

SUN

produce an oscillating force field whose
effect propagates through the vacuum of
space at the speed of light and diminishes
in intensity with the square of the distance
from the sun.

Electrical charges in atoms and molecules
of substances on earth (air, human skin,
rocks, etc.) respond to the oscillating
force by vibrating faster. This increase
in the kinetic energy of the molecules
is sensed as a rise in temperature.

FIGURE 4.3   *Propagation of radiation through a vacuum.*

ground) is not equal to 20, the ratio of these two temperatures, but 160,000, the fourth power of 20! This fourth-power law was formulated before Planck's law and is known as the *Stefan-Boltzman law*.

4.1.3 VARIATION OF WAVELENGTH OF MAXIMUM RADIATION WITH TEMPERATURE

Figure 4.4 shows that the radiation distributions have rather sharp maxima. At certain wavelengths, the radiation is stronger than at either longer or shorter wavelengths. Wien's law describes the relationship between the wavelength at which the maximum amount of radiation is emitted and the temperature of the body (figure 4.4). The wavelength of maximum radiation is inversely proportional to the absolute temperature, so the higher the temperature, the shorter the wavelength of maximum radiation. The outer regions or layers of the sun which emit sunlight have a temperature of about 6000 K. The wavelength of the maximum solar radiation corresponds to visible light (0.4–0.7 μm). Because temperatures on the earth are much lower than the solar temperature, the maximum terrestrial radiation occurs at longer wavelengths. Thus, the earth, the atmosphere, and we ourselves emit the maximum amount of radiation in the wavelengths of the *infrared* (8–15 μm) (figure 4.4). It is important to note that although the maximum earth radiation occurs at around 10 micrometers, and the maximum solar radiation occurs at about 0.5 micrometer, the sun emits more radiation at all

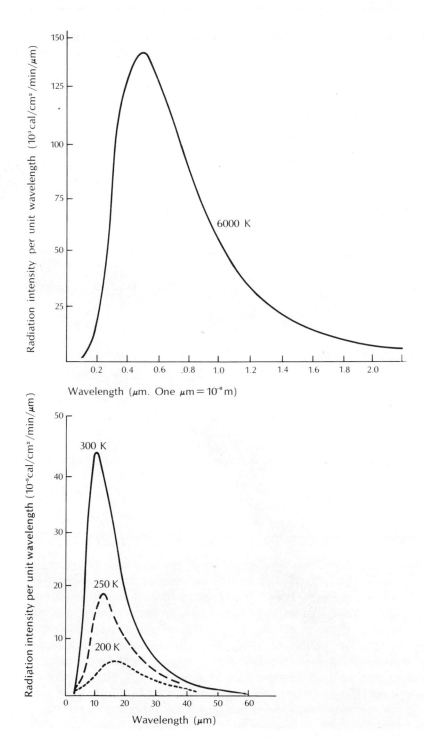

FIGURE 4.4 *Variation of black-body radiation intensity as a function of wavelength for (a) 6000 K, which corresponds to the solar temperature, and (b) 300, 250, and 200 K, which correspond to temperatures found in the earth's atmosphere.*

wavelengths than the earth. (Note the difference of six orders of magnitude—one million—in the scale of the radiation intensity for the solar and the terrestrial radiation curves.)

It is interesting that our eyes have evolved to have maximum sensitivity at the wavelengths of maximum solar radiation. Had early humans been more dependent upon hunting warm-blooded animals than on the diversity (and visual acuity) demanded by cave dwelling, evolutionary changes might have developed for them "eyes" capable of seeing in the infrared, as some snakes have. Although we have not developed a natural visual sensitivity to the infrared wavelengths, we have developed artificially a variety of infrared detectors which enable us to observe the long-wave radiation of the atmosphere. These infrared sensors, carried aloft by earth-orbiting weather satellites, are now used to estimate atmospheric temperatures over the entire globe.

FIGURE 4.5 *Infrared photograph of eastern North America at 12 Noon EST on April 22, 1980.*

An example of the use of infrared sensors on satellites is shown in an infrared photograph over eastern North America during an early season heat wave on April 22, 1980 (figure 4.5). At the time this photograph was taken (12 noon EST), the surface air temperatures over the Great Lakes region ranged from 75 to 90°F (24 to 32°C). At this time of year, water temperatures at the surface of the Great Lakes are about 45°F (7°C). In the infrared photograph, cold objects (lake water and clouds) appear white, and warm objects (ground) appear dark. Thus the cold Great Lakes and the band of clouds across western Pennsylvania, eastern Ohio, and West Virginia are light colored, but the hot ground appears dark.

Figure 4.6 shows a visible photograph for the same time as the infrared photograph of figure 4.5. The visible photograph shows reflected sunlight, providing a measure of the albedo, or reflectivity, of objects below rather than the temperature. Since the albedo of water (about 4 percent when the sun is high in

FIGURE 4.6 *Visible photograph of eastern North America at 11:30 A.M. EST on April 22, 1980.*

the sky) is the lowest of commonly occurring surfaces (table 4.1), the Great Lakes and Atlantic Ocean appear darkest in the photograph. The ground surface, with a typical albedo of 15, appears somewhat lighter, and clouds, with an albedo ranging from 25 to 75 (table 4.1) appear the brightest.

The seemingly esoteric concepts of emissivity and variation of the wavelength of maximum radiation with temperature according to Wien's law have some very important consequences for our weather. For example, snow absorbs very little incoming solar radiation, which is strongest in the visible wavelengths. However, at the temperature of snow, the maximum radiation occurs in the longer wavelengths of the infrared, a region of the spectrum in which snow absorbs (and emits) very well. Thus, snow gains little heat energy during the day by absorption of solar radiation but loses energy rapidly at night by effectively emitting infrared radiation to space. This difference in emissivity at long and short wavelengths is the major reason why snow has such a cooling effect on the earth and atmosphere.

4.1.4 GEOMETRIC RADIATION LAWS

The sun is a large spherical body which emits radiation in all directions into space. As the distance from the sun increases, the spherical area over which a given amount of energy is distributed becomes larger and larger, so the energy received per unit area and per unit time decreases as the square of the radius from the sun (see figure 4.7). Earth, at a distance of 150 million kilometers (km), or 93 million miles, from the sun, receives about 2 calories of heat per square centimeter per minute (the solar constant) at the top of its atmosphere, on a surface perpendicular to the line between the earth and the sun. This constant means that a square measuring 1 centimeter (cm) on a side would receive 2 calories (cal) of heat every minute (min), or enough to raise the temperature of water 1 centimeter deep over this square 2 degrees Celsius every minute. The planet Mercury, located at a distance of only 58 million kilometers from the sun, receives about $(150 \text{ million}/58 \text{ million})^2$, or 6.7 times as much energy per unit area and time as the earth, while Jupiter, at a radius of 778 million kilometers, receives only $(150 \text{ million}/778 \text{ million})^2$, or 0.037 times as much.

The above dependence is known as the *inverse square law* for radia-

TABLE 4.1 *Typical albedo of terrestrial surfaces*

Surface	Percent of visible radiation reflected
Fresh snow	90
Sand	25
Bare ground	15
Forest	7
Water (sun high in sky)	4
Water (sun near horizon)	50
Thick clouds	75
Thin clouds	25
Earth surface (average)	6

Plate 3a   Sea smoke over Lake Mendota, Wisconsin (© Richard A. Anthes).

Plate 3b   Bands of cirrus clouds forming from contrails (© Alistair B. Fraser).

Plate 4a   Early morning ground fog near Stormstown, Pennsylvania (© Dennis W. Thomson).

Plate 4b   Advection fog moving inland from the Pacific off the coast of California (© Dennis W. Thomson).

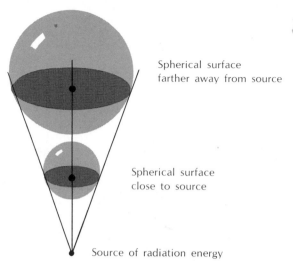

FIGURE 4.7 *Illustration of* ~~*inverse square law.*~~

Spherical surface
farther away from source

Spherical surface
close to source

Source of radiation energy

tion: If we increase the separation of a body from a radiation source from an initial distance r_1 to a final distance r_2, the radiation intensity reaching this body will be reduced by the factor ~~$(r_1/r_2)^2$.~~

The radiation intensity received from a source depends not only on the intensity emitted and the distance from the source but also on the angle at which the radiation strikes the surface. If the radiation arrives at right angles to the surface, the radiation energy received by a given area is a maximum. The more oblique the incident radiation, the smaller the amount of energy received. For this reason, the ground is warmed very little by sunlight near sunrise and sunset and is heated most strongly at local noon. Also, winter is colder than summer because the sun is generally low in the winter sky.

The effect of the sun's elevation on the intensity of radiation received at the ground can be seen in figure 4.8. When the sun is high, a given beam of sunlight is concentrated on a relatively small area, so the radiation energy received per unit area is large. When the sun is low, the same amount of radiation falls on a much larger area, and hence the radiation energy received per unit area is much less. This relationship is called the ~~cosine law~~, which states that the radiation energy received by a unit area depends on the cosine of the angle between the incoming beam and the line perpendicular to the surface.

RADIATION AND THE EARTH'S TEMPERATURE 4.2

The main characteristics of radiation have now been established. Everything emits radiation. The warmer an object is, the more radiation energy it emits, and the shorter is the wavelength that characterizes that radiation. The farther a receiver is from the radiation source, the less radiation is received. The more obliquely the radiation strikes the receiving surface, the less radiation (per unit area) is received. It is now possible to understand the average temperature and temperature distribution on the earth.

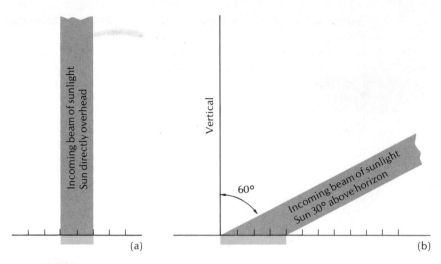

FIGURE 4.8 *Illustration of the cosine law. A beam of sunlight is spread over an area twice as large when the sun is 60 degrees from the vertical as when the sun is directly overhead.*

4.2.1 AVERAGE TEMPERATURE OF THE EARTH

It has already been noted that the earth receives about 2 calories of radiative heat per square centimeter per minute at the top of the atmosphere, on a surface perpendicular to the line between the earth and the sun. This number, however, does not represent the average amount of energy received from the sun, for it does not take into account that nothing is received at night and that much less is received when the sun is low in the sky. These factors lower the average value of received radiation only 0.5 calories per square centimeter per minute. Not even all of this energy is available to warm the earth, though; about 30 percent is reflected back out to space, mainly by clouds. The fraction of the radiation that is reflected is known as the *albedo*, so that for an albedo of 30 percent (the global average), there are only about 0.35 calories per square centimeter per minute available to be absorbed by the earth.

The earth must be losing energy to space in the form of radiation at exactly this same rate for the average temperature of the earth to remain constant. If the earth were not radiating at all, then a great deal of energy would be received each day and none would be lost, so that the earth would get warmer and warmer. If we were to pretend that earth radiated just a little bit less energy than it received, then there would still be an imbalance; with more energy arriving than being lost, the temperature would slowly climb. But as the temperature went up, the amount of radiation that the earth emitted would also go up until a balance between incoming solar radiation and outgoing terrestrial radiation was reached, at which time the average temperature of the earth would remain constant.

Thus on the average, the earth must emit the same amount of radiation that it receives, about 0.35 calories per square centimeter per minute. It is not a difficult task to compute just how warm the earth must be to be able to emit this much energy—about 256°K or about −17°C. This is a little startling at first because common experience tells us that the average surface temperature is much higher than this—more like +15°C. The discrepancy is easily resolved when we realize that the figure of −17°C characterizes the whole atmosphere, not just the part we walk around in; much of the upper atmosphere is, of course, very much colder, so an average value of −17°C is not unreasonable. It does leave us with a problem, however: why are the lower portions of the atmosphere so very much warmer than most of the rest of it.

4.2.2 THE ATMOSPHERIC EFFECT

If the earth had no atmosphere but still had the same albedo as it does now (a contradictory requirement since much of the albedo results from clouds in the atmosphere), then the average ground temperature would be about 256°K or −17°C. With an atmosphere, the surface temperature is much warmer because the ground not only receives radiation from the sun, but also from the atmosphere itself. This is what will be called the *atmospheric effect*.

To see how the atmospheric effect works, let us imagine that the earth's atmosphere is entirely transparent to solar radiation so that, apart from the portion reflected by the clouds, all sunshine penetrates down to the ground and warms it. We will also imagine that the earth's atmosphere is entirely opaque to the long-wave radiation emitted by the surface of the earth, so that all of the earth's radiation is absorbed by the atmosphere. (We will see shortly that neither of these statements is entirely true, but for the moment they serve to simplify the argument.) The atmosphere not only receives radiation from the earth, it also emits the same amount of radiation, because its average temperature is constant. The radiation emitted by the atmosphere is sent in all directions so that some is emitted out to space, but some of it is emitted back down to the earth and is absorbed there; so the ground is receiving radiation not only from the sun but also from the atmosphere and, therefore, has a higher temperature than it would have had in the absence of the atmosphere. This is why we enjoy an average surface temperature of about +15°C instead of −17°C. The fact that the average temperature of the bulk of the atmosphere is about −17°C, whereas the surface temperature is about +15°C, is the basic reason for the decrease of temperature throughout the troposphere.

It is amusing to realize that the atmospheric effect usually goes by another, but erroneous, name, the *greenhouse effect*. It was once believed that a greenhouse stays warmer than the surrounding air because its glass roof allowed the sunlight to pass right through but was opaque to the longer wave radiation emitted by the plants and ground inside. This is the same argument just given for the atmosphere. Actually, the main reason that the greenhouse stays warm is because the glass prevents the warmer air inside from mixing with the cooler air outside. The term *greenhouse effect* is very misleading and should be avoided.

While the basic reason for the decrease of temperature with height in the troposphere has been explained by a very simple and approximate radiation model, it is desirable to describe the behavior of radiation in the atmosphere in somewhat greater detail. We will now abandon the pretenses that all of the solar radiation not reflected reaches the ground and that all of the earth's radiation is absorbed by the atmosphere. These ideas were useful to enable us to establish the basic altitudinal dependence of temperature, but they are simplifications.

4.3 THE INTERACTION OF RADIATION AND THE ATMOSPHERE

Radiation interacts with the atmosphere in a more complex way than suggested up until this point. In this section, this interaction will be discussed in detail sufficient to explain the energy budget of the earth and the atmosphere.

4.3.1 SCATTERING, ABSORPTION, AND REFLECTION OF SOLAR RADIATION BY THE ATMOSPHERE

Radiation does not pass unimpeded through the atmosphere; part is reflected back to space, part is scattered or absorbed by the atmosphere, and part reaches the ground. Both the scattering and absorption of radiation in the atmosphere depend directly upon the electronic properties of the gas molecules, and upon the size of the molecules and any natural or artificial dust particles present in the gas.

Oxygen (O_2) and ozone (O_3) are gases which most effectively absorb, and thus deplete, the sun's radiation as it passes through the atmosphere (figure 4.9). In the ultraviolet wavelengths, the absorption efficiency is so high that virtually no radiation is transmitted. The process by which ozone absorbs dangerous ultraviolet radiation in the stratosphere was discussed in section 3.1.2.

The scattering of solar radiation by gas molecules is the reason for the brightness of the daytime sky; planets without atmospheres have black skies, even during the day. Scattering molecules distribute energy in all directions, forward and backward. The scattering is very strongly dependent upon the size of the scattering particle or molecule. Particles or molecules much smaller than the wavelength of light (0.5 micrometer) produce a scattering of the sunlight that is inversely proportional to the fourth power of the wavelength, which means that shorter wavelengths are scattered much more than longer wavelengths.

Atmospheric molecules are much smaller than the wavelengths of visible light. Therefore, on clear days, the shorter wavelengths of the blue and violet colors are scattered more than the red and yellow colors from the incoming beam of "white" sunshine (figure 4.10). Thus, the clear sky appears blue. Haze,

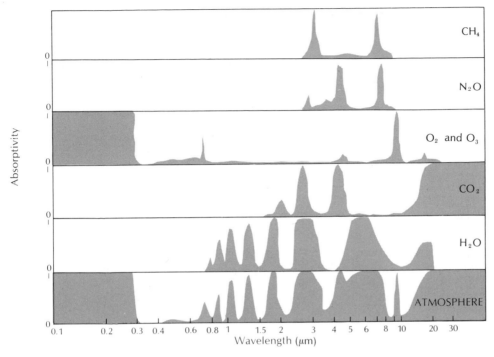

FIGURE 4.9 *Absorptivity as a function of wavelength for methane (CH$_4$), nitrogen oxide (N$_2$O), oxygen (O$_2$), ozone (O$_3$), carbon dioxide (CO$_2$), water vapor (H$_2$O), and the entire atmosphere.*

fog, and cloud droplets, however, have diameters as large as or larger than the wavelength of sunlight, and they do not show a preference for scattering any particular wavelength; all are scattered equally. Thus, when the atmosphere is hazy or cloudy, the sky appears white.

The efficiency of scattering is increased when large amounts of natural or artificial dust particles are in the air. The explosion of the volcano Krakatoa in the East Indies in 1883 produced beautiful sunsets for several years over the entire globe. Frequently now, in highly polluted air, so much scattering and absorption occur that the astronomical sunset is not red or dirty yellow, but, in fact, ceases to exist. For the earthbound observer, the sun may disappear in the murky sky as much as 5 or 10 degrees above the true horizon.

4.3.2 THE ABSORPTION OF TERRESTRIAL RADIATION BY THE ATMOSPHERE

As we have seen, the earth emits very nearly as much radiation to space as it receives from the sun; otherwise, the temperature of the earth and the atmosphere would not remain so constant. The earth receives most of its radiative energy in the visible wavelengths but emits most of the energy in the longer infrared wavelengths (4−70 micrometers).

Figure 4.9 shows the absorptivity as a function of wavelength for some of the important gases in the atmosphere. For example, we have already seen that ordinary oxygen and ozone absorb strongly in the short wavelengths (between 0.1 and 0.3 micrometer) but are relatively transparent (have small absorptivities) in the longer wavelengths. The sum of the absorptivities by the various gases in the atmosphere determines the absorptivity as a function of wavelengths for the entire atmosphere, shown in the bottom of figure 4.9.

While the atmosphere is relatively transparent to radiation in the visible wavelengths, the absorptivity in various bands in the infrared range can be very large. The gases which absorb significant amounts of terrestrial radiation include methane (CH_4), nitrous oxide (N_2O), carbon dioxide (CO_2), and water vapor (H_2O). Most of the important absorption in the infrared range is accomplished by carbon dioxide, a relatively constant component of the atmosphere, and water vapor, which is highly variable. The net effect of these gases is to absorb the radiation in most of the long wavelengths. On the other hand, certain wavelength bands, called *windows*, exist in which essentially all the infrared radiation is transmitted through the atmosphere. These windows occur around 8 and 11 micrometers (see figure 4.9).

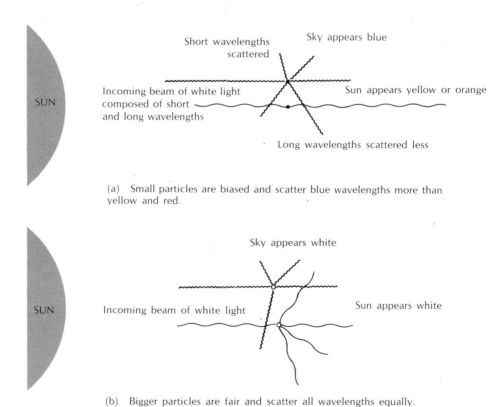

FIGURE 4.10 *Scattering of sunlight by small and large particles. (a) Scattering by molecules in the atmosphere produces blue sky. (b) Scattering by larger dust, haze, or cloud droplets produces white sky.*

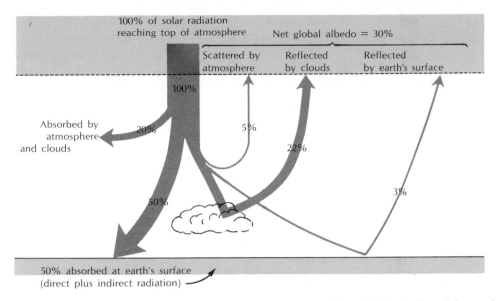

FIGURE 4.11 *Solar radiation budget of the earth and atmosphere.*

The high absorptivity of radiation in the infrared by water vapor and the huge variability of water vapor in the atmosphere have a dramatic effect on the amount of radiation lost to space, and consequently on the nocturnal cooling. Even on perfectly clear nights, a humid atmosphere greatly restricts the net loss of heat to space, and the lowest minimum temperatures occur with very dry air. Thus, the temperature at Miami (which has a very high humidity) during a clear June night may fall to only 20°C (68°F) from a high of 30°C (86°F). At El Paso, Texas, however, where the atmosphere is usually very dry, the temperature may fall from 30°C to 10°C (50°F) by morning.

4.3.3 THE RADIATION BUDGET

The radiation budget of the earth and its atmosphere shows quantitatively the relative amounts of radiation reflected, scattered, or absorbed by the clear skies, clouds, and earth. Figure 4.11 shows approximately what happens to the solar radiation that reaches the top of the earth's atmosphere.* About 20 percent of the radiation is absorbed by the atmosphere and used to warm the air. A large amount, 22 percent, is reflected back into space by clouds and lost, which indicates the very important role clouds play in determining the climate and weather. An additional 5 percent is lost by scattering in clear air, and only

---

*The percentages of the incident solar radiation and the emitted terrestrial long-wave radiation used in the various processes in the energy budgets shown in figures 4.11 and 4.13 are only estimates. The actual percentages may vary by ± 2 percent from the values given. The important points are the relative orders of magnitude and that the budgets balance.

about 3 percent of the incoming radiation at the top of the atmosphere is reflected from the earth's surface, which is relatively dark.

The reflectivity (percentage of the incident radiation reflected), or albedo, of various terrestrial surfaces is listed in table 4.1. The reason for the small overall reflection of radiation of sunlight from the earth is that water, which covers about 75 percent of the earth's surface, has a very low albedo during most of the day when the sun is high above the horizon. Only near the sunrise and sunset does water reflect an appreciable amount of radiation.

The remaining 50 percent of the solar radiation reaches the ground either by direct transmission through the atmosphere to the surface, or indirectly as diffuse radiation from clouds and clear sky. Curiously, because of the significant amounts of diffuse radiation, it is possible for a surface to receive more radiation than the solar constant (2 cal/cm²/min); this is hard to accept at first, because the solar constant is the amount of radiation received at the top of the atmosphere and would appear to be the maximum possible amount that any surface could receive. However, if the sun is directly overhead and the atmosphere is clean and clear directly above the surface, nearly the entire amount of radiation given by the solar constant reaches the ground (figure 4.12). And if there are scattered cumulus clouds, the solar radiation reflected and diffused from the clouds may contribute another 0.4 cal/cm²/min. Thus, the surface at the ground can receive more solar radiation than the same surface at the top of the atmosphere, a situation which would be obviously ideal for rapid suntanning (or, more likely, burning).

The approximate long-wave energy budget is shown in figure 4.13. About 96 units of long-wave radiation from the atmosphere are added to the 50 units of solar short-wave radiation, yielding a total of 146 units absorbed by the ground. In order for the radiation budget at the ground to balance, the same 146 units must be lost. The greatest loss, about 114 units, occurs through the emission of long-wave radiation. About 20 units are used to evaporate water at the earth's surface, adding latent heat to the atmosphere which may later be released in condensation. The remaining 12 units are utilized to warm directly the air near the ground by *conduction* (direct transfer of heat energy in a medium by the collision of rapidly moving molecules) and *convection* (vertical transfer of heat by eddies that are much bigger than the molecular scale).

Most (104 units) of the 114 units of infrared radiation emitted by the ground are absorbed by carbon dioxide and water vapor in the atmosphere. However, part of the infrared radiation is radiated back to the earth (the atmospheric effect), resulting in higher surface temperatures.

The total radiation balance in the atmosphere is achieved by balancing the gains of energy by short-wave absorption (+20%), absorption of long-wave radiation emitted by the ground (+104%), latent heat addition (+20%), and sensible heat addition (+12%) with the loss of energy by emission of long-wave radiation to space (−60%) and to the ground (−96%).

In summary, the disposition of the incoming solar radiation is determined by scattering, reflection, and absorption by the clear atmosphere, clouds, solid particles, and surface of the earth. The infrared radiation which is emitted by the earth is largely absorbed by carbon dioxide and water vapor in the atmosphere. The result of all this scattering, reflection, absorption, and reradiation is a complex equilibrium energy budget which determines the mean temperature of the earth. Because of the importance of the highly variable amounts of clouds and water vapor in this equilibrium, small changes in either or both could have large effects on the climate at the earth's surface.

RADIATION AND LATITUDINAL TEMPERATURE VARIATIONS   *4.4*

At the beginning of this chapter, we noted that the distribution of radiation produced both altitudinal and latitudinal variations of temperature. We have discussed the variation of radiative heating with altitude in sufficient detail to show how it results in warmer surface temperatures and cooler atmospheric temperatures. This, in turn, gives rise to convection, cumulus clouds, and thunderstorms, particularly in the tropics where surface heating is most intense.

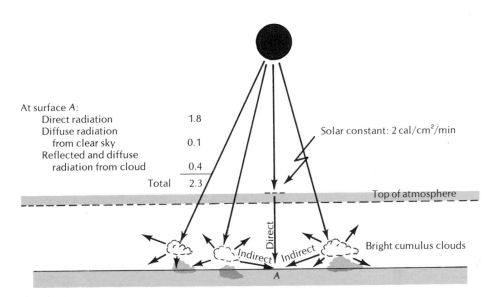

At surface *A*:
Direct radiation   1.8
Diffuse radiation
   from clear sky   0.1
Reflected and diffuse
   radiation from cloud   0.4
         Total   2.3

Solar constant: 2 cal/cm²/min

Top of atmosphere

Direct

Indirect   Indirect

Bright cumulus clouds

A

FIGURE 4.12 *Surface receiving more radiation than the solar constant. Because of indirect radiation from the clouds and scattered radiation from the sky, a surface at the ground can receive more radiation per unit time than the solar constant.*

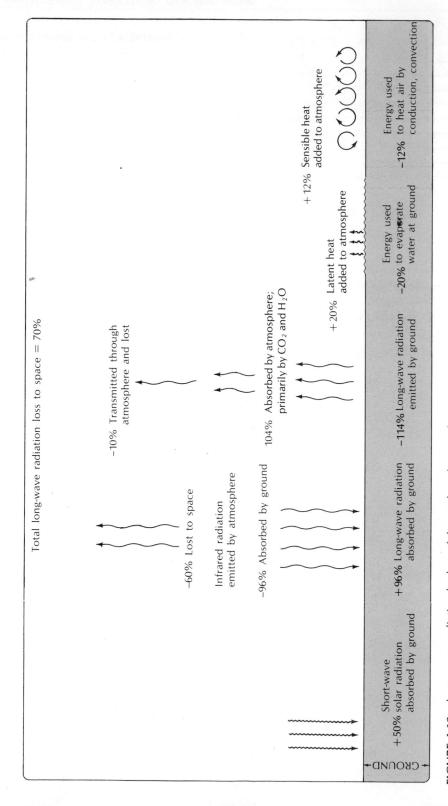

FIGURE 4.13 Long-wave radiation budget of the earth and atmosphere.

94

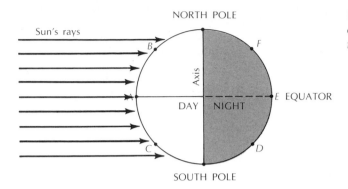

FIGURE 4.14 *Variations in sunlight received on the surface of the earth as a function of latitude.*

We now turn to the issue of the latitudinal variation of temperature—the poles are colder than the tropics. We'll examine the reason for this difference with a hint of the meteorological consequences.

4.4.1 *VARIATION OF TEMPERATURE FROM NORTH TO SOUTH*

The earth is almost a sphere; the diameter across the equator is only 0.3 percent larger than the diameter from the North Pole to the South Pole. Usually, we can ignore the difference. Because the earth is nearly a sphere, the average angle between sunlight and ground changes from latitude to latitude. In figure 4.14, it is noon at places on the earth's surface designated by *A, B,* and *C;* these points are receiving the maximum intensity of sunlight they will receive on this day. The stations on the right, *D, E,* and *F,* are receiving no sunlight at all; they are experiencing local midnight. But do *A, B,* and *C* get the same amount of light and solar energy? Indeed, they do not. At point *A,* the sun is overhead, and the light intensity is great; at *C,* the rays of the sun just graze the surface, and the amount of light received is very small. This figure is drawn for March 21 or September 23, times at which the equator gets most of the light at local noon and the poles receive the least. This distribution of the sun's energy is true only for March 21 and September 23. At other times, as we shall see, other latitudes may get more radiation than the equator. Over the year, however, the equator gets the most radiation, and the poles the least. This simple difference is the basic cause of large-scale atmospheric motions, and, therefore, of the traveling weather systems that course our planet.

The reason that the motion of the atmosphere is linked to these temperature differences can be explained through an analogy with the heating in a room. We know by the drift of smoke across a room, or the rustling of curtains, or the feel of a faint draft on our faces that hot-water radiators produce weak circulations of air in a closed room. The air in contact with the hot surfaces of the radiator is heated, expands, and rises, and cooler air moves in to replace it. The cause of this very small-scale wind circulation is *differential heating,* which means that a portion of air in the room is heated more than the rest, thereby producing temperature and pressure differences that lead to acceleration of the air and the

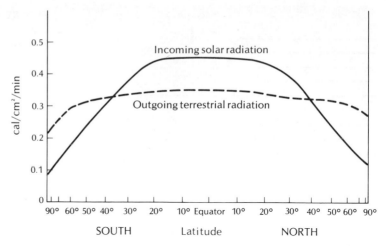

FIGURE 4.15 *Mean annual incoming solar radiation absorbed by the earth and atmosphere and the mean annual long-wave radiation emitted by the earth and atmosphere.*

production of winds. On a much larger scale, the complicated circulations of the atmosphere also owe their existence to differential heating, where the sun is the radiator and the entire atmosphere is the room.

So far in this section we have ignored the role of terrestrial radiation. Earlier we found that the temperature of the atmosphere on the ground depended on a balance between energy gained and energy lost. These two components of the radiation are illustrated in figure 4.15, which shows how the mean annual radiation budget changes with latitude. Obviously, the tropics get much more solar radiation than do the poles. (The radiation level at the poles does not drop to zero because, during its summer, each pole receives quite a bit of radiation.)

The outgoing terrestrial radiation does not show nearly as much variation with latitude as does the incoming solar radiation. Recall that the hotter an object is, the more radiation it emits, and indeed, the tropical regions of the earth are emitting more radiation than the polar regions. The difference is not terribly pronounced, however, because, as we saw in figure 4.13, most of the long-wave radiation that escapes to space is emitted by the atmosphere rather than by the ground. At the levels in the atmosphere from which the radiation can escape out to space there is not a pronounced change in temperature with latitude.

There is a striking discrepancy between the two curves of figure 4.15: The tropics receive more radiative energy than they lose, whereas the poles lose more than they gain. If radiation were the only means of transporting heat, the situation could not long remain as it is; the tropics would warm in response to all the excess energy received, and the poles would cool even further. Clearly there must be some mechanism other than radiation to transport energy from the tropics where there is a surplus to the high latitudes where there is a deficit. This transport of heat is accomplished by both the winds in the atmosphere and by the currents in the oceans. In the atmosphere, the major transport occurs in the cyclones and

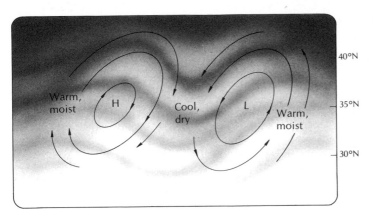

FIGURE 4.16 *Transport of heat and moisture northward from the tropics by anticyclones and cyclones.*

anticyclones, as shown in figure 4.16. These circulatory wind systems act as giant eddies in global circulation, mixing cold, dry air from the polar regions with warm, moist air from the tropics. In the portion of these eddies in which the winds are blowing poleward, the air is warm and moist. On the other side of the circulation, where the winds are blowing equatorward, the air is cool and dry. Therefore, even though there is little net flow of air poleward or equatorward, there is a systematic transfer of heat and moisture from low to high latitudes.

4.4.2 SEASONAL VARIATION OF RADIATION WITH LATITUDE

Even those who have not traveled extensively know that the temperature variation between the tropics and the poles is not as extreme in summer as in winter. Recall that the traveling cyclonic storms that give us much of our weather are one of the important mechanisms for transporting the surplus energy from the tropics to the polar regions. It is everyone's experience that these storms are not nearly as severe in summer as in winter; this points to the decreased temperature variation with latitude in the summer. Thus, if we know why the latitudinal variation of temperature changes with the season, we have a basis for understanding the seasonal variation of storms.

The seasons are caused primarily by the motion of the earth around the sun and the tilt of the earth's axis of rotation. To begin with, the earth revolves around the sun in an ellipse (nearly a circle), as illustrated in figure 4.17. Here, the ellipse is exaggerated to emphasize the characteristics of an elliptical orbit. The sun is at a focus of the ellipse, so both the maximum and minimum distances between the earth and sun occur when the earth is located on the major axis. Minimum distance, *perihelion* (Greek for *around the sun*), occurs around January 3; maximum distance, *aphelion* (*away from the sun*), occurs about six months later. The obvious result of this difference in distance is that July should be colder than January. Yet it is not—at least for people in the Northern Hemisphere. So the changing distance between the earth and sun cannot be the main reason for seasonal change. For many purposes, we can forget about this difference in distances. It is only about 3.5 percent; considering that the intensity of radiation varies as the

97

inverse square of the distance from the source, the entire earth gets about 7 percent more radiation from the sun in January than in July. However, if we are to make accurate calculations, we should take this difference into account. For example, if summer in the Northern Hemisphere occurred at perihelion, the summer would be hotter. Also, in that case, winter would be at aphelion, and winters would be colder than they now are. Therefore, seasons in the Northern Hemisphere would be more severe than they are at present. Actually, this situation did occur 10,000 years ago. Not only that, but astronomers have computed the orbit of the earth for the more distant past and have found that, several tens of thousands of years ago, the orbit was less circular than it is now, and the radiation received at the closest approach was 20 percent greater than at aphelion. Certainly, this would produce a substantially different climate from what we have now.

In figure 4.17, we note another consequence of the elliptical shape of the earth's orbit. The distance the earth has to travel from the beginning of spring (March 21) to the beginning of autumn (September 23) is longer than the distance from the beginning of fall to the beginning of spring. Not only that, but Johannes Kepler showed long ago that the earth travels more slowly when its distance from the sun is greater than when its distance is small. Therefore, in the Northern Hemisphere, the total length of spring and summer is substantially longer (about a week) than fall and winter. If you do not believe this, count the number of days from March 21 to September 23 and compare it with the number between September 23 and March 21. In the distant past, this difference in the length of the seasons has been as much as a month. So, we in the Northern Hemisphere have nice long springs and summers. We must remember, though, that the seasons are reversed in the Southern Hemisphere, which, therefore, has relatively short springs and summers.

Now we come to the real reason for the difference between seasons (figure 4.18): the plane of the earth's equator is not parallel to the plane of the earth's orbit. Instead, there is an angle between the plane of the equator and the plane of the earth's orbit (also called the *ecliptic*). This angle, which has the impressive name *obliquity of the ecliptic,* is now 23.5 degrees. In the last 100,000 years, this angle has changed a bit because the earth rocks back and forth a little.

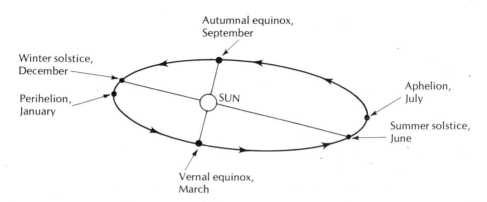

FIGURE 4.17 *Elliptical orbit of the earth (eccentricity exaggerated).*

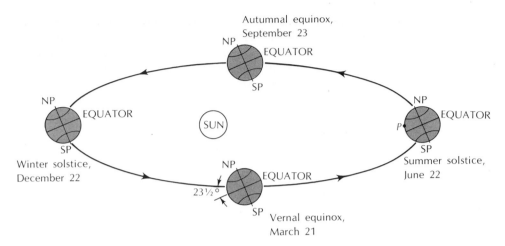

FIGURE 4.18 *Orbit of earth in perspective, showing seasons. Note that the plane of the equator is not parallel to the plane of the earth's orbit.*

Roughly, it has varied between 22 and 25 degrees. We shall see that this angle has important effects on climate, and when it changes, the climate changes accordingly.

As the earth revolves about the sun, its axis points in the same direction in space. Figure 4.18 shows the earth in four positions: on March 21, June 22, September 23, and December 22. On June 22, the Northern Hemisphere is directed toward the sun. The North Pole is in sunlight all day, and the South Pole is shaded all day. Also, the point *P*, where the sun is directly overhead, is in the Northern Hemisphere, and daylight lasts a long time. In the Southern Hemisphere, days are short, and the sun is never far from the horizon. Therefore, all points in the Northern Hemisphere get more sunlight on June 22 than the corresponding points in the Southern Hemisphere. This date is called the *summer solstice,** which means *summer sun-stand-still*. The sun "stands still" in the sense that it ceases its daily northward migration in the noon sky, hesitates, and then heads back toward the southern skies.

The satellite picture of the Western Hemisphere at 4.30 A.M. EST, July 5, 1974, (figure 4.19) illustrates the effect of the tilt of the earth's axis on the length of night and day. Because this date is close to the summer solstice, the North Pole is bathed in sunlight, and the South Pole shivers through its polar night. The satellite is centered over the equator at 45°W longitude, so the line of sunrise on the earth makes an angle of nearly 23.5 degrees with the 45° meridan. Thus, the comma-shaped mass of swirling clouds over the northern Atlantic is illuminated by the early-morning sun. Farther west, dawn is breaking over eastern Canada and northern New England, while locations at the same longitude but farther to the south, such as Miami, have an hour or two of night left.

---

*Strictly speaking, we should say "summer solstice in the Northern Hemisphere." In the Southern Hemisphere, the summer solstice occurs around December 22. In this text, *summer* or *winter solstices* will always pertain to the Northern Hemisphere.

On September 23 and March 21, both hemispheres get exactly the same amount of sunshine. These dates are called the *autumnal equinox* and *vernal equinox,* respectively. *Equinox* means *equal night;* every place has nights and days of equal length. Therefore, at the equinoxes, the time from sunrise to sunset should be twelve hours. It would be exactly twelve hours, too, if the atmosphere did not bend the rays of the sun. Because of this bending, the sun can be seen when it is below the horizon, and the time between sunrise and sunset appears to be a little longer than twelve hours.

On December 22—the *winter solstice*—the Southern Hemisphere gets much more light than the Northern Hemisphere, where the sun rises only a little above the horizon and remains for only a short time.

These changes through the year would not exist if the obliquity of the ecliptic were zero, in which case there would be no seasons. Over tens of thousands of years, this angle has changed, and, as a result, the severity of the seasons has also changed. When the angle is small, the seasons are less harsh than when the angle is large.

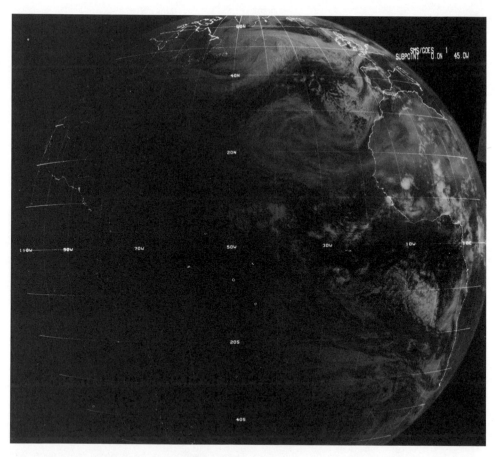

FIGURE 4.19 *Satellite view of the Western Hemisphere for 4:30 A.M. EST on July 5, 1974.*

In most places in the United States, the date of the greatest amount of sunshine (around June 22) is not, on the average, the warmest (see appendix B for examples). Since it takes time to heat the ground, the oceans, and the atmosphere, the period of maximum temperatures follows the period of maximum sunshine. Therefore, the warmest month over most continents outside the tropics is July in the Northern Hemisphere and January in the Southern Hemisphere. Similarly, it also takes time to cool the air, so the coldest date in the Northern Hemisphere is not December 22, but is usually later in the season.

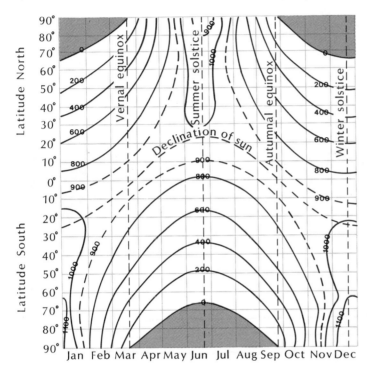

FIGURE 4.20 *Undepleted solar radiation, in cal/cm²/day, as a function of latitude and date. Shaded areas represent latitudes within the earth's shadow.*

Much of the above information is summarized in figure 4.20, showing how much solar radiation is received by the earth at various latitudes and times of the year. In the northern hemisphere winter (upper, right-hand corner of the diagram), we can see that the radiation decreases from about 800 calories per square centimeter per day near the equator down to nothing at about latitude 70°N. This extreme difference vanishes in the summer (top, center) when, for a time, the long polar days actually cause the arctic to receive more radiation than the tropics.

Figure 4.20 presents *undepleted* solar radiation—as if there were no clouds to reflect the radiation and no air molecules to absorb or scatter some of it. The useful solar radiation that actually reaches the ground, therefore, cannot be as high as the values shown in figure 4.20. Because cloudiness varies from place to place, the mean daily radiation measured on the ground shows a more compli-

101

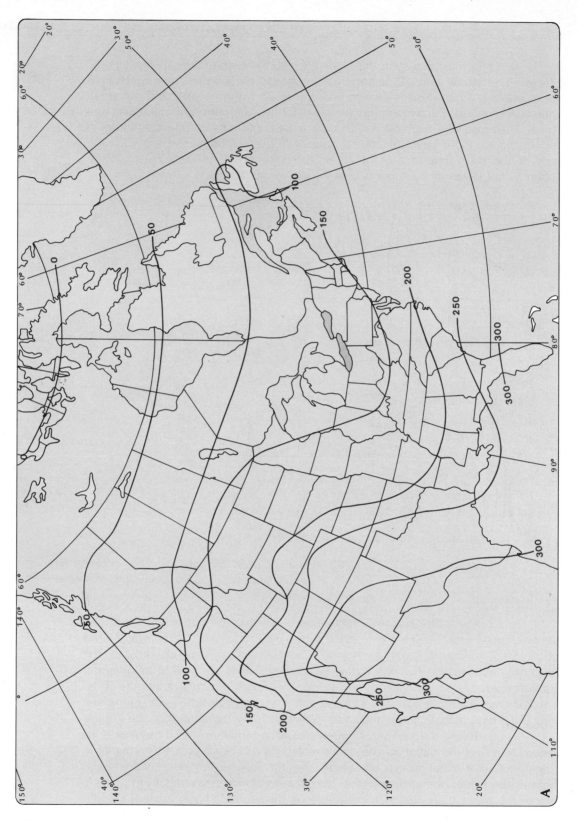

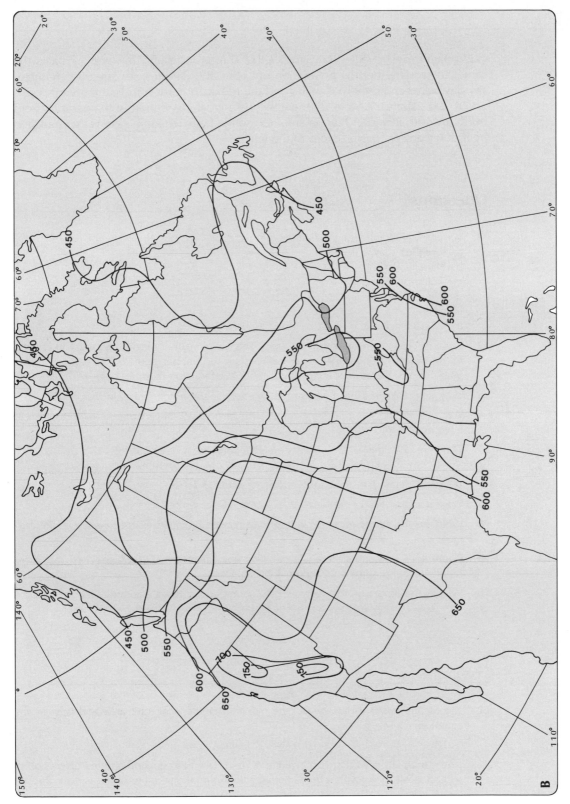

FIGURE 4.21 Mean daily insolation (cal/cm²) in (a) January and (b) July.

cated pattern, as revealed by figure 4.21. The basic result, however, remains the same. In January there is a variation of about 300 calories per square centimeter per day between northern Canada and the Mexican border; in July the variation is about 200 calories. As a result there will be a greater variation in temperature from north to south in winter than in summer, so the large cyclonic storms that transport heat poleward are more severe in the winter.

Questions

1 Summarize the following radiation laws:
   a. Kirchhoff's law
   b. Planck's law
   c. Stefan-Boltzman law
   d. Wien's law
   e. cosine law
   f. inverse-square law

2 What is wrong with the following logic? The wavelength of maximum radiation emitted by an object is longer for cooler bodies than for warmer bodies. The earth is cooler than the sun, therefore it emits more long-wave radiation than the sun.

3 Why does Lake Michigan appear lighter than the surrounding land in figure 4.5?

4 What would the solar constant of the earth be if the distance from the earth to the sun were doubled?

5 Compare and contrast the atmosphere effect and the greenhouse effect.

6 Why is the radiative mean temperature of the planet earth about −17°C, while the mean temperature of the earth's surface is about +15°C?

7 Why is a clean, clear sky blue? Why is a dusty sky white?

8 What are the two main gases that absorb infrared radiation emitted by the earth's surface?

9 Figure 4.15 shows that there is a radiation imbalance at most latitudes. How are these radiative energy surplusses and deficits balanced?

10 Why is the length of time between the spring and autumn equinoxes longer than the time between the autumn and spring equinoxes in the Northern Hemisphere?

Problems

11 The Stefan-Boltzman law relates the radiation emitted per unit area and time by the equation

$$E = \sigma T^4 \ [cal/cm^2/min]$$

where σ equals 8.132×10^{-11} cal/cm^2/K^4/min. Use this equation and the discussion of section 4.2.1 to verify that the radiative equilibrium temperature of the earth is 256°K.

12 If the average albedo of clouds is 50 percent, and the average albedo of the rest of the earth (ground, ocean, and air) is 10 percent, what is the percentage of cloud cover? Use figure 4.11 to obtain net global albedo. (Answer: 50 percent)

13 Describe and sketch the qualitative changes in the radiation curves of figure 4.15 as a function of latitude that would occur if the obliquity of the ecliptic were reduced from $23\frac{1}{2}$ to zero degrees.

FIVE

Behavior of Gases

Gas law • First law of thermodynamics • Continuity equation for air and water vapor • Newton's second law • parcel of air • dry adiabatic lapse rate • Coriolis force • pressure gradient force • absolute and relative vorticity • potential vorticity • friction • eddies • hydrostatic equilibrium

Because air is a mixture of gases, we can better understand the behavior of the atmosphere through a few general physical laws governing the behavior of a gas when it is heated, lifted, or subjected to differences in pressure. We are already aware of many of the consequences of these laws even if we do not formally understand the laws themselves. For example, we know that hot-air balloons rise; that friction slows down moving objects; that winds are usually lighter in a dense forest than on an exposed beach; and that it is harder to set a heavy object in motion than a lighter object, but once in motion, that the heavier object is harder to stop. We also know that mountains are generally cooler than valleys; that boiled eggs take longer to cook in Denver, Colorado, than in Tallahassee, Florida; and that basements are cooler in the summer than attics. The amazing thing is that these, and many more apparently unrelated, phenomena can be explained by a relatively small number of general physical laws or principles. This is the beauty of once learning the general principles—they can be applied over and over again in different contexts to explain phenomena that are new to the observer. Thus, when we understand that expanding gases become colder and compressed gases become warmer, we have the correct explanation for why aerosol sprays, when released from the compressed can, feel cold; why suddenly pressurized air in a hand tire pump feels hot; and why air at a mountain resort is generally cooler than the air in the desert below.

5.1 SEVEN EQUATIONS AND SEVEN UNKNOWNS: THE PROBLEM OF ATMOSPHERIC PREDICTION

A fundamental question of meteorological forecasting can be asked as follows: Given the structure of the atmosphere at a certain time, together with the earth's rotation rate and its orbital characteristics, how will the atmosphere evolve in time? If we could answer this question for all scales of motion, from small eddies to huge storms and for time scales ranging from seconds to centuries, we could make accurate weather and climate forecasts for days, months, years, and even centuries.

The evolution of the atmosphere from a given state to some future state is determined by the laws of physics that the atmosphere must obey. For many, but not all, meteorological problems, the relevant laws of physics can be expressed in terms of seven equations which govern the behavior of seven variables (table 5.1). If the distribution of the variables is given at an initial time, the equations can be solved for the distribution of the same variables at a future time. In principle, this can be done for motion on all scales, from tornadoes to the circulation around the whole earth.

Variable	Equations
1. Pressure *(P)*	1. Gas law or equation of state (relates temperature, pressure, and density: $P = \rho rT$).
2. Temperature *(T)*	2. First law of thermodynamics (relates temperature changes to heating or cooling and changes in pressure).
3. Density (ρ)	3. Continuity equation for air (expresses conservation of mass for air).
4. Water vapor	4. Continuity equation for water vapor (expresses conservation of water vapor).
5. West-to-east component of wind	5–7. Newton's second law (force = mass x acceleration) applied in west-east, south-north, and vertical directions separately. Relates acceleration of air in each of 3 directions to forces.
6. South-to-north component of wind	
7. Vertical component of wind	

TABLE 5.1 *A set of seven atmospheric variables and seven equations to approximate the behavior of the atmosphere.*

If we know the laws and variables, why can't we make perfect forecasts, even short-term ones? In the first place, we would have to know the value of all the meteorological variables everywhere in the atmosphere, an obvious impossibility. If we are interested in large-scale motions, for example, we have information only at weather stations, which are typically several hundred kilometers apart. To obtain data between the weather stations, we assume that the variables vary smoothly from one station to the next. Actually, the variables do not vary smoothly, because there are "eddies" of all sizes between stations. These eddies affect the larger-scale motions by transferring heat, water vapor, and velocity from one part of the atmosphere to another. We often ignore such motions, and the penalty for the approximation and the resulting errors is an inaccurate prediction.

Another trouble arises from the nature of the lower boundary of the atmosphere—the earth's surface. This surface is complex, with varying elevations and physical characteristics. Land, lakes, forests, farms, cities, and mountains occur in irregular patterns, often on much too small a scale to be treated properly in a problem of large-scale meteorology. Usually the ground characteristics are smoothed and simplified, and, occasionally, ignored entirely. In any case, the real world is approximated, and more errors are introduced.

There are additional problems. One is that certain physical processes in the atmosphere cannot be handled very well in the mathematical equations. For example, heating by radiation depends in a complex fashion on the moisture distribution, the temperature, and cloudiness. Approximations are used to represent these effects. Also, the equations are so complex that exact solutions are impossible. Again, error-producing approximations are used.

For all these reasons, then, the behavior of the seven basic meteorological variables is not understood or predicted perfectly, although much work is being done to improve the current state of affairs. However, even if we could predict perfectly these seven variables, we would not know the complete state of the atmosphere, because there are other variables which are not specified by the equations. For example, knowledge of the seven variables does not tell us the amount of liquid water, ice, or ozone present. These and other variables affect the weather, too. If we want to understand the behavior of ozone, we have to add another variable, the ozone concentration, and another equation to predict ozone changes. But ozone can be affected by concentrations of other trace gases, for example chlorine or oxides of nitrogen. So we need to introduce more equations and more variables. It has been estimated that a complete study of the ozone problem requires at least fifty variables and equations in addition to the original seven. Similar situations exist for other pollutants. Generally, air-pollution meteorology (for chemically active pollutants) requires large numbers of equations. Still, many important problems of the atmosphere have been attacked quite successfully using the seven fundamental equations. We shall now consider the individual equations in more detail.

5.2 BEHAVIOR OF GASES—CHANGES IN TEMPERATURE, PRESSURE, AND DENSITY

In the range of temperatures and pressures in the atmosphere, the gases which make up the atmosphere (mainly nitrogen and oxygen; see table 2.1) behave as *ideal gases*. Ideal gases are by definition those which obey the relatively simple relationship between pressure, density, and temperature known as the *gas law,* discussed in section 5.2.2. Although necessary in quantitative calculations, this law has only limited use in qualitatively explaining how temperature changes are caused by expansion or compression because of the presence of three variables, all of which may change simultaneously. To explain many important atmospheric phenomena, we need a relationship between pressure and temperature alone. We find this more-revealing and useful dependency by making use of the *first law of thermodynamics,* which tells us what happens to the temperature of a gas if it is heated, cooled, or if the pressure is changed by processes such as the lifting of air over a mountain.

Before discussing the behavior of gases that make up the atmosphere, let us introduce the useful concept *parcel of air* and the environment of this parcel.

5.2.1 AIR PARCELS AND THEIR ENVIRONMENT

Meteorologists are always talking about rising parcels of air, squeezing the water vapor out of parcels of air, or the acceleration of a parcel of air. What is this nebulous parcel of air? A parcel of air usually refers to a very small volume of the atmosphere with horizontal dimensions of a few meters within which the temperature, pressure, humidity, density, dustiness, etc., are uniform. This homogeneous parcel is viewed as a sort of invisible balloon which can be followed as it moves up, down, or sideways over a period of time (figure 5.1). The parcel may expand, stretch, shrink, or even twist, but the air molecules which made up the original parcel always stay together; that is, the parcel does not break in two, with half of it heading south and the other half north. We might imagine coloring all the air molecules in a volume the size of a classroom green and then observing what happens when this green volume of air is subjected to lifting, heating, cooling, condensation, or any other process we might wish to consider. We can visualize parcels of air in other ways. By riding along in a weightless balloon, we follow the sample of atmosphere in the immediate vicinity of the balloon as it drifts around the earth. A smoke puff rising from a chimney identifies a parcel of air, as does a cloud of tiny gnats, rising and falling with each turbulent eddy near the ground.

We find it useful to talk about parcels rather than about the entire atmosphere when we want to isolate certain processes or forces. A parcel is simply a convenient increment of the atmosphere which behaves as a unique element during the process under discussion.

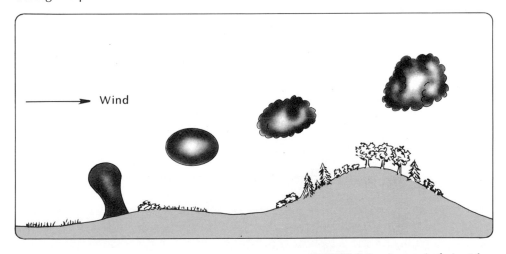

FIGURE 5.1 *A parcel of air at four successive times.*

The air-parcel concept is very useful when considering the formation of cumulus clouds (to be discussed in detail in chapter 10), for the condensation of water vapor into cloud droplets makes visible this particular parcel of air (figure 5.2). Imagine a spherical parcel of warm, moist, buoyant air near the ground. Because of its high temperature and moisture content, it is slightly less dense than the surrounding air, which is called the *environment*. It is important to realize that the properties of the parcel and the environment are not usually the same, although they could be. As the parcel of air rises through its environment (much like a bubble rises through water), its temperature decreases, but so does the temperature of the surrounding air (which is not moving). Therefore, the parcel may remain warmer than its environment and continue to rise. Figure 5.2b shows a plot of

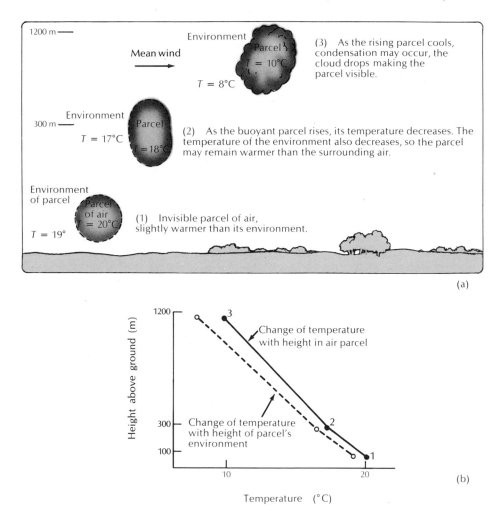

(a)

(b)

FIGURE 5.2 *(a) Rising parcel of air and its environment. (b) Graph of temperature* vs. *height in a parcel of air, compared with the temperature of the parcel's environment.*

temperature versus height above the ground in this example and emphasizes that the rising parcel and its environment are quite different. Finally, as the parcel continues to rise, condensation may occur, making the parcel visible.

5.2.2   *THE GAS LAW: EQUATION OF STATE*

The *gas law,* or *equation of state,* says that pressure of a gas is proportional to the product of the absolute temperature and the density (mass of the gas per unit volume); that is,

$$\text{Pressure} = \text{constant} \times \text{density} \times \text{temperature} \qquad \textbf{(5.1)}$$

Thus, in a closed container holding a fixed amount of gas (so that the density is constant), increasing the temperature increases the pressure, and decreasing the temperature decreases the pressure, as we can demonstrate by heating the air in an open metal can on a stove and then removing and quickly capping the can. As the gas in the sealed container cools and the inside pressure falls, the can will be crushed by the higher atmospheric pressure outside.

In equation (5.1), the temperature is the absolute temperature and must be expressed in Kelvins; otherwise, we could calculate a negative pressure at temperatures below zero, which of course would be nonsense. If density is expressed in kilograms per cubic meter (kg/m^3), a numerical value of 2.87 for the constant in equation (5.1) gives the pressure in millibars (mb). (See problem 15 at the end of the chapter.) For example, typical values of temperature and density near the earth's surface are 280° K and 1.250 kg/m^3 respectively, yielding a pressure of 1004.5 mb from equation (5.1).

The gas law, by itself, has only limited direct applicability in explaining the behavior of the atmosphere. We know, for example, that at the earth's surface, pressure and temperature are not directly proportional. In fact, during the winter over the continents, the lowest temperatures frequently occur with the highest surface pressures, and the temperature usually rises as the pressure falls. However, these observations do not invalidate the gas law; the difficulty lies in the fact that the density of the atmosphere is not constant because the atmosphere is compressible. Thus, when the temperature falls as the pressure rises on a cold winter day, the density also rises, and the gas law is still satisfied.

Let us try to apply the gas law in one more way. We hear repeatedly how the temperature of rising air decreases, whereas the temperature of sinking air increases. Can the gas law be used to explain this important observation? If air rises to a greater elevation, its pressure must fall, and if the density were constant, the temperature would have to fall also. Thus, the gas law seems to explain the phenomenon of rising air becoming cooler. However, the density of rising air does not stay constant; in fact, it decreases, as does the pressure. Looking back at the equation of state, we see that if both the pressure and the density fall, we cannot say whether the temperature will rise, fall, or remain the same; it could do any of these things, depending on which experiences a higher percentage of decrease, the pressure or the density. Again, the problem is that we are working with three variables.

To explain what happens as air is lifted and pressure falls, we would much prefer to have a relationship between temperature and pressure alone so that we do not have to worry about what the third variable, density, is doing.

5.2.3 TEMPERATURE CHANGES OF A GAS UNDER HEATING, COOLING, OR CHANGES IN PRESSURE

One of the most useful laws in explaining the behavior of a gas (or a mixture of gases such as the atmosphere) under various conditions of pressure, temperature, and heating or cooling is the empirical (experimentally determined) first law of thermodynamics, which says that the temperature of a gas may be changed by addition (or subtraction) of heat, a change in pressure, or a combination of both:

$$\text{Change in temperature} = \text{constant} \times \text{heat added (subtracted)} \qquad \textbf{(5.2)}$$
$$+ \text{ another constant} \times \text{pressure change}$$

The effect of adding or subtracting heat is simple to understand; some examples are given in figure 5.3. When the atmosphere is heated by solar radiation, by long-wave radiation from the earth, or by contact with the warm ground, its temperature rises. When the atmosphere loses radiation to space, or is cooled by evaporation of rain falling through a dry layer, the temperature decreases.

In spite of the ultimate importance of heating and cooling processes (called *diabatic* processes), there are many atmospheric processes, usually involving time scales of a day or less, in which the amounts of heat added or subtracted are small. In these cases, we may neglect the heating or cooling term in the first law of thermodynamics and are left with a very simple relationship between small changes in temperature and pressure:

$$\text{Change in temperature} = \text{constant} \times \text{change in pressure} \qquad \textbf{(5.3)}$$

Because this latter relationship is valid if no heat is added or subtracted, it is called the *adiabatic* form of the first law of thermodynamics. Note that the temperature may change without heating or cooling; thus, heating a gas is not synonymous with a temperature increase, nor is cooling (removing heat) synonymous with a temperature decrease. An adiabatic process is simply one in which no heat energy (calories) is added or subtracted from the gas during the process. Thus, the process of a parcel of dry air flowing over a mountain can be adiabatic, but the temperature will decrease as the parcel flows up the mountain slope and the pressure falls, and will rise again as the parcel descends the opposite side and the pressure increases.

It is difficult to overemphasize the importance of the adiabatic form of the first law of thermodynamics—that temperature changes are directly proportional to pressure changes when no heat is added or subtracted from the gas. Because the pressure changes so rapidly over small vertical distances, large temperature changes result as the atmosphere sinks or rises. For example, the adiabatic form of the first law of thermodynamics predicts a 10C° temperature fall for a decrease in pressure corresponding to an increase in height of 1 kilometer (or 5.5F°/

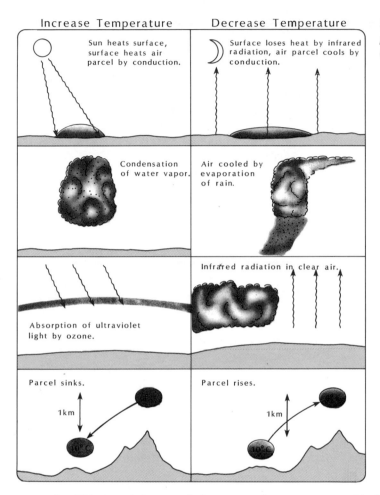

Increase Temperature

Sun heats surface, surface heats air parcel by conduction.

Condensation of water vapor.

Absorption of ultraviolet light by ozone.

Parcel sinks.

1km

Decrease Temperature

Surface loses heat by infrared radiation, air parcel cools by conduction.

Air cooled by evaporation of rain.

Infrared radiation in clear air.

Parcel rises.

1km

FIGURE 5.3 *Diabatic and adiabatic processes leading to temperature changes in air.*

1000 ft). This special rate of decrease of temperature with height (10C°/km) is called the *dry-adiabatic lapse rate*.

Air flowing over tall mountains, rising in vigorous thunderstorms, or being lifted in the complex cyclones of middle latitudes may change elevation by thousands of meters. Thus, if a parcel of dry air at the surface, with a comfortable temperature of 25°C (77°F) was lifted to 6 kilometers (about 20,000 feet), the temperature would be a frigid −35°C (−31°F). Conversely, if air at a typical temperature of −20°C (−4°F) at 6 kilometers sank to the surface, its temperature would be a roasting 40°C (104°F).* All of these changes in temperature could be produced without any externally produced heating or cooling whatsoever, so powerful is the effect of compression or expansion on temperature changes.

---

*The warming effect of compression has a surprising consequence. Air conditioners, not heaters, must be used in aircraft flying at low pressures (high altitudes) even though temperatures of the air outside may be −35°C (−31°F). A compression of this air at a height of 9 km (30,000) feet to sea-level pressures in the aircraft would produce a temperature of 55°C (131°F) if the air conditioners were not used to extract heat from the air.

5.3 FORCES, ACCELERATIONS, AND THE PRODUCTION OF MOTION

The previous sections have considered the radiative and thermodynamic properties of the gaseous atmosphere. Now, we turn briefly to the dynamics of gases—the laws that explain the motion of the ever-restless atmosphere.

Starting the wind in motion, changing its direction, or causing it to cease requires acceleration (change in velocity), and producing acceleration requires a net force. The basic law relating force to acceleration is Newton's second law:

$$\text{Net force} = \text{mass} \times \text{acceleration} \qquad (5.4)$$

TABLE 5.2 *Important forces in the atmosphere*

Force	Remarks
1. Gravity	Acts downward; is nearly balanced by the vertical pressure gradient force (see 2a).
2. Pressure gradient force	Acts toward low pressure.
a. Vertical pressure gradient force	Component of the pressure gradient force in the vertical; acts upward; is nearly balanced by gravity.
b. Horizontal pressure gradient force	Component of the pressure gradient force in the horizontal; acts from high to low pressure; is much smaller in magnitude than the vertical pressure gradient force but is very important in producing horizontal air motions (winds). On large scales, tend to be balanced by the Coriolis force (see 3).
3. Coriolis force	Force caused by the rotation of the earth; acts at right angles to the direction of motion; is proportional to the speed of motion; is strongest at the North Pole; vanishes at the equator. For large-scale motions, balances the horizontal pressure gradient force.
4. Friction	Generally acts in direction opposite to the direction of motion and slows the speed of the moving air. Friction is usually important only near the ground, in the lowest kilometer of the atmosphere.

Thus, changes in velocity are produced by unbalanced forces acting over time; a balance of forces (zero net force) means no acceleration and, therefore, constant velocities.

The important forces that accelerate (or decelerate) the lower atmosphere are summarized in table 5.2.** Usually, forces are nearly, though not exactly, balanced. Therefore, the net forces (and, consequently, the accelerations) acting on air parcels are small. Note that, though there is little or no acceleration, there can still be rapid motion. For example, if you drive a car at a constant speed of 80 km/h and in a constant direction, there is no acceleration, but the speed is considerable. According to Newton's law, there must be zero net force in this case. The accelerating force of the motor is exactly balanced by the decelerating force of friction. If you wish to accelerate, you must step on the gas to increase the forward force in order to exceed the friction. Before Newton, people had always believed that a constant motion required a net force, and that the absence or cancellation of forces implied zero motion.

Another point about equation (5.4) is that it is a vector equation: it governs not only changes in speed but also changes in direction. If your car is traveling in a given direction and you wish to turn right, a force is required. In general, a force at right angles to the direction of motion will change the direction, not the speed. A force along the direction of motion will change the speed but not the direction.

As pointed out earlier, equation (5.4) is equivalent to three equations, one for each dimension. For example, Newton's law applies in the vertical regardless of what is happening in the horizontal. Thus, no matter whether you run or merely step out of an upper-story window, the force of gravity will produce the same downward acceleration and you will hit the ground at the same time.

5.3.1 GRAVITY

The first force listed in table 5.2 is gravity. Gravity always points downward because the downward direction is defined as the direction of gravity. Gravity varies with distance from the center of the earth, but for most meteorological purposes it may be considered constant. The force of gravity produces an acceleration on a body of about 9.8 meters per second per second. This acceleration is a little larger at the poles than at the equator for two reasons: the surface at the poles is a little closer to the earth's center than the surface at the equator; and the earth rotates, resulting in some centrifugal force at the equator (less than 1 percent of the gravitational attraction) but none at the poles.

5.3.2 PRESSURE GRADIENT FORCE

The next force in table 5.2 is the pressure gradient force. This force always acts from high pressure toward low pressure. Because pressure changes near the ground over a given vertical distance are about 10,000 times greater than typical pressure changes over the same horizontal distance (figure 5.4),

---

*At very high levels (above 100 km), electrical and magnetic forces also become important.

FIGURE 5.4 *Pressure changes over vertical distances are about 10,000 times greater than over horizontal distances.*

the vertical pressure gradient forces are typically 10,000 times larger than horizontal pressure gradient forces. It is for this reason that we consider separately horizontal and vertical forces. In the vertical, we have a balance or near balance of huge forces; in the horizontal, we have a near balance of tiny forces.

The vertical pressure gradient force is almost the same everywhere because it balances (or nearly balances) gravity, which is almost the same everywhere. However, the horizontal pressure gradient forces vary from place to place. In stormy regions, pressures change rapidly in the horizontal, and pressure gradient forces are relatively large; in quiet regions, pressure changes slowly and the horizontal pressure gradient forces are small.

5.3.3 CORIOLIS FORCE

The next force in table 5.2 is the Coriolis force. Coriolis forces arise whenever the coordinate system of interest is rotating and motion of objects occurs relative to the moving coordinate system. The wind is air moving relative to the earth, which is rotating about an axis connecting the North and South poles. However, before discussing the Coriolis force on the rotating spherical

FIGURE 5.5 *Illustration of the Coriolis effect on the movement of a dart (. . . .) fired from the center of a turntable.*

  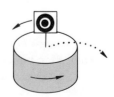

earth, let us consider the simpler case of a rotating plane such as a turntable (figure 5.5). Suppose a turntable is rotating in a counterclockwise direction as viewed from above. Furthermore, suppose a small dart gun is located at the center of the turntable and fires a dart at a target on the rim. As soon as the dart leaves the gun, it has no net forces on it except gravity, which tries to pull it downward. Therefore, the dart travels in a straight horizontal line relative to the nonrotating coordinate system of the room. However, the turntable rotates under the target, and by the time the projectile reaches the rim of the turntable, the target has rotated out of the way. To the nonrotating observer, it is obvious what has happened—the only force affecting the projectile was gravity, and so no horizontal deflection occurred.

FIGURE 5.6 *Illustration of the Coriolis effect on the movement of a dart (. . . .) fired from a location off the center of a turntable.*

However, consider the view of an observer on the turntable and ro-tating with the turntable. Whether or not he is aware of the rotation, the target appears fixed in space. He sees the projectile turn and miss the target to the right. In his coordinate system, the projectile changed its horizontal direction. The ob-server, wishing to keep Newton's laws satisfied in his coordinate system, ascribes this acceleration to a force—the Coriolis force, named after the French scientist Coriolis who studied it. We note that the apparent deflection and the associated Coriolis force are proportional to the rate of rotation of the plane about the vertical.

The apparent deflection to the right of the direction of motion does not depend on where the gun is positioned or on the direction of fire. Figure 5.6 shows the effect when the gun is located away from the axis of rotation and fires at a target in a direction not along a radius from the center of the turntable. Again, a rightward deflection is sensed by an observer on the turntable.

Before applying the Coriolis force to the wind, we need to remind ourselves that the earth, the trees, our rooms, and we ourselves are moving at all times. Wind is air movement relative to the earth, but the earth, together with its atmosphere, is traveling eastward at high speeds at low latitudes, and more slowly at higher latitudes, causing the local horizon to rotate about a vertical axis (picture a flagpole) once per day at the poles, and somwhat more slowly away from the poles (figure 5.7). This effect vanishes at the equator, where the north and south horizons move eastward at the same speed. A flagpole at the equator tumbles over once per day as it races eastward at about 1600 km/h (1000 mi/h), but neither it nor the horizon rotates about a vertical axis. Hence, for the horizontal winds, Coriolis deflective effects due to the rotation of the local horizontal plane are sig-nificant at middle and high latitudes, but not at the equator.

Now imagine wind (or any moving object, such as an airplane) trav-eling in a straight line with respect to a motionless observer in outer space (figure 5.8). Since the object is moving in a straight line, there are no real accelerations or forces. However, the earth rotates under the moving object, and to an earth-bound observer, the object appears to deflect toward the right of its direction of

FIGURE 5.7 *Rotation of the horizon about the local vertical axis.*

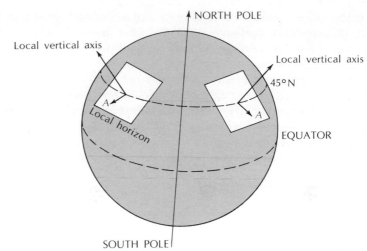

motion. This deflection is very real to the observer on earth. As shown in figure 5.8, the object (airplane) crosses central Florida instead of southern Georgia. Again, the apparent force responsible for this deflection is the Coriolis force. Note that there would be no horizontal deflection at the equator, and that the deflection would be a maximum at the North Pole.

As in the case of the projectile moving over the turntable, these deflections of the winds are not limited to air flowing from west to east. Test your own understanding by picturing the apparent deflection caused by the rotation of the underlying ground when the winds is from the south. The ground is rotating counterclockwise, so the air appears to turn off to the opposite direction, that is, to the right.

Coriolis deflections usually need to be considered when objects or fluids move substantial distances relative to the earth, either by traveling rapidly or by traveling for long times. Some cases for which Coriolis forces are too important to neglect include ballistics problems, flow in rivers, ocean currents, and the situation of interest to us—large-scale flows of air.

The properties of the Coriolis force are summarized in table 5.2. The Coriolis force acts at 90 degrees to the right of the flow in the Northern Hemi-

FIGURE 5.8 *Apparent deflection of an airplane over a rotating earth.*

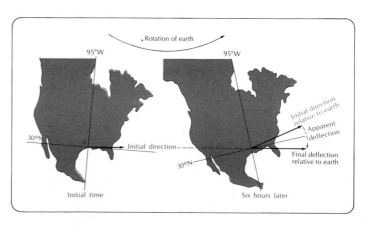

sphere, increases with latitude from zero at the equator to a maximum value at the North Pole, and is proportional to the wind speed. For large-scale motions, we will see that the Coriolis force, which acts to the right of the flow, nearly balances the horizontal pressure gradient force, which acts to the left of the flow. The result of this balance is that large-scale wind currents blow parallel to the isobars (lines of equal pressure), with low pressure to the left of the flow. In the Southern Hemisphere, the Coriolis force acts to the left of the direction of flow because the local horizon is rotating in a clockwise rather than counterclockwise direction.

5.3.4 SPINNING PARCELS OF AIR AND VORTICITY

A fundamental property of such atmospheric vortices as dust devils, tornadoes, hurricanes, and extratropical cyclones and anticyclones is the rotation or spin of the air about the vertical axis. A measure of this rotation which meteorologists find useful is called the *vorticity,* which is twice the angular rate of rotation about a vertical axis. Its units are radians per time, (1 revolution $= 2\pi$ radians) so that the vorticity of a 45 rpm record is $2 \times (2\pi \times 45)$ rad/min or 9.4 rad/s. Radians have no dimensions so vorticity is often written in units of s^{-1}.

In the atmosphere, there are two important types of vorticity. Because the earth is rotating, even air which is not moving relative to the earth is spinning about its vertical axis, except directly on the equator (figure 5.7). The vorticity associated with this spin due to the Earth's rotation is represented by the symbol f, which is given by

$$f = 2\Omega \sin \phi \qquad (5.5)$$

where Ω is the angular velocity of the earth (2π rad/24 hours $= 7.27 \times 10^{-5}$ rad/s), and ϕ is the latitude.

If the air is also spinning relative to the earth, as in a tornado, hurricane, or extratropical cyclone, this relative spin or *relative vorticity* adds to the earth's vorticity to give the *absolute vorticity.* If the air is rotating in the same sense as the earth's rotation (counterclockwise when viewed from above, over the Northern Hemisphere), the relative vorticity is positive. If it is rotating in the opposite sense, the relative vorticity is negative. Thus the relative vorticity associated with anticyclones is negative. Only rarely does the negative relative vorticity associated with anticyclonic flow exceed the positive Earth's vorticity, so the absolute vorticity of large scales of motion is almost always positive.

The advantage of introducing the concept of vorticity is that the formation and decay of cyclones and anticyclones can be studied by considering the changes in vorticity. For example, cyclogenesis is associated with increases in vorticity. The two major physical properties which change the relative vorticity of a disc of air are horizontal divergence and change of latitude (figure 5.9). If we consider a disc of air with absolute vorticity $(\zeta + f)$ and depth D, both effects can be seen in the statement

$$\frac{\zeta + f}{D} = \text{constant} \qquad (5.6)$$

FIGURE 5.9 *(a) Effect of horizontal divergence on relative vorticity, if potential vorticity is conserved. If horizontal divergence reduces D by ½, the rate of spin will also decrease by ½. If horizontal convergence doubles D, then the spin will also double. (b) Effect of latitude on relative vorticity. If absolute vorticity (ζ + f) is conserved, ζ must increase as f decreases. In this example the direction of spin relative to earth reverses as parcel moves from 40°N to 30°N.*

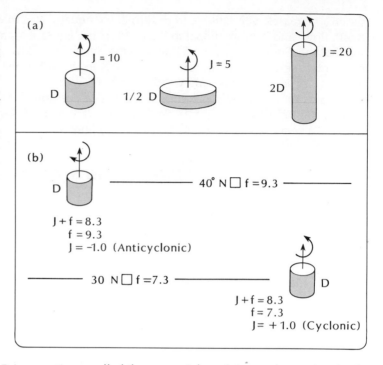

The quantity $(\zeta + f)/D$ is sometimes called the *potential vorticity,* and equation (5.6) states that the potential vorticity of the parcel of air is approximately conserved. Now we can consider the effects of horizontal divergence and latitude changes on the relative vorticity, ζ. Horizontal divergence (figure 5.9a) results in a decrease of D. At a constant latitude, f is constant, and so the relative vorticity must also decrease to satisfy (5.6). Conversely, horizontal convergence represents an increase in D and consequently in more rapid cyclonic spin (increase of ζ).

The latitude effect is quite simple. For a constant D (no horizontal divergence or convergence) a disc of air moving southward over the Northern Hemisphere will experience a decrease in f and an increase of equal magnitude in relative vorticity in order to keep $\dfrac{\zeta + f}{D}$ constant. Thus, air moving southward will tend to spin more cyclonically relative to the earth. On the other hand, as air moves northward, f will increase and ζ must decrease. Thus, air flowing northward will tend to spin more anticyclonically relative to the earth.

5.3.5   FRICTION

The final force listed in table 5.2 is friction. Friction is important only in the lowest kilometer or so, a region that is therefore often called the friction layer. Friction usually retards the flow. When you step on the brake of a car, you retard the car by friction. But if you put your palm on a book and move your hand, you can accelerate the book. As a result of friction, the book begins to share the motion of your hand and is accelerated.

The mechanism for this type of friction is that the molecules in one object collide with those of another. If one object (in this case, your hand) moves (has momentum), the other object will receive some of this momentum and will move. In general, molecular friction accelerates or decelerates a body that is in contact with another body moving at a different speed.

On a larger scale, "eddies" can exchange momentum and therefore cause frictional forces. Consider figure 5.10. Here a wind is blowing at a height of 1 kilometer. Between the ground and 1 kilometer, the air is "turbulent." That is, below 1 kilometer, there exist eddies commonly driven by heating of the ground. These eddies try to exchange the momentum at the ground (which cannot move if it is solid) with the momentum of the wind. The result is that the wind experiences a retarding force, which we call *eddy friction, turbulent friction,* or simply *friction.* The ground, on the other hand, experiences an accelerating force. If the surface is free to move (as in the case of a lake surface), this force can start waves and currents. If the ground is loose, blowing dust or snow can occur.

Thus, the frictional force in the lower layer of the atmosphere is dependent on the existence of turbulent eddies. Such eddies are well developed and vigorous on clear days when the ground is warm but tend to be suppressed on calm nights when the surface is cold. Therefore, the force of friction varies strongly from day to night, and the friction layer tends to extend much higher in the daytime than at night.

5.3.6 BALANCE OF FORCES IN THE ATMOSPHERE

We have now discussed all of the important forces which cause the air to move faster, slow down or change direction. It is important to note that each force given in table 5.2 is normally counteracted by another force of opposite direction but nearly the same magnitude. Therefore, the net forces (and, therefore, accelerations) acting on parcels of air are small, a condition we call *balanced* (figure 5.11). In the balanced large-scale flows above the friction

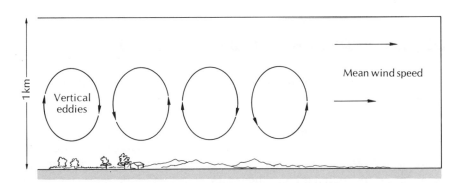

FIGURE 5.10 *Turbulent eddies exchange horizontal momentum in the vertical by bringing fast winds downward and slower winds upward.*

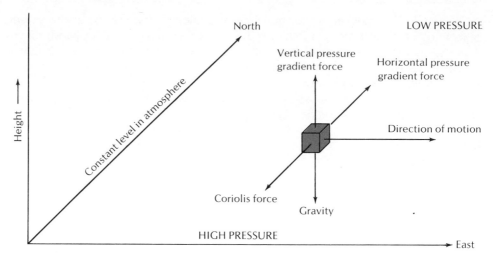

FIGURE 5.11 *Balance of forces on an air parcel traveling eastward. The parcel is experiencing no acceleration because all the forces are balanced. The vertical pressure gradient force (directed upward) is balanced by gravity (directed downward). The horizontal pressure gradient force (directed northward toward low pressure) is balanced by the Coriolis force (directed southward, at right angles to the direction of motion). The horizontal balance is called* geostrophic balance.

layer, the vertical pressure gradient force balances gravity, and the horizontal pressure gradient force is balanced by the Coriolis force. Note that even though there is little acceleration, there can still be high winds. Exceptions to the nearly balanced conditon of large-scale motions occur in violent weather such as tornadoes, or on small scales such as in the gusts and eddies that form close to objects on a windy day.

5.4　HYDROSTATIC EQUILIBRIUM AND STABILITY

One of the most important balances of forces in meteorology is the balance between gravity and the vertical pressure gradient force. A balance of forces in a given direction is called *equilibrium* (in that direction). In particular, the balance of gravity (pulling down) and the vertical pressure gradient force (pushing up) is called *hydrostatic equilibrium*. We say that scales of motion larger than about 10 kilometers are in hydrostatic equilibrium. For systems on a smaller scale, such as thunderstorms, or for individual eddies, there may be a small, but important difference between gravity and the pressure gradient force. This difference makes possible large vertical accelerations.

The balance of vertical forces on a parcel of air in hydrostatic equilibrium is depicted in figure 5.12. The pressure on the bottom of the parcel is p_1,

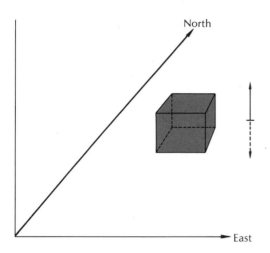

FIGURE 5.12 *Balance between the vertical pressure gradient force and gravity in hydrostatic equilibrium.*

and the pressure on the top is p_2. Because the pressure is force per unit area, the net upward force on the parcel by the difference in pressure is $(p_1 - p_2) \times A$ where A is the area of the top and bottom of the parcel. In hydrostatic equilibrium, this upward force is balanced by the weight, the downward force of gravity, which is the mass (M) of the parcel times the acceleration of gravity (g). Hence, we have

$$(p_1 - p_2)A = Mg \qquad \textbf{(5.7)}$$

The mass is equal to the density (ρ) times the volume of the parcel, which is $A \times \Delta z$. Thus, equation (5.7) may be written

$$(p_1 - p_2) = g\rho\Delta z \qquad \textbf{(5.8)}$$

This is called the *hydrostatic equation*. The hydrostatic equation tells us how rapidly pressure decreases over a given vertical distance Δz. It also says that the pressure difference across the layer Δz is proportional to the weight of the layer. The denser (heavier) the air, the greater is the pressure decrease. Since cold air is denser than warm air, pressure decreases faster with height in cold than in warm air. This important result has many applications in meteorology, as discussed in section 11.3.2.

Any equilibrium may be stable, unstable, or neutral. If a stable equilibrium is disturbed, the forces act to return the system toward the equilibrium state. If an unstable equilibrium is disturbed, the forces act to increase the effect of the disturbance.

Consider, for example, a ball resting in the bottom of a smooth bowl. The ball is in equilibrium. If we move the ball slightly and release it, it accelerates back toward its equilibrium position. Therefore, the ball is in stable equilibrium. A ball perched precariously on top of a smooth mound is also in equilibrium. In this case, a small push results in acceleration of the ball away from its original position; so the ball is in an unstable equilibrium. Finally, if the ball rests on a horizontal plane, an initial push results in constant speed with zero acceleration. This is known as neutral equilibrium.

125

Hydrostatic equilibrium also may be described as stable, unstable, or neutral. If air is pushed upward, but the forces act to force the air down, we say that we have *hydrostatic stability*. If air is pushed up and the forces cause a further upward acceleration, we have *hydrostatic instability*. These expressions are rather cumbersome, so for simplicity, the terms *stable air* and *unstable air* are usually used to mean hydrostatic stability and instability, respectively.

The vertical temperature (and, to a lesser extent, moisture) distribution characterizes the stability of the atmosphere. The details of the vertical temperature structure of the atmosphere are important in many meteorological processes. For example, the change of temperature in the vertical may determine whether morning cumulus clouds will grow into afternoon thundershowers, or whether pollution from a smokestack will rise thousands of meters and disperse over a wide area instead of turn back toward the ground.

The relationship of the vertical temperature structure to stability is illustrated in figure 5.13. When no sources or sinks of heat are present, the temperature of a rising parcel of air decreases at a rate of 10C°/km (the *dry adiabatic lapse rate*). If the rate of temperature decrease with height in the environment is less than 10C°/km (sounding *A* in figure 5.13), the parcel becomes colder (and denser) than its environment and sinks back toward its original level, a condition termed *stable*. If the environmental rate of temperature decrease exceeds 10C°/km, however (sounding *B* in figure 5.13), the parcel remains warmer than its environment and accelerates away from its original position, a condition termed *unstable*.

The vertical distribution of moisture also affects the stability of the atmosphere. Because water vapor is less dense than dry air, moist air underlying dry air favors instability. If condensation occurs in the rising parcel of air, the latent heat of condensation is added to the parcel and reduces its rate of cooling to approximately 6C°/km in the lower troposphere. If the rate of decrease of the environmental temperature is more than this *wet adiabatic lapse rate*, the saturated

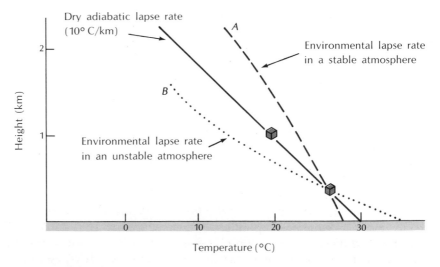

FIGURE 5.13 *Illustration of stability in the atmosphere.*

(100 percent relative humidity) parcel of air may continue to rise. In this case, the atmosphere is unstable with respect to vertical motions of saturated air parcels. This process, which leads to the development of cumulus clouds and thunderstorms, is discussed in chapters 6 and 10.

The ground typically gets heated strongly in the daytime and cools rapidly at night due to loss of heat by infrared radiation. These temperature changes are much greater than those in the air above. As a result, the lowest 100 meters of the atmosphere are usually stable at night and unstable in the daytime. Strong turbulence occurs in the unstable lower atmosphere in the daytime, with much smoother conditions prevailing at night. This is why flying a plane at low levels on a sunny day is bumpier than on a clear night.

CONSERVATION OF MASS 5.5

The sixth basic equation of meteorology states that a given mass of air can be neither created nor destroyed—a given kilogram of air will always be a kilogram of air. However, the shape of the kilogram of air may change because the atmosphere is compressible.

The continuity equation gives us an important relationship between horizontal and vertical motions (figure 5.14). Consider a parcel containing 1 kilogram of air (parcel A in figure 5.14). If the horizontal winds squeeze the parcel, as shown, it expands vertically to conserve mass. The result after some time might be parcel B, which still contains 1 kilogram of mass but is more squashed in the horizontal and elongated in the vertical than parcel A. Thus, horizontal convergence of air has been compensated by a vertical expansion. The equation of continuity can be expressed

$$
\begin{aligned}
\text{Change of} & = \text{amount of mass flowing into} \\
\text{mass in a volume} & \quad \text{(or out of) the volume by} \\
& \quad \text{horizontal motions} \\
& + \text{amount of mass flowing into} \\
& \quad \text{(or out of) the volume by} \\
& \quad \text{vertical motions}
\end{aligned}
\tag{5.9}
$$

From equation (5.9), we see that the continuity equation relates the vertical motions in the atmosphere to horizontal motions, so if we know how the horizontal wind varies, we can compute the vertical motion.

A common application of the equation of continuity is to the wind field in a low or a high. For example, in the Northern Hemisphere near the surface, the wind spirals clockwise and outward from a high, as shown in figure 5.15. Because air is moving away from the center, there is "horizontal divergence" or expansion; hence, there must be vertical compression. But since there is no vertical motion at the ground, the air above the ground must sink, and so we have subsidence (sinking) above highs. As will be seen, this subsidence is responsible for the fair weather in high-pressure areas and for air pollution problems in these regions. Conversely, lows are characterized by horizontal convergence at the ground and rising air above the ground.

5.6   CONSERVATION OF MOISTURE

The seventh basic equation in meteorology is the continuity equation for water, which states that the total amounts of water vapor, liquid, and ice must be conserved in a closed system. This relationship forms the basis for precipitation forecasting. The moisture variables and the condensation and precipitation processes are discussed in detail in chapter 6.

FIGURE 5.14 *Illustration of the continuity equation. Horizontal convergence in (a) squeezes the volume and results in a vertical stretching, as shown in (b).*

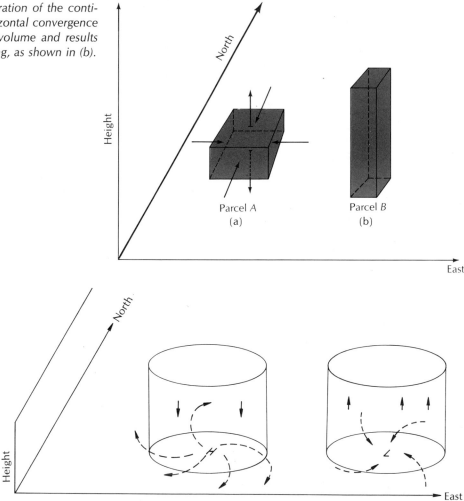

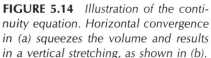

FIGURE 5.15 *Application of the continuity equation to relate vertical motions to horizontal motions in high- and low-pressure systems.*

Questions

1 List the seven basic atmospheric variables that can be used to describe atmospheric flow.

2 List at least three reasons we cannot make weather forecasts as accurately as forecasts of astronomical events such as eclipses.

3 Calculate the mass (in kg) of air in a room measuring $10 \times 10 \times 4$ m³, if the temperature is 20°C and the pressure is 1013 mb. (Answer: 480 kg)

4 What are the two forces that generally balance the flow of air at 500 mb?

5 Sketch the wind flow around circular highs and lows in the Southern Hemisphere.

6 If the earth were not rotating, what would be the relationship between isobars and wind direction?

7 What is the vorticity of a $33\frac{1}{3}$ RPM record? Express answer in s^{-1}.

8 Suppose a tube of air of depth 100 m has an absolute vorticity of $10^{-4}s^{-1}$. If this tube is stretched to a depth of 1 km, what will its absolute vorticity be?

9 If the density is 1 kg/m³ and the acceleration of gravity is 9.81 m/s², how much lower will the pressure be on the fourth floor of a building compared to the ground floor? Assume the distance between the ground and fourth floors is 10m. Note 1 mb equals 100 kg/m/s². (Answer = 0.98 mb)

10 Why does the pressure decrease more rapidly with increasing height near the ground than it does in the stratosphere?

11 How long would it take a parcel of air to make one complete turn if its vorticity is $10^{-4}s^{-1}$? (Answer: 34.9 hours)

12 If a parcel at 45°N (f = $1.03 \times 10^{-4}s^{-1}$) has a *relative* vorticity of 0, what will its relative vorticity be at a latitude of 30°N (f = $0.73 \times 10^{-4}s^{-1}$) if its *absolute* vorticity is conserved. (Answer: $0.27 \times 10^{-4}s^{-1}$)

Problems

13 The balance between the Coriolis force, fv where v is the wind speed, and horizontal pressure gradient force, $\dfrac{1}{\rho}\dfrac{\Delta p}{\Delta n}$ where ρ is density and Δp is the difference in pressure over the horizontal distance Δn (measured perpendicular to the isobars), is

$$fv = \frac{1}{\rho}\frac{\Delta p}{\Delta n}$$

In this formula, p must be expressed in kg/m/s² and ρ in kg/m³. If ρ is 1 kg/m³, and the distance between isobars drawn every 4 mb is 400 km, what is the wind speed at the following latitudes: 90° (North Pole), 45°N, and 5°N? (Answers: 6.9 m/s, 9.7 m/s, 78.7 m/s)

14 If the rotation rate of the earth doubled and the typical wind speeds remained as they are now, would you expect the intensity (maximum and minimum pressures) of highs and lows to increase or decrease? Explain.

15 The gas constant for dry air in the equation of state (5.1) is 287 Joules/kg/K, where a Joule is a unit of energy and is equal to 1 kg m/s². What are the units of pressure using this value for the constant? If 1 mb equals 100 kg/m/s², what should the constant be in order to give the pressure in mb? (Answer: 2.87)

129

SIX

Moisture and Clouds

latent heat of fusion • latent heat of vaporization • vapor
pressure • mixing ratio • saturation • relative humidity •
dewpoint • hygroscopic • solute effect • curvature effect •
sublimation nuclei • freezing nuclei • condensation trails •
wave clouds • pileus • anvil • entrainment • mamma • collision •
coalescence • fall streak • Bergeron-Findeisen process • riming

In regions of plentiful annual precipitation, we frequently underestimate the importance of water—yet water, next to air, is most vital to us. Without water, there would still be weather, but it would bear little resemblance to the weather as it is now. Cyclones and anticyclones would still exist, spawned by the north-south temperature difference. However, without the tempering effects of the oceans, these arid storms would probably be nightmarish whirlwinds of dust, first scorching the earth with torrid heat from the tropical deserts, then chilling the hostile landscape with glacial cold.

6.1 UNUSUAL PROPERTIES OF WATER

Water is a much more complicated substance than its simple formula, H_2O, might indicate. Composed of two hydrogen atoms combined with one atom of oxygen (figure 6.1), water has some very unusual properties in comparison with other commonly occurring materials. First, water is the only substance on earth that appears in all three forms—gas, liquid, and solid—at temperatures commonly found on the planet. In each of these states water still consists of two hydrogen and one oxygen atom; the only difference is the organization of the molecules. In the ice phase (figure 6.1b), water molecules are arranged compactly in groups of six molecules. Because of the low temperatures, the water molecules have less energy and are not free to move around. As the temperature increases, the molecules become more energetic and the bonds of the ice phase are broken. In the liquid state, water molecules are still close together, but they are free to slide around, and thus liquid water flows readily. In the gas or vapor phase, the water molecules are highly energetic and fly about in a totally disorganized manner (figure 6.1d). A popular misconception is that the visible plume rising from a teakettle or the mist hanging over a pond on a cool morning is water vapor. In reality, water vapor is invisible; the teakettle's steam or the mist over the lake are actually clouds and consist of tiny drops of liquid water.

Unlike most substances, water expands just before it freezes and during the freezing process. As a result of this expansion, ice floats on water rather than sinks, a fortunate property for water life. If ice behaved as a typical substance and became denser as it froze, lakes would freeze from the bottom rather than the top and it would be much more difficult for water plants and animals to survive the winter. An unfortunate consequence of this peculiar property of water is the possibility of broken pipes, caused by the enormous pressure exerted by the freezing water.

Hey Seafoodies,
When it comes to your fish we
know you want something as good
for you as it is for the Earth.

Learn more about our SMART
mission, find recipes, coupons,
or just say hi!

www.seacuisine.com

Facebook & Instagram:
@seacuisinemeals

BE SMART, SPEND LESS!

...e caution when consuming.

& committed to **Making waves in...**
SMARTer Tomorrow.

SEA CUISINE

$1.00 OFF

YOUR NEXT
SEA CUISINE
PURCHASE.

0035493-011178

Encode: 8101000354930111783010110002200601498070
SYMBOL HEIGHT 0.65676 INBAR 0.01004 BWA -0.0020
011178 SMART INGREDIENTS ON-PACK BOUNCEBACK 2018

20 GRAMS PROTEIN*

SUSTAINABLY WILD-CAUGHT

NO PRESERVATIVES

ART CHOICE. *SMART INGREDIENTS.*

Serving Si...

Calories	Per serving 270		Per Cu... 530	
		% Daily Value*		% Daily Value*
Total Fat	8g	11%	17g	22%
Saturated Fat	1g	5%	2g	11%
Trans Fat	0g		0g	
Cholesterol	45mg	15%	90mg	29%
Sodium	610mg	26%	1220mg	53%
Total Carb.	27g	10%	54g	20%
Dietary Fiber	<1g	3%	2g	7%
Total Sugars	<1g		1g	
Incl. Added Sugars	0g	0%	<1g	2%
Protein	20g	28%	40g	56%
Vitamin D	0.7mcg	4%	1.3mcg	6%
Calcium	40mg	4%	80mg	6%
Iron	0.9mg	4%	1.8mg	10%
Potassium	380mg	8%	760mg	15%

* The % Daily Value (DV) tells you how much a nutrient in a serving of food contributes to a daily diet. 2,000 calories a day is used for general nutrition advice.

RESPONSIBLY CHOSEN SEAFOO...

THIS PACKAGE IS 97% RECYCLABLE

...ENTS: WILD COD, BREAD CRUMB COATING: WHEAT FLOUR, WATER, BREAD (WHEAT FLOUR, YEAST, SUGAR**, VINEGAR, MALTED BARLEY FLOUR, SEA ...T, BAKING SODA), CORN STARCH, POTATOES*, MINCED ONION, YELLOW ...R, SALT, PARMESAN CHEESE (PASTEURIZED COW'S MILK, CHEESE CULTURE, ...ZYMES), CHEDDAR CHEESE (PASTEURIZED MILK, CHEESE CULTURE, SALT, ANNATTO), BEEF STOCK, EGG WHITES, SUGAR**, RICE STARCH, ONION YEAST, GARLIC POWDER, PARSLEY, BLACK PEPPER, CELERY SEED, WHITE ...AR-FRIED IN CANOLA, COTTONSEED, AND/OR SOYBEAN OIL. *DRIED **ADDS ... AMOUNT OF ADDED SUGARS

...S: FISH (COD), WHEAT, MILK, EGGS

...NING: While every effort has been made to remove bones from this product, ...bones may remain. Please use caution when consuming.

SEND QUESTIONS OR COMMENTS TO
feedback@highlinerfoodsusa.com
or call 1-888-860-3664.

PRODUCT OF CANADA

...BUTED BY
...INER FOODS, PORTSMOUTH, NH 03801
...seacuisine.com

...gistered trademark of High Liner Foods Incorporated.
...s a trademark of High Liner Foods USA Incorporated.

PLEASE RECYCLE TRAY & CARDBOARD SLEEVE

0 35493 29440 7

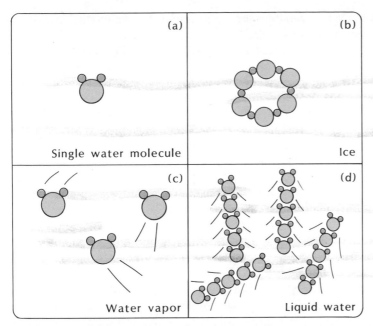

A third unusual property of water is of its heat capacity, which is higher than all natural substances except ammonia. A substance with a high heat capacity requires more heat energy to increase its temperature by a given amount than does a substance with low heat capacity. Thus, raising the temperature of 1 gram of water 1 C° requires 1 calorie of heat input. In contrast, the same calorie of heat would raise the temperature of a gram of air by more than 4 C°. The high heat capacity of water is one reason that oceans and large lakes are slow to warm in spring and cool in autumn.

A fourth remarkable property of water is the amount of heat required to melt ice *(latent heat of fusion)* or evaporate water *(latent heat of vaporization)*. To melt 1 gram of ice requires about 80 calories of heat; to evaporate water re-quires an amazing 597 calories of heat at a temperature of 0°C and typical surface pressures. These values are much higher than the latent heats of most substances. Conversely, when water condenses (freezes), it releases 597 (80) calories per gram. When the condensation or freezing occurs in the atmosphere, this heat is trans-ferred to the air and represents an important energy source. Thunderstorms, torna-does and hurricanes all depend on the release of latent heat. Finally, the evapora-tion of water at low latitudes and the subsequent poleward transport and recondensation at high latitudes represents a significant transfer of energy from low to high latitudes.

MOISTURE VARIABLES 6.2

In discussing the role of water in meteorological processes, it is necessary to define the amount of water in its three states present in a given parcel of air. There are over thirty variables currently used in meteorology that are related to the quantity of moisture in the air. Certain variables are more useful than others in describing specialized meteorological processes. Indeed, the large number of variables suggests the complicated ways moisture affects the behavior of the atmosphere. For our purposes, it is sufficient to know and understand the meaning of five moisture variables: the vapor pressure, the mixing ratio, the relative humidity, the dew-point temperature, and the wet-bulb temperature.

The air is a mixture of gases; the exact composition varies with time and place. The most important variable constituents are carbon dioxide and water vapor. The percentage of water vapor in a parcel of air varies from nearly zero over cold deserts to a maximum of about 4 percent in the warmest, most humid tropical rainforests.

According to Dalton's law, each gas in the mixture we call air exerts a pressure independent of the other components. The pressure we measure with a barometer is the sum of all the pressures of the individual gases. The water-vapor pressure is the pressure that would be exerted by the water-vapor molecules alone and is therefore a measure of the amount of water-vapor molecules present in a given volume at a certain temperature.

If we introduced dry air into a closed container with a flat surface of pure water, evaporation would begin as water molecules escaped the liquid surface. These molecules would have enough energy to escape the surface tension of the liquid, which represents the energy holding the water molecules together in liquid form. As the number of water-vapor molecules increased, so would the vapor pressure. As the vapor pressure, which is a measure of the kinetic energy of the molecules, increased, some of the faster-moving molecules would return to the water surface and the liquid state. Ultimately, an equilibrium would be reached in which the number of molecules escaping the liquid surface (evaporation) would equal the number of vapor molecules recondensing on the liquid surface. The vapor pressure in this equilibrium is called the *saturation vapor pressure*, and air containing the vapor in this equilibrium state is said to be saturated.

Because the water-vapor pressure depends on the temperature as well as upon the mass of water present in a given parcel of air, it is not a unique measure of the amount of water vapor present. The amount of water vapor in a parcel of air is given by the mixing ratio, which is the mass of water vapor per unit mass of dry air. The mixing ratio is usually expressed as grams of water vapor per kilogram of dry air. Typical values of the mixing ratio range from less than 0.1 gm/kg on cold winter days to 30 gm/kg on sultry, humid days in a tropical jungle. The mixing ratio of saturated air is the *saturation mixing ratio*.

One of the properties of water most important to the formation of clouds and precipitation is that the saturation vapor pressure and the saturation mixing ratio increase exponentially with increasing absolute temperature (figure 6.2). In the range of temperatures normally encountered in the atmosphere ($-20°C$ to $+30°C$), the saturation mixing ratio approximately *doubles* for every 10C° increase in temperature. Thus, saturated air at a summertime temperature of 20°C contains roughly four times more water vapor than saturated air at the freezing point. This property is often described by the statement, "Warm air can hold more water vapor than cold air." And because the more water vapor present, the more is available for precipitation, this property explains why copious rains are likely in the summer when the air is warm and why heavy precipitation is unlikely when the temperature is low ("It's 'too cold' to snow!").

Figure 6.2 illustrates another important property of water. The saturation mixing ratio (and the saturation vapor pressure) is *less* over ice than over water. This means that the saturation mixing ratio (and vapor pressure) over a surface of ice in a closed container is less than that over a surface of supercooled water. (*Supercooled water* is water that exists in liquid form at temperatures below freezing.) This property plays an important role in the precipitation process, as we will see in section 6.5.2.

Although the mixing ratio tells us how much (in an absolute sense) water vapor is present in a given parcel of air, it does not tell us how close the air is to being saturated. Therefore, it does not tell us the likelihood of clouds (liquid waterdrops or ice crystals) or precipitation. The proximity to saturation is given by the *relative humidity,* which is the ratio of the actual amount of water vapor present in the parcel of air to the amount that would be present at saturation. Mathematically, the relative humidity may be defined as *either* the actual vapor pressure divided by the saturation vapor pressure—

$$\text{Relative humidity} = \frac{\text{actual vapor pressure}}{\text{saturation vapor pressure}} \qquad \textbf{(6.1a)}$$

or the actual mixing ratio divided by the saturation mixing ratio—

$$\text{Relative humidity} = \frac{\text{actual mixing ratio}}{\text{saturation mixing ratio}} \qquad \textbf{(6.1b)}$$

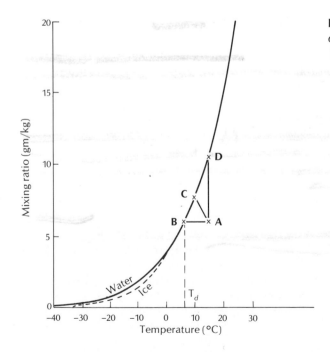

FIGURE 6.2 *Saturation mixing ratio over water and ice at an air pressure of 1000 millibars.*

Both definitions are used in meteorology. Although they are not exactly equivalent, the difference is small enough to be ignored.

Two additional moisture variables that have great significance in meteorology are expressed as temperatures. The *dew-point temperature*, or more simply, the *dew point,* is the temperature to which a parcel of air must be cooled to reach saturation. The definition of the dew point can be illustrated in figure 6.2. A parcel of air with a given temperature and water-vapor content, as expressed by the mixing ratio, can be represented by a point in figure 6.2; for example, suppose a parcel of air has a temperature of 15°C and a mixing ratio of 6 grams of water per kilogram of air. It would be represented by point *A.* Now, the cooling of the air without addition or loss of water vapor would be represented by the line *A-B.* When the temperature reaches 6°C (point *B*), the saturation mixing ratio becomes equal to the actual mixing ratio (which is constant in the process), and saturation occurs. This temperature is therefore the dew point.

For a given temperature and pressure, the dew point is a unique, but complicated, measure of the amount of water vapor present. The dew point has great forecasting significance—if the air is cooled to the dew point, condensation is probable. When the cooling occurs near a solid surface such as a car top or grass, condensation occurs on the surface of the object and forms dew. When the air is cooled to its dew point away from the surface of the earth, condensation occurs as cloud droplets. Therefore, knowledge of the dew point and the amount of cooling experienced by the air is crucial to the prediction of dew, fog, clouds, and precipitation. Furthermore, the dew point of the surface air often serves as a lower limit on the temperature decrease at night, because as the air reaches the dewpoint, the latent heat released by condensation retards further cooling. The dew point is therefore a reasonable forecast for the minimum nocturnal temperature.

At temperatures below freezing, the *frost point* may be defined as the temperature to which air must be cooled for ice crystals to be formed directly from water vapor by sublimation (deposition). Because the saturation vapor pressure over ice is less than it is over water, the frost point is slightly higher than the dew point.

The dew point expresses the temperature at which saturation will occur in a parcel of air in which the water vapor content remains constant. On the other hand, the *wet-bulb temperature* provides an estimate of the cooling of air that can occur as water is evaporated into the air. If water is evaporated into initially dry air, the temperature falls, and the vapor pressure, mixing ratio, relative humidity, and dew point all rise. This process is illustrated by line *A-C* in figure 6.2. At saturation (point *C*), the air will have reached a certain minimum temperature called the *wet-bulb temperature,* or more simply, the *wet bulb.* The wet-bulb temperature is also the temperature to which a thermometer can be lowered by evaporating water at the bulb of the thermometer. Such an instrument is called a *psychrometer.*

An important application of the wet-bulb concept to meteorology occurs fairly often when precipitation falls from an upper layer of clouds into an unsaturated layer below. As the waterdrops or ice crystals evaporate, they cool the

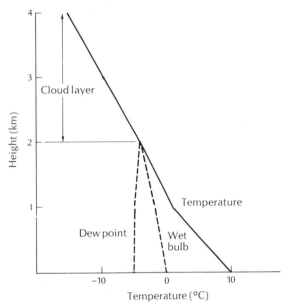

FIGURE 6.3 *Temperature, wet bulb, and dew-point sounding (temperature plot versus height) at the onset of precipitation. A saturated cloud layer exists above 2 kilometers. As precipitation falls into the lower dry layer, evaporation cools the layer to the wet-bulb temperature.*

layer and add water vapor. Both processes increase the relative humidity of the lower layer. Often in wintertime, dry layers of air that are initially slightly above freezing can be lowered below freezing by the evaporation. If this wet-bulb effect is not considered, forecasts of rain can turn out quite badly indeed. This situation is illustrated in figure 6.3. The initial temperature in the layer below 1 kilometer is above freezing, which would seem to indicate rain if precipitation were to fall. However, this layer is quite dry, with the wet-bulb temperature below freezing. Thus, as snow and rain fall into the layer and evaporate, the temperature in the entire layer may fall below freezing as the wet-bulb temperature is approached. The wet-bulb effect is the major reason that the temperature often drops sharply as precipitation begins.

THE CONDENSATION PROCESS 6.3

In section 6.2, we noted an extremely important property of warm air—that it can contain more water vapor than an equal mass of cold air. We also showed how cooling air below its dew point produces clouds, and then precipitation. However, to understand how the various types of precipitation—rain, sleet, graupel (soft hail), hail, and snow—actually form, grow, and fall to the earth, it is important to understand the microphysical processes (physical processes that occur on a very small scale) by which these varied forms of liquid and solid water are produced.

To describe the formation of waterdrops and ice crystals, we set aside the large-scale motions of the atmosphere and center our attention on the tiny world of the cloud, where relevant distances are measured in micrometers rather than in kilometers. The typical sizes of the components of the precipitation process

FIGURE 6.4 *Typical sizes and terminal velocities of particles and forms of precipitation in the atmosphere.*

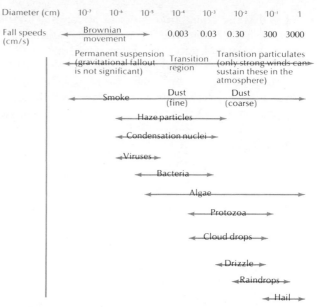

are illustrated in figure 6.4. We first discuss the condensation process when temperatures are above freezing, and then introduce the complicated, but extremely important effects of subfreezing temperatures.

All visible waterdroplets in nature start with microscopic (typical radius 0.1 micrometer) particles called *cloud nuclei*, which are simply any of the abundant particles of dust that are always present even in the purest arctic air. Although condensation may occur on any speck of dirt, if the humidity becomes high enough (it sometimes must exceed 100 percent if *hydrophobic*—having no affinity for water—particles are present), certain particles are much preferred for condensation. These particles with a special affinity for converting water vapor to liquid drops, even when the relative humidity *(RH)* is less than 100 percent, are called *hygroscopic* particles.* Ordinary table salt is hygroscopic, turning wet with moisture on humid summer days, in spite of advertising claims to the contrary. Garden fertilizer, which includes nitrates of ammonia, is also very hygroscopic.

The perceptive reader will sense in the preceding discussion that there is nothing particularly magical about the 100 percent relative humidity figure when condensation of water vapor into waterdroplets is considered. It is entirely possible to have tiny liquid droplets in equilibrium with water vapor in air with 98 percent relative humidity, just as it is possible, though not as likely in the atmosphere (which always seems to have enough hygroscopic "dirt" in it), to have perfectly clear air with a relative humidity of 101 percent. To understand these paradoxes, we note that relative humidity is defined as the ratio of the actual mixing ratio to the saturation mixing ratio (equilibrium amount of water vapor that would occur in a closed volume of air over a flat plane of pure water). The subtle

*Strictly speaking, *hygroscopic* means "having an attraction for water vapor. Particles that actually dissove and become liquid are *deliquescent*. However, virtually all meteorologists mean *deliquescent* when they say *hygroscopic,* and we will follow this convention.

point is that the equilibrium amount of vapor at saturation is defined with respect to a flat plane of pure water, but that newborn cloud droplets are neither flat nor pure. The effect of curvature alone is to require higher relative humidities—perhaps 101 percent—before condensation can occur. On the other hand, impurities in the form of hygroscopic chemicals reduce the ambient relative humidity required for the onset of condensation, to as low as 80 percent for large sea-salt particles. This latter effect, called the *solute effect*, is an important basis for warm cloud seeding, in which hygroscopic particles such as ordinary table salt are scattered into a cloud, increasing the rate of condensation.

Sea-salt particles, abundant near and over oceans where the evaporation of the salty water in the spray of breaking waves leaves behind giant salt nuclei (as large as 10 micrometers), are very important for the initial condensation process. These particles are responsible for the bluish haze over the oceans, which sometimes extends kilometers out to sea. Other hygroscopic particles that are efficient cloud nuclei are produced by combustion and are therefore abundant over cities.

The curvature and solute effects are illustrated in figure 6.5 which gives curves for the equilibrium relative humidity for a pure waterdroplet and a waterdroplet formed by the deliquescence of a salt particle of a given mass. The curvature effect is illustrated by the curve for a pure waterdroplet—relative humidities must be greater than 100 percent for pure waterdroplets to exist in an equilibrium state. For example, a pure waterdroplet with a radius of 0.4 micrometer will evaporate unless the relative humidity exceeds 100.4 percent (point A in figure 6.5). On the other hand, a drop of the same radius obtained by dissolving 10^{-15} gram of salt (NaCl) in pure water will be in equilibrium at a relative humidity of approximately 99.6 percent (point B) because of the solute effect. Note, however, that for drops to be in equilibrium at humidities less than 100 percent (for this concentration of salt solution), the droplets must be very small (less than 0.5 micrometer). This size corresponds to haze particles rather than visible cloud drops, and so the presence of a cloud in equilibrium necessitates large drops and requires that humidities be greater than 100 percent.

The growth or evaporation of waterdroplets can be illustrated by plotting points on figure 6.5. Let us consider what will happen to a drop containing 10^{-16} gram of salt with a radius slightly greater than 0.05 micrometer in a relative humidity of 99.8 percent (point C in figure 6.5). Because the equilibrium relative humidity for a drop of this size is much less than 99.8 percent (it would be represented by a point off the scale of this graph), the environment appears supersaturated with respect to this drop, and the drop will grow. When its radius reaches approximately 0.15 micrometer, the environmental humidity will equal the equilibrium humidity for that size drop, and the drop will stop growing (point D).

Now, consider a drop in the same environment with an initial radius of 0.2 micrometer (point E). The equilibrium relative humidity for this drop is over 100 percent (point F), and so the environment appears unsaturated to this drop. The drop therefore evaporates until its radius reaches the equilibrium value of 0.15 micrometer (point D again). The droplet represented by point D is in a stable equilibrium—droplets that are either slightly too large or too small for this equilibrium relative humidity evaporate or grow until they reach the equilibrium value.

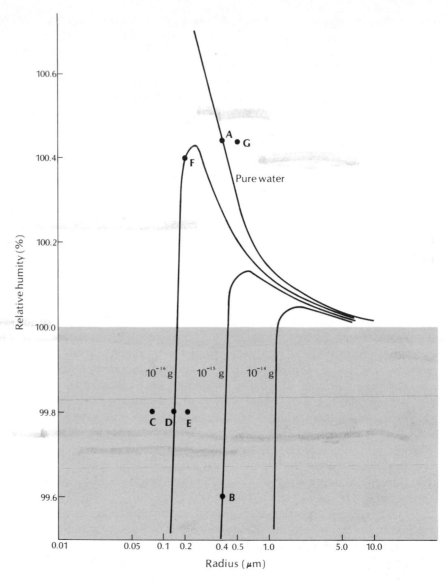

FIGURE 6.5 *Equilibrium relative humidity over a pure waterdroplet and over droplets of a dilute salt solution. The curves for the salt-solution droplets are labeled in terms of the mass of the original salt particle.*

The radii for this part of the curve are too small for cloud-drop radii (10–100 micrometers); therefore, droplets cannot grow to cloud-drop size for humidities less than 100 percent.

Unstable equilibrium can occur if the relative humidity exceeds 100 percent. For example, if a pure waterdroplet with a radius of 0.4 micrometer exists in a relative humidity of 100.42 percent (point *A*), it will be in equilibrium with its environment. However, if it is ever so slightly larger than 0.4 micrometer (point

140

G), the equilibrium relative humidity of the drop will become smaller than the relative humidity value for the environment, and the drop will grow rapidly. Thus, a droplet represented by point *A* is in an unstable equilibrium. The rapid growth of such drops helps explain why a cloud may form very suddenly when the relative humidity exceeds 100 percent.

If temperatures are below freezing as saturation is approached, water-vapor molecules may be converted directly to ice crystals, a process called *sublimation*. Sublimation requires tiny particles called *sublimation nuclei*. In contrast to cloud nuclei, which are abundant, sublimation nuclei are almost completely absent in the atmosphere, and so the initial formation of clouds by sublimation is rarer than by condensation, even in clouds at temperatures well below freezing. However, once supercooled waterdroplets form, a rapid freezing occurs if microscopic particles called *freezing nuclei* are present. Freezing nuclei, which have a structure resembling ice crystals, are much more common in the atmosphere than sublimation nuclei. Ice crystals themselves are excellent freezing nuclei. Thus, ice clouds form first as supercooled water clouds and then freeze if freezing nuclei are present.

MECHANISMS THAT PRODUCE CLOUDS   *6.4*

The descriptions of condensation, sublimation, and freezing in the preceding section do not account for the remarkable variety of cloud forms. The kaleidoscopic pageant of clouds delights perceptive observers and provides them with unmistakable clues concerning atmospheric structure and motion. Appendix A presents a glossary and illustrations of common clouds.

The shape of a cloud reveals much more about the motion of the atmosphere and the stability (vertical temperature structure) in the vicinity of the cloud. The texture of a cloud depends on whether the cloud is composed of ice crystals or of waterdrops. The brightness varies with the number and size of waterdrops (or ice crystals) present, as well as with the relative positions of the sun (or moon), the cloud, and the observer. Reflection, refraction, scattering, and diffraction of light may produce a cornucopia of special optical effects (see chapter 15).

Condensation in an initially unsaturated parcel of air can be described by again considering figure 6.2. For a cloud to form, an unsaturated parcel of air (point *A* in figure 6.2) must be cooled and/or moistened so that the temperature and mixing ratio of the parcel fall on the saturation-mixing-ratio curve. This can occur in two ways— *mixing* the original parcel with warmer, more humid air or *cooling* the parcel to its dew point without mixing. Of course, the cooling and mixing processes also may occur simultaneously.

6.4.1   CLOUDS PRODUCED BY MIXING

Let us first consider how a cloud can be produced by mixing of two unsaturated air parcels of differing temperatures and mixing ratios. Suppose that parcel *A* with a temperature of $-10°C$ and a mixing ratio of 1.6 gm/kg (point *A* on figure 6.6) mixes with a parcel of equal mass, a temperature of

10°C, and a mixing ratio of 7.6 gm/kg (point *B*). The temperature and mixing ratio of the mixture would be 0°C and 4.6 gm/kg (point *D*), respectively, if condensation did not occur. However, as shown by point *D*, this mixture would be supersaturated: the actual mixing ratio (4.6 gm/kg) would exceed the saturation mixing ratio (3.8 gm/kg). Therefore, waterdroplets form and a cloud is created by the mixing of the two initially unsaturated air parcels. Clouds or fog produced by horizontal mixing can never condense sufficient quantities of water to produce rain, but they do restrict visibility.

The production of clouds by mixing occurs often in nature when cold air flows over warmer water. As the cold, dry air aloft is mixed with moist, warm air immediately in contact with the water, clouds form (plate 3a). These clouds are often called *sea smoke* or *steam fog*. Sea smoke is common in middle latitudes in autumn or early winter when cold air flows across unfrozen lakes. It also occurs in all seasons when cold air drains over warmer lakes at night.

Condensation trails (or contrails) from jet aircraft are also clouds produced by horizontal mixing. A major component of jet exhaust is water vapor that exists at a very high temperature. When this hot, extremely moist air mixes with the cooler environmental air, a cloud is produced. The time required for the proper amount of mixing for saturation to occur explains the lag of the visible cloud behind the jet. As the mixing continues, the dry air ultimately dominates the mixture and the contrail evaporates. However, if freezing nuclei are present, the contrail may freeze. Once freezing occurs, the ice crystals may persist or even grow because of the difference in saturation mixing ratios for ice and water. As will be discussed in section 6.5.2, the environmental air may be saturated with respect to

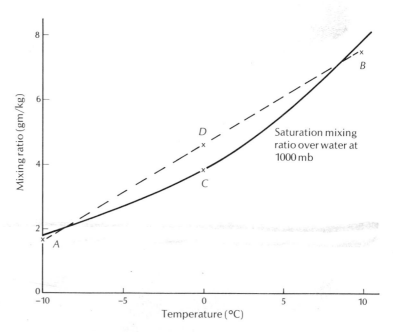

FIGURE 6.6 *Formation of a cloud by mixing at constant pressure of parcels* A *and* B.

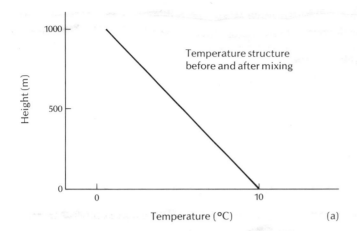

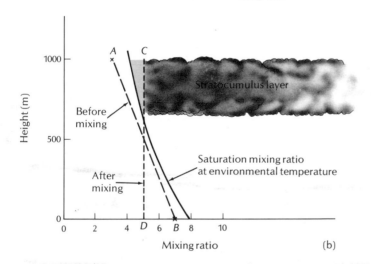

FIGURE 6.7 (a) Temperature sounding before and after vertical mixing. (b) Mixing ratio before vertical mixing (line A-B) and after vertical mixing (line C-D). Also shown is saturation mixing ratio at the environmental temperature.

water but supersaturated with respect to ice. Plate 3b shows an extensive band of cirrus clouds forming from contrails under just these conditions.

Under ideal conditions, vertical mixing may produce a layer of stratocumulus clouds near the earth's surface. Suppose that the temperature and mixing ratio of the lower kilometer of a calm atmosphere in which no vertical mixing is occurring are given in figure 6.7. Here the temperature lapse rate (figure 6.7a) is nearly dry abiabatic. The saturation mixing ratio for this sounding is depicted in figure 6.7b. As shown by the actual mixing-ratio profile (line A-B in figure 6.7b), the air is close (but not equal) to saturation at all levels. Now, if moderate winds develop and vertical mixing occurs, the temperature structure will not change appreciably because each rising and sinking parcel of air will cool or warm at the dry adiabatic rate. However, the moisture in the lower levels will be mixed upward, producing a more uniform mixing ratio in the vertical (curve C-D in figure 6.7b).

143

A layer of clouds will form in the layer where saturation is reached. This process can produce an overcast from a cloudless sky in a matter of minutes. It usually occurs after sunrise on calm, humid mornings as the vertical mixing associated with the heating of the ground begins.

In a more general situation in which the original temperature lapse rate is not dry adiabatic, appreciable temperature changes will also occur as the air is mixed vertically. In this case, the upper part of the layer will cool while the lower parts warms. The cooling in the upper part of the layer will then contribute to the cloud formation.

6.4.2 CLOUDS PRODUCED BY COOLING

The cooling of air to its dew point is the most common mechanism by which clouds are produced. Cooling of air may occur as a result of *diabatic* processes in which heat is removed from the air or *adiabatic* processes in which no heat is removed. The diabatic processes include radiation and conduction. Adiabatic cooling occurs when air expands as it is *lifted* and decreases in pressure.

Clouds produced by diabatic cooling are most common near the ground. Cooling by conduction when moist air comes in contact with a colder surface (such as snow, cold water, or cooler ground) can occur only at the surface. Cooling by radiation can occur anywhere in the atmosphere. However, the typical radiative cooling rates in the troposphere away from the ground are about 1 C° per day—too slow to produce clouds under most conditions. In the lowest kilometer, however, radiative cooling can exceed several Celsius degrees per hour, a rate that is sufficient to cool a layer of air to its dew point and produce a cloud.

Radiation-produced clouds are most likely on calm, clear nights with high humidities near the surface but lower humidities aloft. These conditions favor strong, radiative cooling of a shallow layer near the surface. If condensation occurs at the ground, the cloud is called *fog*. Ground fogs (plate 4a) and valley fogs (plate 20b) are formed in this way. If the cloud does not touch the ground, it is termed a *stratus cloud*.

Stratus clouds and fogs are also produced when warm, moist air is cooled as it flows over a colder surface. The sea fogs and stratus off the coast of the Pacific Northwest and Newfoundland are examples (plate 4b). Over land, such fogs and clouds often occur in spring as warm moist air from the south is advected (transported) over cold ground. A fog formed in this way is called an *advection fog*.

The fogs and stratus clouds formed by radiative and conductive cooling are layered clouds and do not show appreciable vertical development because the air is stable (cold air underlies relatively warm air). Also, because of the relatively slow cooling rates, these clouds and fogs are not usually more than a few hundreds of meters thick. At most, they are capable of producing light drizzle.

The mechanism that can produce the most rapid cooling rates in the atmosphere is the cooling of rising air by expansion. A parcel of air in a thunderstorm updraft, for example, can cool 40 C° in 15 minutes as it is carried from the lower troposphere to a height of 8 kilometers or more. Even the gentle, large-scale

144

upward motion of air associated with middle-latitude winter storms can produce a cooling rate in a parcel of air of more than 20 C° per day. Thus, upward motion can cause rapid cooling and a great amount of condensation in a relatively short time. Therefore, thick, precipitating clouds are always associated with upward motion, and any process that produces upward motion in the atmosphere can generate clouds.

A wide variety of clouds are produced by lifting. The nature of the cloud produced depends on the stability of the atmosphere (which is primarily determined by the temperature lapse rate) and on the horizontal scale of the lifting. In a stable atmosphere, the vertical development of clouds is restricted, and clouds tend to have a layered appearance with smooth tops and bottoms. Layered clouds produced by lifting over horizontal distances of 100 kilometers or more in a stable atmosphere include *cirrostratus* (high, thin, whitish clouds comprised of ice crystals—figure 15.7), *altostratus* (light, grayish clouds comprised of waterdrops through which the sun may be dimly visible—Figure 6.8), *nimbostratus* (thick, gray clouds that are producing precipitation), and sometimes *stratus clouds*. As we shall see in later chapters, cirrostratus, altostratus, and nimbostratus clouds are usually produced by large-scale upward motion associated with extratropical storms. A common mechanism that produces an extensive layer of low stratus clouds is the lifting of low-level air when east winds blow over the sloping Great Plains (figure 6.9).

In addition to the large-scale mechanisms for lifting the air, there are several common small-scale processes that produce clouds with a typical horizontal scale of a few tens of meters to several kilometers. Lifting on this scale is often a result of the wavy nature of stable atmospheric flow. Perfectly horizontal laminar flow is rare in a stable atmosphere; any perturbation or irregularity at the earth's

FIGURE 6.8 *Sun through altostratus clouds.*

145

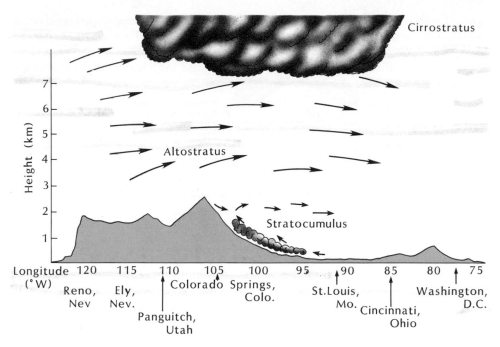

FIGURE 6.9 *Cross-section across the United States at 39° N. A stratocumulus layer is formed by east winds blowing up the sloping Great Plains. Altostratus and cirrostratus layers are formed by upper-level air rising over a large horizontal scale.*

surface or within the atmosphere itself will induce waves in the flow. The most common example is the perturbation to the winds induced by air flowing over a mountain (figure 6.10 and plate 5a). The trajectories of air parcels are depicted in figure 6.10. The horizontal size and vertical thickness of the mountain-induced cap cloud depends critically on the humidity of the air. If the air approaching the mountain is already saturated, a distinct cloud associated with the mountain will not be present. As the humidity of the approaching air is reduced, fewer parcels will experience sufficient lifting and cooling to produce condensation, and the cap cloud will shrink. It is obvious that the shape and thickness of the cap cloud is determined by the initial relative humidity of the air parcels that pass over the mountain and by the amount of lifting experienced by each parcel. If the parcel has been cooled to its dew point, cloud material will be present. Alternating layers of high and low relative humidity can produce two or more cap clouds (figure 6.10c). In this example, the parcels of air following the middle trajectory were initially too dry to allow condensation.

Mountains are not the only source of waves in the atmosphere. Rapidly growing cumulus clouds may act as a partial barrier to the winds aloft, causing the air to flow over the cumulus top (figure 6.11). If the ambient humidity is high enough, a cap cloud, also called a *pileus,* can form (figures 6.11 and 10.4).

Waves of very short horizontal wavelength (tens of meters) also can occur in the atmosphere, sometimes extending for tens of kilometers downwind

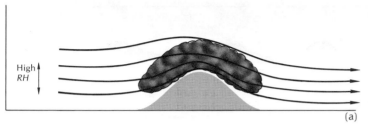

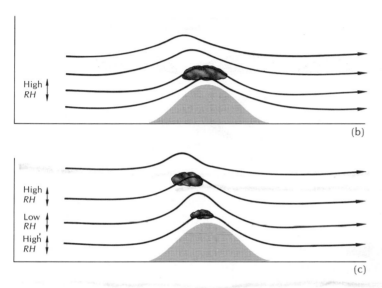

FIGURE 6.10 *Clouds produced by air of varying relative humidity flowing over a mountain. (a) High relative humidity through a deep layer. (b) High relative humidity through a shallow layer. (c) Two layers of high relative humidity.*

from an obstacle. When they occur in a layer of air that is close to saturation, the small upward and downward motions associated with the waves may be sufficient to noticeably increase the cloud thickness in the upward-moving portion of the wave and to evaporate the cloud in the downward-moving portion of the wave (figure 6.12). The horizontal spacing of the clouds depends on the mean relative humidity as well as on the horizontal wavelength of the wave (compare figures 6.12a and b). An example of waves in a layer of altocumulus clouds is shown in plate 5b.

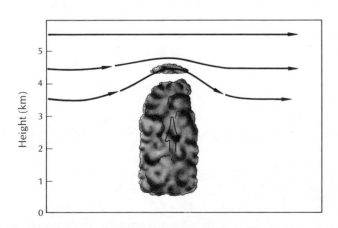

FIGURE 6.11 *Cap cloud (pileus) formed by air flowing over a growing cumulus cloud.*

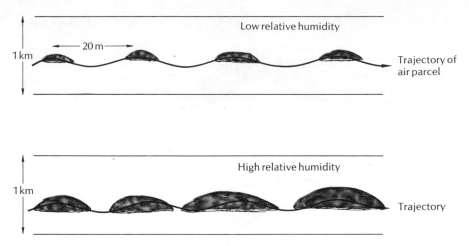

FIGURE 6.12 *Wave clouds produced in air of slightly different relative humidities.*

When a stable layer of air underlies a less-stable layer, as in an inversion, waves can move along the top of the inversion. If clouds exist at this level, the vertical displacements of the air that occur with the moving waves become visible. An example of such a series of waves moving along the top of a cirrus layer is shown in figure 6.13. The wavelength of these waves is about 11 kilometers, and the vertical extent of the cloud layer is 1.5 kilometers. The clouds are approximately 10 kilometers above the surface.

Small-scale vertical motions are often responsible for the development of isolated or detached cirrus clouds that resemble tufts of fine hair. Cirrus clouds can form in the upward-moving part of short-wavelength waves in the upper troposphere. If enough condensation and subsequent freezing occur, the release of latent heat can make the cirrus clouds buoyant. Recent observations have found small-scale upward velocities of over 1 m/s in detached cirrus clouds. The temperature in these clouds was about 0.1°C warmer than the temperature of the surroundings. Detached cirrus clouds may also be the remnants of the top (anvil) of a thunderstorm (cumulonimbus) cloud.

The stable atmosphere resists vertical displacements and produces smooth, layered-type clouds. Very different types of clouds are produced when the atmospheric structure is unstable with respect to saturated vertical motions. We have seen that because of the latent heat given to the air by condensation, a parcel of saturated air will cool at a rate of approximately 6 C°/km.*

When an environmental lapse rate greater than 6 C°/km exists, saturated particles of air that are displaced upward will no longer become colder than their environment and be forced back to their original level. Instead, they will remain warmer than their environment and will accelerate upward. Under these

---

*Unlike the dry adiabatic lapse rate, which is a constant 10 C°/km, the wet adiabatic lapse rate varies somewhat with temperature and pressure.

Plate 5a Wave clouds over Mount Rainier, Washington (© Alistair B. Fraser).

Plate 5b Wave in altocumulus clouds over Florida (© Richard A. Anthes).

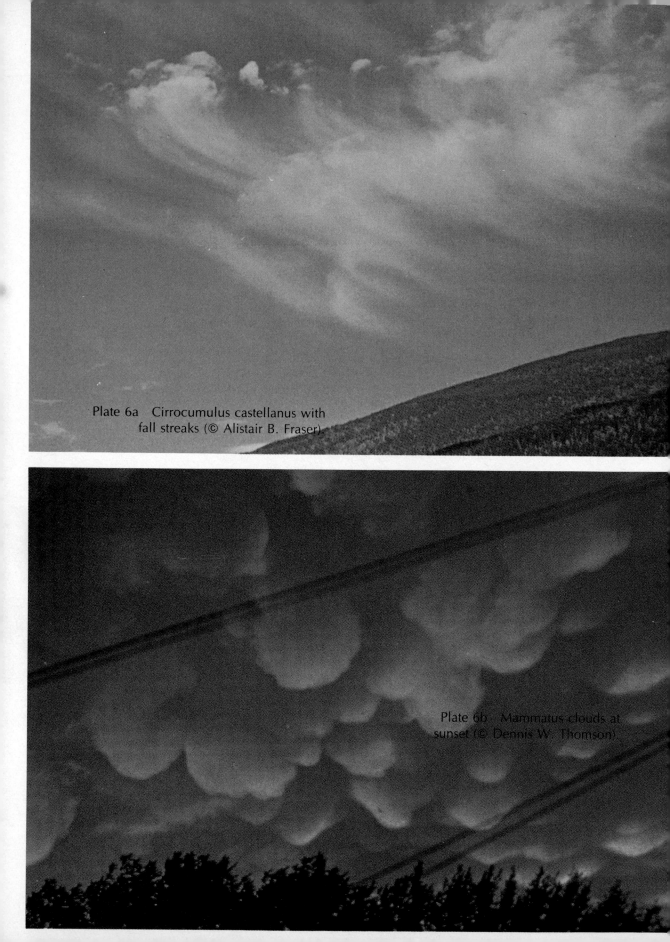

Plate 6a Cirrocumulus castellanus with fall streaks (© Alistair B. Fraser).

Plate 6b Mammatus clouds at sunset (© Dennis W. Thomson).

FIGURE 6.13 *Waves on cirrus clouds.*

conditions, clouds of small horizontal extent have a greater upward acceleration than larger clouds; therefore, the preferred scale of clouds in an unstable environment is always small—tens of meters to several kilometers.

Unstable layers of air (with respect to saturated motions) can occur at any level in the troposphere. However, because most of the incoming solar radiation heats the ground, which in turn heats the low-level air, unstable air is most likely in the lower half of the troposphere. Almost every day during the summer in middle latitudes there is sufficient low-level instability to permit the growth of cumulus clouds—small isolated clouds with flat bases and vertical scales comparable to their horizontal scale (typically 100 meters).

The vertical development of clouds in an unstable layer is limited by two factors, the depth of the unstable layer and the amount of moisture in the unstable layer. The limiting effect of the depth of the unstable layer on the growth of the cloud top is obvious. Once the cloud penetrates into a stable layer, the cloud soon becomes colder than its environment and experiences a downward acceleration. In figure 6.14a, a parcel of air that is slightly warmer than its environment rises from the ground. Condensation occurs when the temperature is lowered enough that the saturation mixing ratio equals the actual mixing ratio. The parcel of air then cools at the wet adiabatic rate indicated by the dashed line in figure 6.14a. In the unstable layer, the parcel remains warmer than its environment, and the upward acceleration continues. When the cloud encounters a stable layer, it becomes colder than its environment, and the acceleration becomes downward. If the cloud has sufficient upward motion, its momentum can carry it part of the way into the stable layer. In fact, if the stable layer is shallow enough (figure 6.14b) and the cloud's upward velocity is great enough, the cloud may

149

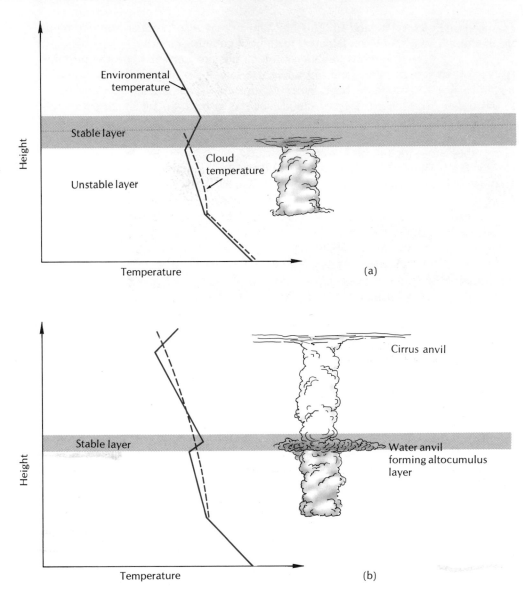

FIGURE 6.14 *Effect of the temperature structure on the development of thunderstorms. In (a), the stable layer stops the rise of the thunderstorm, whereas in (b), the thunderstorm penetrates the thin, stable layer.*

"coast" through the entire stable layer and redevelop upon emerging into an unstable layer.

When the cloud encounters a stable layer, it tends to spread out horizontally (figures 6.14a and b). Thus, thin stable layers in the atmosphere are often revealed by patchy layers of clouds that formed by the lateral spreading of convective clouds. The cirrus veil associated with the thunderstorm anvil is a common example (see plates 10a and b). This mechanism may also produce stratocumulus or altocumulus layers by the lateral spreading of cumulus clouds.

Cumulonimbus (thunderstorm) clouds are formed when the unstable layer is deep (greater than approximately 4 kilometers). Examples are shown in plates 1a and 10 and figure 10.4. When the unstable layers are shallow and occur in the upper troposphere, layered clouds such as cirrostratus or altocumulus may break into discrete convective elements. When these convective clouds consist of ice, they are called cirrocumulus castellanus clouds (plate 6a); when composed of water, they are altocumulus castellanus.

The depth of convective clouds is also limited to some degree by the amount of moisture in the surrounding air. Rising clouds mix with their environment, a process known as entrainment. If the environmental air is dry enough, the entrainment of dry air into the rising cloud may be sufficient to evaporate all of the liquid water in the cloud. With mixing, therefore, the rising parcel cools at a rate between the wet and dry adiabatic rates. The greater the mixing and the drier the environmental air, the closer is the cooling rate to the dry adiabatic value. Consider the situation depicted in figure 6.15. The environment cools at a rate of 8 C°/km, and so the layer is unstable with respect to wet adiabatic motion. If the humidity of this layer is high (figure 6.15a), a rising parcel will cool at nearly the wet adiabatic rate because the entrainment of moist air will not evaporate very much water in the cloud. This parcel will remain warmer than its environment and extend throughout the entire depth of the unstable layer.

In contrast, if the environment is very dry (figure 6.15b), the dilution of the rising parcel by entrainment can evaporate so much water that the parcel cools at a rate close to the dry adiabatic rate, say 9 C°/km. This parcel becomes colder than the environment at a level far below the top of the unstable layer. Therefore, in this case, the humidity of the environment, rather than the depth of the unstable layer, controls the cloud top.

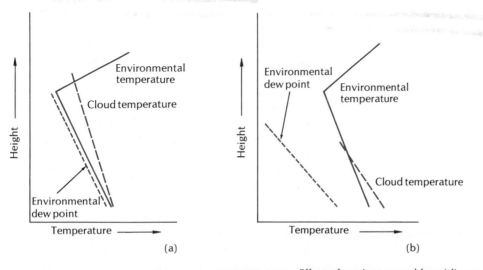

(a)   (b)

FIGURE 6.15   *Effect of environmental humidity on cloud temperature.*

FIGURE 6.16 *Downward-moving convection.*

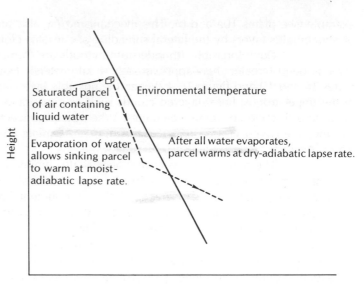

Although the most common form of convective cloud is associated with rising parcels of air that remain warmer than their environment through the release of latent heat, convection caused by cold parcels of air sinking relative to the warmer environment also occurs (figure 6.16). The difficulty in producing downward-moving clouds lies in maintaining saturation. As the cloud warms by compression, a source of liquid or frozen water must be present for evaporation to maintain the cold temperature of the sinking parcel relative to the environment. The amount of liquid water present in a cloud is typically 1 gram per kilogram of air. This amount of water will allow the parcel to warm by about 2.5 C° before all of the liquid is evaporated. Since with evaporation the parcel warms at a rate of approximately 6 C°/km (the wet adiabatic rate), this amount of liquid water will evaporate in about 400 meters. Thus, downward-moving convection tends to be rather shallow, since it is limited by the amount of liquid water that is initially present or falls into the cloud. Exceptions occur in the downdrafts of thunderstorms that are kept near saturation by precipitation falling into the downdrafts.

The most commonly observed type of downward-moving convection are *mamma clouds,* which appear as hanging protuberances on the undersurface of a cloud (plate 6b). Mamma clouds can occur on the undersides of altocumulus, altostratus, cirrus, or stratocumulus clouds, but they are most commonly associated with cumulonimbus. The environment of cumulus clouds is unstable, and the large condensation rates in the cumulonimbus clouds provide abundant liquid water. Mamma clouds are sometimes observed at the base of the cumulonimbus, but they are more frequent on the undersurface of the projecting anvil.

6.5 PRECIPITATION

Thus far, we have discussed the condensation process and the types of clouds that form under various atmospheric conditions of stability and vertical motion. The condensation process, of course, only explains the formations of liquid cloud drops; we have not yet discussed in detail the formation of ice crystals. Although ice crystals probably do form directly from vapor on rare occasions if sublimation nuclei are present, it is far more common for ice crystals to form from the freezing of liquid water. Thus, the production of cloud drops is the first stage in the formation of most clouds and precipitation.

However, the formation of cloud droplets does not water the corn-field, for the terminal velocity of a cloud drop with radius 10 micrometers is about 1 centimeter per second; at this rate of fall, it would take eighty-four hours for the water to reach the ground from the middle cloud level, a distance of about 3 kilometers. Furthermore, raindrops must usually survive a fall through an unsaturated layer between the base of the clouds and the ground. A droplet of radius 10 micrometers would survive for a distance of only a few meters through a layer of 90 percent humidity near the surface. Therefore, some process must increase the size of the thousands of millions of cloud droplets to a size that will make them heavy enough to fall to the ground before they evaporate or are carried away by the horizontal winds.

6.5.1 *PRECIPITATION IN CLOUDS ABOVE FREEZING*

Although it might seem plausible that the droplets would continue to grow by the condensation process, the growth rate by continued condensation on existing waterdroplets is just too slow. As a drop grows in size, the hygroscopic solution responsible for the condensation in the first place becomes more diluted and, therefore, less efficient in condensing water vapor. Indeed, most clouds never produce rain, since larger-scale changes in the synoptic-scale flow (such as subsidence) dry the cloud before sufficient growth can occur through simple condensation process. The processes of collision and coalescence are necessary in the occasional active clouds that do exhibit rapid growth and fallout of precipitation. Figure 6.17a illustrates these two processes in warm clouds.

Collision is simply the bumping together of two cloud droplets. Collision is favored between drops of different sizes and, therefore, different terminal velocities. The large drops overtake the slower-moving small drops.

In the laboratory, drops that collide usually bounce off each other unless an electric field is present. In the latter case, the electric field causes the electrical charges on the drop to separate. *Coalescence*, or the merging of two drops after they collide, is favored if the drops have opposite charges or if the charges are separated, because opposite charges attract each other. Therefore, the atmosphere's electric field favors the coalescence process in natural clouds. Thus, atmospheric electricity plays a role in the growth of cloud droplets.

While collision and coalescence are sufficient to produce copious rainfall in the tropics, where relative humidities are high and hygroscopic nuclei are present, another process, involving the growth of ice crystals rather than raindrops, is efficient in middle latitudes. Even in summer, much of the rain that falls over the United States begins as snowflakes at elevations of 3 to 5 kilometers, melting before reaching the ground in the warm air near the surface. Thus, during

153

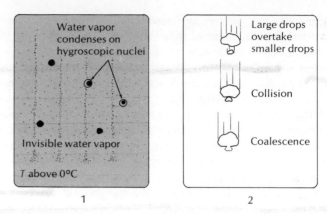

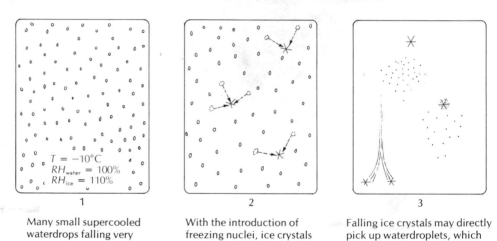

(a) Growth of waterdroplets by collision and coalescence

1	2	3
Many small supercooled waterdrops falling very slowly.	With the introduction of freezing nuclei, ice crystals form, which then grow at the expense of waterdroplets.	Falling ice crystals may directly pick up waterdroplets, which freeze upon contact, producing riming. Others may fracture, producing additional small ice crystals, which serve as freezing nuclei.

(b) Growth of ice crystals at the expense of waterdroplets

FIGURE 6.17 *Precipitation process in (a) warm and (b) cold clouds.*

a warm, muggy thundershower, it may be refreshing to know that a raging blizzard is occurring a scant 5 kilometers away—directly overhead.

6.5.2 ROLE OF ICE IN PRECIPITATION PROCESS

The ice phase of water complicates the precipitation process enormously because direct transition from each of the three phases of water to any of the other phases is possible. Thus, frost is not frozen dew, nor snow frozen rain, but they are instead results of a direct conversion of vapor to solid (called *sublimation* or *deposition*) at subfreezing temperatures.

We have seen that one of the most important properties of water is that the saturation vapor pressure is less over ice than over water, which means that air of 100 percent relative humidity (calculated with respect to liquid water) has a relative humidity considerably greater than 100 percent when calculated with respect to ice. The difference, which depends on temperature, is shown in figure 6.18. Thus, if the temperature is −10°C and the humidity is 100 percent with respect to supercooled water, the relative humidity with respect to ice is 110 percent.

The difference in saturation vapor pressure means that water droplets and ice crystals cannot peaceably coexist in the same cloud, for the air will always "appear" more saturated to the ice crystals than to the liquid droplets. Consequently, the ice crystals will continue to grow at the expense of the water droplets, which evaporate and yield the resulting vapor to the "hungry" ice crystals. The huge supersaturations (up to 150 percent) that are possible with respect to ice mean that the growth of ice crystals by this process can be very rapid, and so sizable ice crystals (large enough to fall to the earth before melting or evaporating) can form. There is one catch, however; ice crystals seem reluctant to form, even at temperatures as low as −20 or −30°C, unless given a start by freezing nuclei. When the temperature reaches −40°C (−40°F), even pure water freezes without the benefit of freezing nuclei. Once the freezing process is initiated, a feedback process begins, with the first ice crystals growing, fracturing, and serving as freezing nuclei for future ice crystals (figure 6.17b).

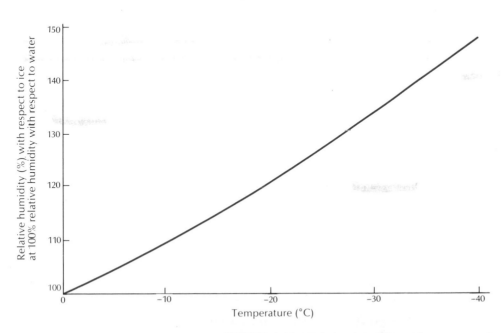

FIGURE 6.18 *Relative humidity with respect to ice when the air is saturated (100 percent relative humidity) with respect to water.*

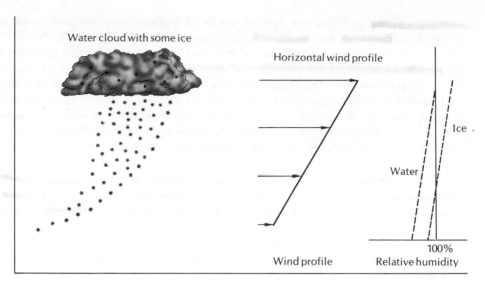

FIGURE 6.19 *Production of a fall streak.*

The growth of ice crystals and the subsequent precipitation through the process of robbing Peter (water) to pay Paul (ice) is known as the *Bergeron-Findeisen process,* after the scientists who discovered it in the 1930s. Before World War II, this process was believed to be essential to the formation of all precipitation.

The difference in saturation vapor pressure over ice and water has another important consequence in the cloud and precipitation processes. Clouds in the upper troposphere often are formed as water drops, even when the temperature is as cold as −30°C. Air in the vicinity of these supercooled water clouds may be slightly unsaturated with respect to water but supersaturated with respect to ice. The introduction of freezing nuclei (either by natural or artificial means) into the water cloud will induce a freezing and subsequent growth of ice crystals through the Bergeron-Findeisen process discussed above. As these ice crystals grow and become heavier, they fall out of the water cloud, forming a fall streak. This process is illustrated in figure 6.19 and plate 6a. Now, if the layer under the water cloud is saturated with respect to ice, as indicated in this example, the crystals continue to grow through sublimation. Furthermore, if the horizontal wind varies with height (wind shear), the parent cloud and the falling ice crystals move at different velocities. Thus, the curvature of the fall streaks (plate 6a) reveals the vertical shear of the horizontal wind. In later chapters, we will see that this shear is an indication of horizontal temperature variation and can be used to help make a temperature forecast.

6.5.3 TYPES OF PRECIPITATION

The growth of raindrops through collision and coalescence or ice crystals through the Bergeron-Findeisen process produces precipitation that is heavy enough to fall to the ground. The form of precipitation that actually reaches the ground depends on the history of the raindrop or ice crystal after

156

it is formed. The most important aspect of this history is the temperature structure of the layers through which the precipitation particles, or *hydrometeors*, fall. In this section, we discuss the temperature structures associated with rain, snow, graupel (soft hail), sleet (ice pellets), freezing rain, and hail.

Rain reaching the ground in middle latitude is usually melted snow. Even in the summertime, the layers of the atmosphere where precipitation forms (at heights of 3 to 5 kilometers) are cold enough to produce snow. The freezing level in summertime is typically 3 kilometers. Snow that forms above this level melts in a deep layer of above-freezing temperature and reaches the ground as rain (figure 6.20a).

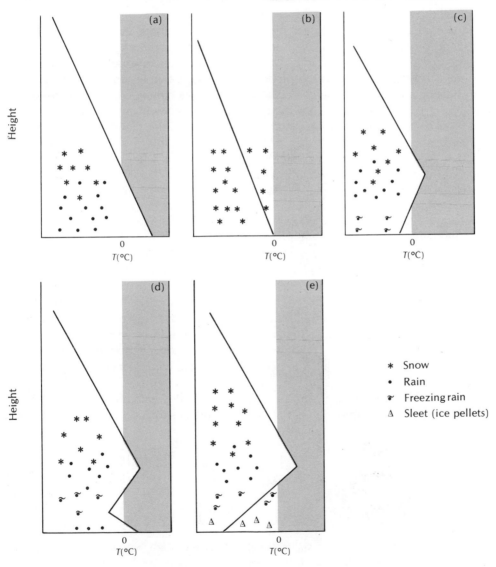

*	Snow
•	Rain
~	Freezing rain
Δ	Sleet (ice pellets)

FIGURE 6.20 *Temperature soundings that lead to various types of precipitation.*

157

Although most of the rain that falls in middle latitudes begins as snow, rain in the humid tropics, especially in the vicinity of tropical oceans, often does not go through an ice stage. Because of the high mixing ratio and the abundance of giant sea-salt nuclei, collision and coalescence are very efficient, and precipitation can form in cumulus clouds that extend to a height of only 3 kilometers or so.

If the entire troposphere is below freezing, snow can reach the surface (figure 6.20b). The variety of snow crystal types is legendary—ice crystals in the form of hexagonal plates, starlike crystals, columns, needles, or amorphous graupel are common (figure 6.21). The type of crystal that forms is determined by the temperature and relative humidity. The relationship, however, is complicated. In general, needles and columns form at low temperatures and relative humidities that are greater than 100 percent with respect to ice but less than 100 percent with respect to water. As the relative humidity increases, plates and stellar crystals become common. As the humidity increases still further, to exceed saturation with respect to water, graupel is the predominant type. Graupel forms when small ice crystals pass through a supercooled water cloud. The falling crystal captures liquid waterdrops, which freeze immediately, a process known as *riming*. The riming process also occurs on subfreezing solid objects in water clouds such as airplane wings or trees (plate 7a).

A wide variety of precipitation is possible when layers of air that are above freezing occur. Freezing rain occurs when a shallow layer of subfreezing air exists at the ground (sounding c in figure 6.20). Rain falling into this layer from the warmer air aloft freezes upon contact with subfreezing objects. The accumulation of weight can be devastating. For instance, a wire 100 meters long coated with 2 centimeters of ice must support 126 kilograms.

The occurrence of freezing rain can vary tremendously from place to place in mountainous regions. If the coldest air occurs in the valleys, the severest condition will be in the lower elevations. In fact, some of the higher mountains may penetrate into the warmer air and escape the freezing rain altogether. However, the reverse can also occur. When an elevated inversion exists, as depicted in figure 6.20d, a layer of above-freezing air can exist at the lowest levels as well as aloft, with a narrow layer of below-freezing air sandwiched in between. Under these conditions, the freezing rain occurs only on the higher elevations; the valleys receive only rain.

A sequence of surface temperatures and weather conditions during a typical freezing rain event in central Pennsylvania is shown in figure 6.22. Precip-

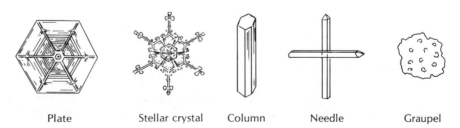

| Plate | Stellar crystal | Column | Needle | Graupel |

FIGURE 6.21   *Examples of snowflake types.*

158

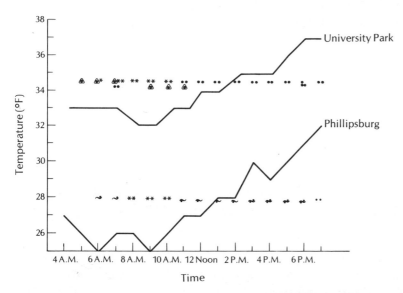

FIGURE 6.22 *Surface temperature and precipitation at University Park and Philipsburg, Pennsylvania, on December 30, 1975.*

itation began around 6 A.M. EST at University Park and Philipsburg in the form of sleet and freezing rain. Shortly after the precipitation began, the sleet and freezing rain changed to snow, probably as a result of evaporational cooling in the unsaturated air near the ground. However, continued warm advection restored the temperature aloft to above the freezing point, and the snow ended at both stations around 10 A.M. EST. In the valley, the temperature started out slightly above freezing and rose gradually throughout the afternoon and evening. However, on the mountaintop at Philipsburg, the temperature remained below freezing until 7 P.M. EST, when it finally reached 0°C. During this period, light, and sometimes moderate, freezing rain occurred. Thus, a relatively harmless precipitation event in the valley was a major ice storm on the nearby mountaintops.

Figure 6.23 shows the effects of an elevated inversion on the morning after an ice storm. The freezing level of the night before is clearly visible about halfway up the slopes of Mt. Nittany, which is located about 10 kilometers northeast of State College. Above this level, which is about 90 meters (295 feet) above the valley floor, trees glisten in their coat of fresh ice. Below this level, the trees and ground are free of ice.

Sleet or ice pellets reach the ground when a deep layer of air that is well below freezing occurs near the ground (figure 6.20e). Partially melted snow falling through such a layer can freeze into pellets of ice, arriving at the ground with considerable noise as they rattle against the roof or windowpane.

The formation of hail is associated with the updraft and downdraft structure of thunderstorms and is discussed in detail in chapter 10. Here, we simply note that hail is formed when ice particles are lifted several times by strong updrafts through layers of supercooled waterdrops. With each passage, the ice pellets accumulate a coating of ice. The hailstones grow in this way until they become too

159

FIGURE 6.23 *Mt. Nittany, Pennsylvania, on the morning after an ice storm.*

heavy to be supported by the rising air. The large size of hailstones enables them to fall through deep warm layers of air near the ground, so hail in summer is common.

6.6 WHEN LEPRECHAUNS PLAY, RAIN, SNOW, AND THUNDERSTORMS ALL IN ONE DAY

To see how some of the concepts concerning moisture, clouds, and precipitation act together to produce important weather, let us consider the situation on the evening of St. Patrick's Day, March 17, 1977, in western Pennsylvania. In spite of the tranquil spring day, an approaching storm from the Ohio Valley pretty much guaranteed precipitation later that night and the next day. The forecasting problem was to determine the form of precipitation that would fall. Because afternoon temperatures on March 17 were generally in the low 50s (°F) under bright, sunny skies, rain seemed to be a good bet. However, a look at the temperature and wet-bulb sounding for 7 P.M. EST, March 17 showed some ominous features (figure 6.24). Although the surface air temperature had cooled slightly from the daily high of +8°C, it was still well above freezing at +6°C. The temperature decreased rapidly with height in the layer from 250 to 1200 meters, cooling at the dry adiabatic lapse rate of 10C°/km, and fell below freezing at a height of about 1 kilometer above the ground. Thus, a layer of above-freezing temperatures approximately 1 kilometer

160

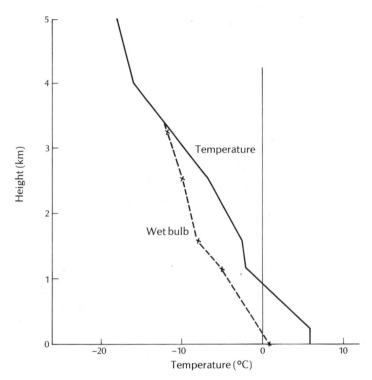

FIGURE 6.24 *Temperature and wet-bulb sounding for 7 P.M. EST on March 17, 1977, in western Pennsylvania.*

thick was next to the ground. Under many conditions, this layer would be thick enough to melt any snow falling through it.

However, the layer of air near the ground was not only warm; it was also very dry, as shown by the wet-bulb temperature curve in figure 6.24. The wet bulb of the entire lower atmosphere was below freezing, with the exception of a very thin layer next to the ground. If precipitation fell into this layer, evaporation could lower the temperature to below freezing everywhere except right at the surface. Thus, in spite of the mild surface temperatures, snow was a distinct possibility if the air remained dry until the precipitation began.

Later that night, precipitation did reach western Pennsylvania. The sequence at Altoona, Pennsylvania, was typical of the region. At 2:20 A.M. EST, March 18, the first raindrops reached the ground. At this time, the temperature was +5°C (40°F). In twenty minutes, evaporation had chilled the air to +2°C (36°F), and wet snow had mixed with the rain. By 3 A.M., the precipitation was entirely wet snow, and the temperature was at the wet-bulb value of +1°C. Heavy snow fell for the next six hours, accumulating 8 to 15 centimeters (3 to 6 inches) by 9 A.M. EST, and Pennsylvania blamed leprechauns for the sudden return to winter. However, the leprechauns were not quite finished.

The temperature and wet-bulb sounding for 7 A.M. EST, March 18, 1977, is shown in figure 6.25. In contrast to the sounding twelve hours earlier (figure 6.24), the lower troposphere is nearly saturated. In the lowest 2 kilometers, the temperature is within 1°C of the freezing point, the cooling coming from the evaporation of water into the layer.

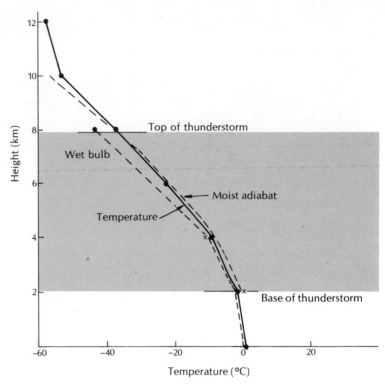

FIGURE 6.25 *Temperature and wet-bulb sounding for 7 A.M. EST on March 18, 1977, in western Pennsylvania.*

While cooling was occurring in the lowest kilometer, permitting snow to reach the ground, important changes were also happening in the layer between 2 and 4 kilometers. Warm, moist air from the lower Ohio Valley swept in at these levels under the colder air aloft. For example, at 3 kilometers, the temperature rose from −10°C to −3°C, producing a slightly unstable layer between 2 and 8 kilometers (figure 6.25). A parcel of air that was 0°C (1C° warmer than its environment) at 2 kilometers would remain slightly warmer than its environment if it rose and cooled at the *wet* adiabatic rate up to a height of about 8 kilometers (see dashed line labeled *wet adiabat* in figure 6.25). The nearly saturated environment in this layer guaranteed that any drying effects by entrainment would be negligible. Thus, thunderstorms were a real possibility even with the cold surface temperatures. And sure enough, in the middle of the wet snowstorm, Pennsylvanians were awakened by the crashing of thunder and the rattle of hail on the morning of March 18. Obscured by the snow and fog, cumulonimbus clouds reached from their base at 2 kilometers to heights of almost 8 kilometers (as measured by weather radar)—more leprechauns at play—giving a mixture of the worst of summer and winter weathers.

Questions

1   List at least four unusual properties of water, and give one important consequence of each property.

2   Explain why there is some truth to the adage, "It's too cold to snow."

3   What is the saturation mixing ratio over water at 0°C, 10°C and 20°C? (Use figure 6.2.)

4   Use figure 6.2 to find the dewpoint of a parcel of air with a temperature of 20°C and a mixing ratio of 10 g/kg. (Answer: 14°C)

5   Explain the difference between the dewpoint and wet bulb.

6   Why does the temperature usually fall at the beginning of precipitation?

7   How can condensation occur at relative humidities below 100 percent?

8   What effect does the curvature of water drops have on the equilibrium relative humidity?

9   Explain how stratocumulus clouds form over the Great Lakes in early winter when very cold, dry air flows across the water.

10 Why is cooling by radiation not an important mechanism for producing precipitation? What is the important cooling mechanism?

11 Contrast the appearance of clouds which form in stable and unstable layers.

12 Briefly describe the difference in precipitation production in warm clouds (above 0°C) and cold clouds (below 0°C).

13 Why would seeding a supercooled cloud with ice crystals enhance the formation of precipitation?

14 Sketch a vertical temperature sounding that would give freezing rain on mountain tops but not in the valleys.

Problems

15 A 100-watt light bulb puts out 100 Joules of energy per second. How long would it take a 100-watt bulb to completely melt all of the ice in an ice tray holding 800 g of water? How long would it take to then evaporate the water? (1 calorie = 4.187 Joules. Assume all of the energy from the bulb is used to melt the ice or evaporate the water.) (Answers: 44.7 minutes, 5.4 hours)

16 Suppose the temperature and relative humidity in a classroom measuring 10 m × 10m × 5m are 20°C and 75 percent, respectively. If all the water vapor were condensed, how deep would the layer of water on the floor be? (Assume an air density of 1.2 kg/m³, and use figure 6.2 to obtain the actual mixing ratio.) (Answer: 0.007 cm)

17 Explain physically why the dewpoint is always less than the wet bulb (except at saturation, when they are equal).

18 Why do wave clouds usually occur in stable atmospheric layers?

SEVEN

The Changing Weather Outside the Tropics

cyclone • polar air mass • arctic air mass • comma cloud pattern • comma head and tail • front • cyclonic rotation • anticyclonic rotation • stratocumulus clouds • stratus sheets • cumulonimbus clouds • cumulus convection • cirrus outflow • water vapor blanketing • cold anticyclone (high pressure) • potential temperature • maritime or continental tropical air mass • frontal zones • jet stream • jet streak • trough aloft • Norwegian cyclone model • sea-level pressure • pressure gradient force • Coriolis force • Buys Ballot's law • geostrophic wind • vertical motion effect on clouds • buoyancy • overrunning • orographic lifting • cold or warm advection • veering and backing wind • convergence and divergence • frictional convergence • warm, cold, stationary front • occlusion • steering principle • warm sector • stable and unstable air • surge region

Weel, do ye ken, sir, that I never say in a' my born days what I could wi' a safe conscience hae ca'd bad weather? The warst has aye some redeemin' quality about it that enabled me to thole it without yaumerin. [*Shepherd*, from Richard Inwards, *Weather Lore* (London: Elliot Stock, 1893), 190 pp.]

7.1 RHYTHMS IN THE WEATHER

We are accustomed to certain rhythms in the weather. The daily cycle of the earth's rotation warms the air from morning to afternoon, stirring the wind and drying the morning fogs; fluffy daytime cumulus clouds, afternoon showers in the mountains, clearing skies at sundown, and the nighttime thunderstorms of the Midwestern summer. Annual rhythms mark seasonal temperature swings at the high latitudes of Leningrad and Montreal, the wet winters and dry summers of San Francisco and Rome, winter clouds and brighter summer skies at New Orleans and London, and clean, refreshing summer lake breezes at Cleveland and Buffalo, accelerating to a howling blizzard in late fall.

Rhythms aside, the weather also changes from day to day or after a few days, albeit irregularly. Why is it sunny today, rainy tomorrow? The answer has little to do with seasonal or daily cycles; instead, it is related to traveling storm systems that cover hundreds of kilometers and have typical lifetimes of about four days. Spatially, the weather that we observe is highly variable, even over very short distances. Who has not noticed rain at one end of town and not at another, deeper snow lying a few kilometers distant, or cooler temperatures literally within walking distance? Nevertheless, consistent weather patterns do occur on larger scales: beneficial rains water the entire Corn Belt, snow may blanket several states, and cold waves engulf half of the United States.

Thus, it is appropriate to consider the larger patterns of weather, recognizing that there may be considerable variation of actual weather within them. Implicit in this approach is the idea that the larger weather pattern usually controls the framework or range of the smaller events "permitted" within it. For example, over 60 percent of all tornadoes in the United States move from southwest to northeast; these tornadoes are very small, but the larger-scale winds steer them preferentially northeastward. Plainly, the condition for tornadoes often involves southwesterly flow on a large scale. The characteristic time between cloudy or rainy periods at most locations in the United States, Canada, and Europe is a few days; the cause lies in sporadic weather disturbances, now lifting the air to form clouds, now compelling it to sink, dissipating them.

In what follows, we study these large regions of disturbed weather, called *cyclones,* concentrating more of our attention on them than on the even larger areas of fair weather in between. However, as we emphasize the cyclones outside the tropics, we will learn that they form and move near the boundaries of great volumes of cold air that are generated by radiative cooling, mainly at high latitudes. Such cold air masses are usually quite free of clouds; clouds and rain are favored near the edges, sometimes called *fronts.* But the existence of these cold masses and the air motions within them play a vital role in the formation and life of the cyclone. Indeed, the wind energy of the storm must come from another energy source—the sinking motion of the cold air. As the cold air gives up the potential energy associated with its elevation above ground, that energy becomes available to the winds. Thus, the alert stormcaster keeps watch for the buildup and movement of cold air masses as an early sign of the exciting cyclonic things to come.

MOVING WEATHER SYSTEMS VIEWED FROM SATELLITES 7.2

Cyclones can move across the earth's face as rapidly as 100 km/h or almost imperceptibly, but typically at 30 to 60 km/h. Interestingly enough, they move faster in winter than in summer, but are farther apart in the winter, so the characteristic time between them is comparable throughout the year. However, it is never so regular that we can say that it will rain every three days or four, but only that the weather will change sooner or later.

Figures 7.1–7.5 illustrate the migrations of clouds for the period of May 1–May 5, 1973. These are composite daytime satellite pictures produced by the NOAA–2 satellite, which orbits over an approximately north-south path. The picture is built up by strips as the earth rotates eastward. Thus, it is always midmorning underneath the satellite. Although the pictures are not taken at one instant, they do depict the general cloud conditions on a given day. The clouds, being good reflectors, appear white from sunlight reflecting back up to the satellite. By contrast, in the dark regions of the pictures, the skies are more nearly clear. One exception is the strong reflection from snow, notable over Greenland and the polar regions. We can differentiate the deep clouds from snow by looking at an *infrared* picture. Figure 7.1b shows the infrared view for the same time as figure 7.1a. Infrared sensors read the temperatures of objects viewed rather than their reflective properties, so this picture is white for only the high clouds, whose tops are very much colder than the snow (perhaps as low as $-40°C$). Throughout early May, there is snow at high latitudes and broken bands of tropical clouds north of the equator, but we will concentrate on the changing middle latitudes.

On May 1 (figure 7.1), extensive clouds cover Japan, with a distinct band extending southwestward to China. Much of the Pacific Ocean is clear, ex-

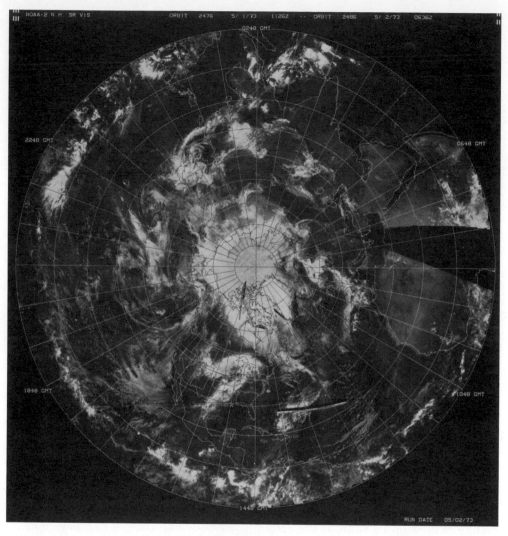

FIGURE 7.1(a)   *NOAA—2 satellite view of the Northern Hemisphere for May 1, 1973.*

cept for a sprawling disturbance near the dateline (180th meridian) and broken cloud sheets off lower California. A band of clouds is invading British Columbia and swirls off southwestward to nothingness north of Hawaii. The western states of the United States are clear; then heavy clouds run from east Texas to the Great Lakes, curling back to northeastern Colorado. Eastward from a clear east coast, cyclones can be seen in the western Atlantic and along the coast of Norway.

By May 2 (figure 7.2), partial clearing has occurred over Japan, but we can see a great cloud mass over China. Thus, the clearing over Japan may not persist for long. The mid-ocean storm is approaching the Aleutians now, and clouds have penetrated the Rockies in British Columbia. The storm in the central United States has spread clouds northeastward to Labrador, but clearing is very

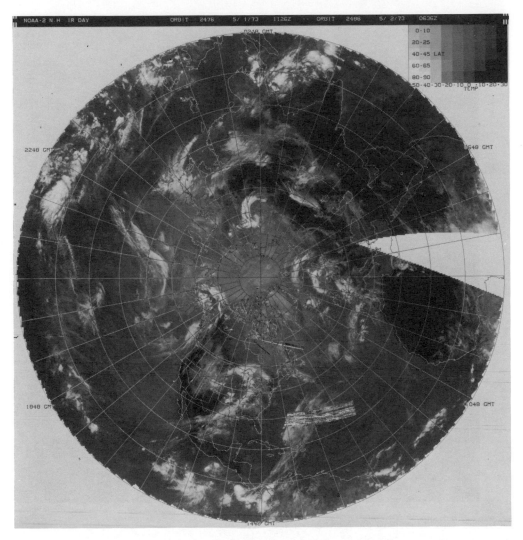

FIGURE 7.1(b) *Infrared view of the Northern Hemisphere for May 1, 1973.*

slow in the Plains. To the east, two cyclones over the Atlantic are spewing clouds into regions that were clear the previous day, including southern Ireland.

On May 3 (figure 7.3), a vigorous cyclone northeast of Japan produces the characteristic cloud pattern that looks something like a huge "comma." The head of the comma is off the Siberian Coast, and the long thick tail extends back to China, giving Japan another cloudy day, as well. Further eastward in the mid-Pacific, another comma pattern has pushed its head northward into the Aleutians, with a long, thin elongated tail trailing all the way southward to about 30°N. For the mid-Pacific cyclone, the comma head has lost some of its sharpness since yesterday (figure 7.2), but the tail has sharpened and pushed southward about 10 degrees of latitude (1000 km). Could we guess that a cold air mass is pushing

FIGURE 7.2 *NOAA–2 satellite view of the Northern Hemisphere for May 2, 1973.*

southward (and eastward) toward the Hawaiian Islands? And could that "comma tail" cloud band be the front, the leading edge of the cold air? Probably so, but we will learn that all cloud bands and comma tails are not fronts, and some fronts don't have cloud bands. Nevertheless, often enough, the comma-shaped cloud forms along the edges of air masses, and the cloud bands of the tail are near enough to the edge to permit us to make such a rough guess.

     Further east, clouds infest the northern Rockies, while the Plains have finally cleared as the Midwestern storm heads for the East Coast. The mid-Atlantic storm lacks coherence, but the comma shape rewards the imagination. Farther east, the slow-moving storm south of Ireland has spread its clouds farther north and east, so now only Scotland is clear.

     One day later, on May 4 (figure 7.4), Japan is finally clear, and two commas ride the Pacific waves, moving inexorably eastward. Clouds in the north-

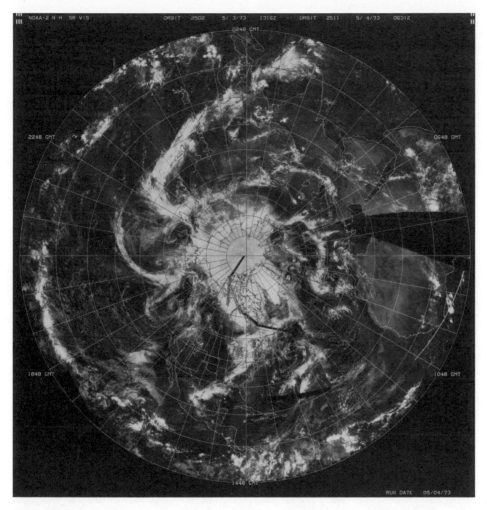

FIGURE 7.3 *NOAA–2 satellite view of the Northern Hemisphere for May 3, 1973.*

ern Rockies and near the Great Lakes stand out in contrast to an otherwise clear United States. The mid-Atlantic comma is still poorly formed, but it is also moving eastward; the cyclone near the British Isles and Ireland looks complicated and is very slow moving. In fact, look carefully—those clouds are moving westward. The western edge of the clouds was near 17° longitude on May 3 but now appears to be backed up to 20°W.* So much for the old forecasters' rule, "If it's west, bring it east," although most weather systems outside the tropics do move eastward. They do not simply move, however; they spread northward or southward, expand, shrink, intensify, or weaken.

---

*The zero longitude passes through Greenwich, England. It is the longitude of Greenwich mean time; 1200 GMT (noon of Greenwich) is the mean time at which the zero meridian passes under the sun. It is often abbreviated Z; for example, 1200Z.

By May 5 (figure 7.5), the cyclone that originated near Japan has moved far off to the northeast and is now approaching the Aleutian Islands. But notice that a bump has appeared east of Japan on the long comma tail. A new storm is being born, as the increased cloudiness suggests. Here, in one part of the world, we see the life cycle of cyclones. The old storm that formed on May 2 near Japan is, a few days later, already past its prime, heading for extinction in the Gulf of Alaska. But a new cyclone, called a *wave cyclone* (because the new storm appears as the crest of a wave rippling along the cloud band), is just beginning to form. In a few days, it will probably dissipate off the coast of Washington State, but not before it stirs up the Pacific.

Amazingly, a group of Norwegian meteorologists, just after World War I, described the basic cyclone structure and this "family" behavior of cyclones

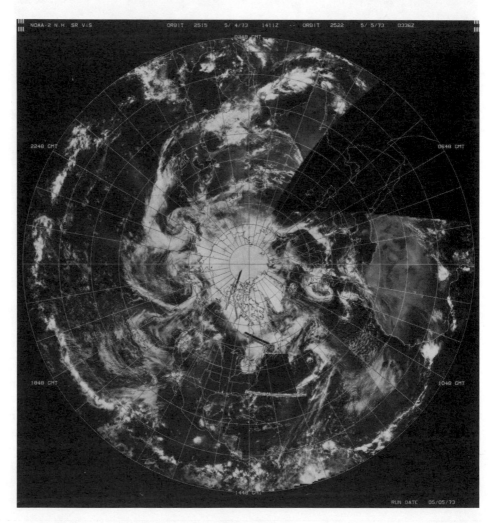

FIGURE 7.4 *NOAA–2 satellite view of the Northern Hemisphere for May 4, 1973.*

from surface observations taken from only a few ship and land stations. They even predicted that each successive track would be farther southeast of its parent.

Figure 7.6 depicts surface wind flows associated with the comma and its offspring, the wave. Note the association of the cloud bands with the fronts. Note also that the second member of the family is traveling along a path south and east of its parent. The winds depict a counterclockwise rotation about the cyclone center; indeed, the northwesterly flow to the rear of the parent has moved the cold front southeastward, ordaining the future track of the new cyclone.

Returning to figure 7.5, the other Pacific storm is plowing into the northern Rockies, and extensive cloudiness covers the West, in contrast to three days earlier. Even so, there are plenty of clear places. Dry-weather regions are distributed uniformly over large areas, with the result that the atmosphere has a

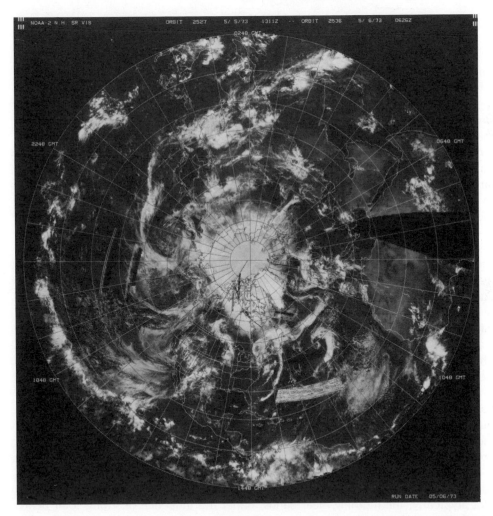

FIGURE 7.5 *NOAA–2 satellite view of the Northern Hemisphere for May 5, 1973.*

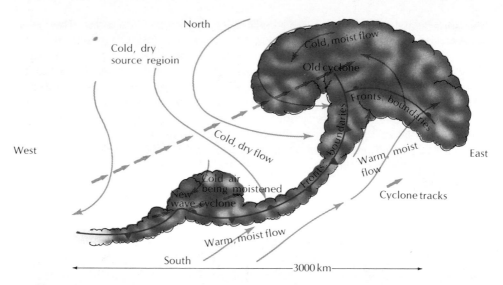

North

Cold, dry
source regioin

Cold, moist flow

Old cyclone

Fronts, boundaries

West

Cold, dry flow

Fronts, boundaries

Warm, moist
flow

East

Cold air
being moistened

New
wave cyclone

Cyclone tracks

Warm, moist flow

South

3000 km

FIGURE 7.6 *Schematic diagram showing relationship of clouds, wind, and fronts in an extratropical cyclone.*

strong bias toward good (dry) weather. Thus, forecasts of good weather are easy to make and frequently correct, whereas forecasts of bad (wet) weather are more challenging and less likely to be perfectly accurate. The pictures tell the story: the dimensions of the fair weather systems, anticyclones and dry flows, are much larger than those of the cloud bands within cyclones. Forecasters can often afford to make errors in locating anticyclones because they are so big; the same error in locating a cyclone can mean a completely incorrect forecast.

Finally, over the Atlantic, a comma cloud near Newfoundland and near the Azores mar an otherwise quiescent ocean scene. Some observations should be made here, however. First, the quiescent weather scene is not necessarily clear; pebbly sheets of clouds appear west of Africa, as well as in the eastern Pacific. These stratus and stratocumulus cloud sheets are the result of the trapping of moist air that has been weakly heated at the ocean's surface, causing it to expand and cool to its dew-point temperature. The trapping is caused by a persistent warm layer aloft, called an *inversion,* that acts as a lid to the mixing process. Though shallow and inert, these clouds are quite persistent and extensive. A steamer trip to Hawaii can be depressingly cloudy. The bright spot north of Haiti is a cumulonimbus cloud, so there can be random showers as well as organized bands of rain. We can see that cumulus convection (that is, showers) can break out at any point where the lapse rate is sufficiently large to permit unstable vertical motions due to buoyancy. Such unstable conditions are especially likely to occur over the continents during daytime, during the summer, and in lower latitudes.

Although apparently random showers do occur, there are favored spots for convection, and these are within the portions of cyclones that get heated. Not atypically, the morning clear bands, under the influence of strong solar heating, become the location of afternoon showers within the large storms. On the ground, we sense the buildup of swelling cumulus in the humid, sticky air, as it is

174

heated. The wind freshens, and we watch for the almost inevitable showers or thunderstorms.

DEAD AND ALIVE CLOUDS 7.3

Mist is the residue of the condensation of air into water, and is therefore a sign of fine weather rather than of rain; for mist is as it were unproductive cloud. [Aristotle, *Meteorologica*]

Of the many clouds we have seen in the previous pictures, only a few were actually producing precipitation. Some clouds are building in upward-moving currents of air; these are the active clouds that are likely to drop snow or rain. Other clouds are dead; they did not necessarily form where they are seen, but may be debris from old upwind showers, or simply shallow clouds that never bore their aqueous fruit. Indeed, most clouds are barren. Even rain clouds may be raining themselves to extinction, especially if they are of the cumulonimbus variety (showers and thunderstorms), with a lifetime on the order of an hour. Nevertheless, within the masses of old, dead clouds in a cyclone, there are usually spots where new ones are forming and generating precipitation. Occasionally, a large cloud mass produces rain or snow for hours, or even days. In these cases, the cloud system must be regenerated constantly by inflow of moist air near its base, upward motion within, condensation of water vapor to rain, and the dessicated air flowing out aloft. Figure 7.7 illustrates this process, which is valid for any size of cloud

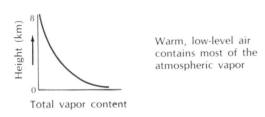

Warm, low-level air contains most of the atmospheric vapor

FIGURE 7.7 *Vertical cross section of a precipitating cloud system.*

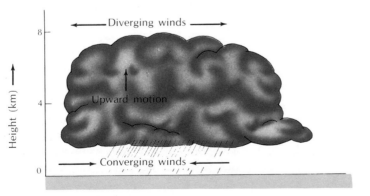

175

system, whether an individual cumulonimbus or a vast rain shield of a great winter storm. We have called the outflowing air near the top of the precipitating system *dessicated* because much of its water vapor has been condensed out to liquid or solid form and precipitated to the ground. However, this high-level air is often still saturated; high aloft, the air is very cold and easily saturated, even by a minor amount of remaining vapor. Thus, we can often see the outflow at the top of a thunderstorm, hurricane or even a great winter snowstorm, in the form of high, cold, cirrus clouds blowing away from it near its downwind edges. Sometimes those advancing cirrus are a precursor of things to come.

Because so many clouds are dead, the change to a cloudy tomorrow may be more certain than to a rainy one, for the active clouds occupy smaller volumes than the entire cloud system. But rain or snow is certainly favored where clouds are observed. So, we will study the parents of clouds, the cyclones, for they have the environments that are favorable for some of the more interesting weather events—thunderstorms, blizzards, tornadoes, and wild snowstorms. Nevertheless, we must recognize that not all cyclones will actually produce these severe phenomena; many will pass with no more than a sprinkle or a subtle shift in wind direction. The vigor of the cyclones often depends on the strength of the upper air "support," so we begin by considering the upper level winds.

7.4 COLD AIR AND MEANDERING JET STREAMS

In chapter 4, we learned that the high latitudes receive less incoming sunlight than low latitudes. On the outgoing side, however, the earth's surface and the air above it are like some families (or even a few students) and "spend" more radiant energy than they receive. This is partly a consequence of the rather low values of water vapor pressure in the cold polar atmosphere. Cold air cannot hold much vapor, and the vapor is a major factor in the atmospheric effect, letting sunlight in but retarding earth radiation losses. Thus, the higher latitudes, and especially the dry continental expanses such as Russia and Canada, are net losers of heat on an annual basis. Furthermore, the losses are staggering in the winter, when very little income is received.

Accordingly, we should not be surprised to see the large volumes of cold air that we sometimes call *polar,* or even *arctic,* air masses building up over Canada and Siberia, especially during winter. Similar masses form over Antarctica, but the greater amount of open ocean surface at the high latitudes of the Southern Hemisphere supplies heat and water vapor to the air, ameliorating the effect of cold air masses on populated lands of the Southern Hemisphere.

Cold air is heavy; it hugs the ground. The sea-level pressure is usually high near the center of the mass, so cold, high-pressure systems (anticyclones) are characteristic of these air masses. Because cold air is heavy, the arctic and polar air masses are rather shallow, extending upward only a few kilometers; above them, the air may be cold ($-40°C$). but it is potentially* much warmer than the

---

*Potential temperature is the temperature the air would have if brought down to 1000 mb. Increasing the pressure raises the temperature. (See section 5.2.3.)

air near the surface, which may boast a temperature of $-30°C$ or so. Furthermore, this cold air cannot have much water vapor in it. It formed, in part, because the water vapor content was small, and such cold air cannot hold any significant amount of vapor anyway. So these cold air masses are dry in the sense of absolute water vapor content. However, they may, and often do, have high *relative humidity*, exactly because cold air saturates so easily. This is especially true if the air mass formed over the ocean or if a large portion of it has passed over a body of water. Then it might be called *maritime polar* or *maritime arctic air*.

7.4.1 WHEN THE EAST IS COLD, THE WEST IS USUALLY WARM

Because the cold air masses form preferentially over the high latitude continents, and because the continents are broken by oceans and mountain chains, the distribution of cold air is very irregular along any given latitude circle (in the east-west direction). For instance, on any winter day, you could fly along the 38th parallel and find the Washington, D.C., area freezing in a arctic blast from Canada and San Francisco enjoying a sunny, pleasant 15°C (60°F) day, walled-off from Canadian air masses by the Rockies and with the sting of Siberian ones removed by the Pacific. Farther west, at the border of North and South Korea, a bitter Siberian air might hold full sway, while at Athens, Greece, a warm pleasant Mediterranean breeze might be blowing.

We can visualize that the cold air sits in huge, but finite, domes; from time to time, the high-level winds shift their patterns and these masses move away from the cold surfaces over which they formed, invading the lower latitudes and downwind places (usually toward the southeast of the cold sources) that may have to brace for cold waves. In summer, the migration of cool Canadian air is most welcome, however, in hot U.S. cities. It's hard to have it both ways, though; Miami, Florida, or Cairo, Egypt, get few really cold outbreaks in winter, but precious little relief in summer. Boston and New York may look northwestward for summer relief, but they pay the price in winter.

What of the air displaced by these moving cold air masses? Can we speak of warm air masses? The answer is a guarded yes. Often it is convenient to tag a large volume of warm air that, like the polar or arctic air masses, has taken on the characteristics of the type of surface it has overlain for several days or even weeks. More than half the earth is tropical or semitropical, and the air between about 30°N and 30°S is often called *tropical* or *equatorial*. The temperature characteristics are obvious, but when the air has resided over warm oceans for several days, it can contain much water vapor in the lowest kilometer or two. Such masses are called *maritime tropical**; comparable dry masses are *continental tropical*.

In middle latitudes, which are zones of primary interest, there are invasions of maritime tropical air, and invasions of polar or arctic air. The former are more common in summer and can mean a heat wave; the latter, more common

---

*True maritime tropical air masses are characterized by dry air above the lowest kilometer or so; equatorial masses often have deeper moisture. However, there are many variations encountered in nature. A few thunderstorms will moisten a maritime tropical air mass to great depths in a hurry.

in winter. But the cold air gets warmed as it moves equatorward and the warm air cooled as it moves poleward. On any given day the air over the United States or Europe may be neither tropical nor polar, neither maritime nor continental. On those occasions it is a bit silly to try to classify the air mass, so we don't. But the air mass concept has value on occasion, and that is when we use it.

Finite cold air masses must have boundaries, or fronts. Fronts are zones where the temperature changes rapidly in the horizontal as one travels from one air mass into the other. That is, fronts are the regions of the strongest horizontal temperature gradient. We will see in the next section that frontal zones with their horizontal temperature variation are linked inexorably to the *jet streams* of fast-moving air above them.

7.4.2 JET STREAMS

During World War II, American military aircraft on bombing missions to Japan flew from Pacific Ocean island bases toward the west, into the wind. These planes had a maximum air speed of about 175 knots (320 kilometers per hour). On one mission, the airplanes encountered head winds of 175 knots and found themselves virtually standing still over the ocean. These were the first detailed observations of jet streams, which are relatively narrow (300 to 500 kilometers) bands or ribbons of rapidly moving currents of air characteristic of the upper-level wind patterns that circle the earth.

The best-defined jet stream in the atmosphere is associated with the polar front. Because of this relationship, the jet migrates daily and seasonally northward and southward, following the polar front. Its intensity varies with the strength of the polar front, with the fastest winds (on occasion exceeding 250 knots) occurring with the strongest fronts. Because the temperature contrasts associated with the polar front are strongest in winter, the jet reaches its maximum intensity during this season. Over the United States, the average jet-stream speed varies from about 50 knots in the summer to 100 knots during the winter.

Figure 7.8 shows a jet stream for 7 P.M. EST, February 22, 1975, at the 300-millibar level (about 9.2 kilometers). The direction of flow along the jet stream is depicted by the solid arrow. Lines of equal wind speed, called *isotachs*, are dashed. Over the northeastern Pacific Ocean, two branches of the jet stream merge. The combined jet enters western British Columbia, then turns sharply southward over the Great Plains. The jet plunges all the way into northern Mexico before it recurves northeastward over the Mississippi River valley. Finally, the jet turns eastward and leaves Canada over Nova Scotia. The wind speeds along the jet axis over the United States are everywhere greater than 90 knots. Superimposed on the generally fast wind current are even faster-moving pockets of air. For example, a maximum of over 110 knots occurs along the United States–Canada border, and maxima of over 120 knots are present over Texas and New Brunswick. These local maxima, called *jet streaks,* are often indicators of developing storm systems.

The rapid variations in wind speed and direction across the jet stream are quite remarkable. Imagine flying from New Orleans, Louisiana, to Boise, Idaho. You would cross the jet stream twice. The first time over Texas, you would

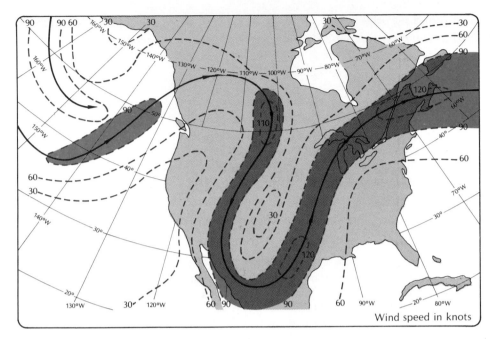

FIGURE 7.8 *Jet stream at 7 P.M. EST on February 22, 1975, at the 300-millibar level.*

observe 120-knot winds from the southwest. The second time, about an hour later, you would cross the jet coming from the opposite direction, northeast, at over 90 knots. In between, the winds would drop to a comparatively light speed of 30 knots. Such strongly curved, intense jet streams are always associated with strong fronts in the lower atmosphere and are very favorable zones for developing low-pressure systems at the surface. Next, we will show why the upper-level winds tend to be concentrated into narrow bands of rapidly moving air and why the jet streams are associated with strong horizontal temperature differences across fronts in the lower troposphere.

The generally westerly winds aloft in the middle latitudes are a result of cold air at high latitudes and warmer air in the tropical regions. Because cold air is denser than warm air, it is associated with low pressure aloft. Conversely, warm air, being less dense, is associated with high pressure aloft (figure 7.9). On the average, therefore, the pressure on a constant-height surface (or similarly, the height on a constant-pressure surface) is low to the north and high to the south throughout the middle and upper troposphere.

Figure 12.6a shows the mean height contours for the 700-millibar surface for January, 1974. Although irregularities in the height patterns exist (due mainly to land-sea contrasts), the mean heights are much lower to the north than over the tropics. The center of the lowest heights, 2580 meters, occurs close to the North Pole. In contrast, the height of the 700-millibar surface is high (over 3150 meters) in a broad zone around the world between latitudes 20 and 30 degrees north.

179

FIGURE 7.9 *Relationship of temperature and pressure aloft.*

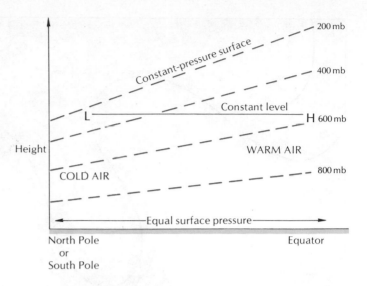

Using the geostrophic relationship between pressure and wind (figures 5.11 and 7.14), we know that the winds aloft tend to follow the height contours, with low pressure to the left of the direction of motion over the Northern
Hemisphere. Thus, in the middle latitudes, westerly winds are the rule throughout
the middle and upper troposphere.

If the temperature in the troposphere decreased uniformly from the
equator to the poles, the pressure aloft would decrease uniformly, and variations
in the speed of the westerly winds would be caused only by latitudinal variations
in the Coriolis force. However, the temperature does not decrease uniformly toward the poles, because as we have seen, the temperature tends to remain constant
over great distances and to change abruptly over the narrow zones called *fronts*.
These fronts occur where great converging wind currents bring together air from
warm and cold source regions, thereby producing large temperature contrasts.
Thus, the major decrease in temperature from the equator to the poles occurs
across one or two frontal zones. The temperature is relatively constant away from
these zones. Therefore, the north-south differences in pressure aloft are also concentrated in narrow zones, coinciding with the large horizontal temperature contrasts that occur in frontal regions. Because of the strong relationship between pressure gradients and winds, the fastest winds aloft occur over fronts. The stronger the
front in terms of horizontal temperature differences, the faster the westerly winds
aloft.

Figure 7.10 shows a vertical cross section through a cold front at 7
A.M. EST, October 16, 1973. This figure depicts the temperature and wind speed
variations through a slice of the atmosphere from Ottawa, Canada (MW), to Athens, Georgia (AHN). Imagine flying from Athens to Ottawa at the 700-millibar level
(about 3 kilometers). The temperature would gradually fall from the +5°C over
Athens as you flew northward until you reached the frontal zone over Huntington,
West Virginia (HTS). As you flew through the front, the temperature would decrease rapidly, and by the time you left the frontal zone north of Pittsburgh, Pennsylvania (PIT), the temperature would have fallen to about −11°C. Beyond the

Plate 7a Riming on sticks and branches (© Alistair B. Fraser).

Plate 7b (below) Pollution-induced fog near Tyrone, Pennsylvania (courtesy of Alistair B. Fraser).

Plate 8   Hurricane Gladys, taken by
Apollo 7 astronauts in October, 1968
(photo supplied by Robert Sheets, National
Hurricane Research Laboratory).

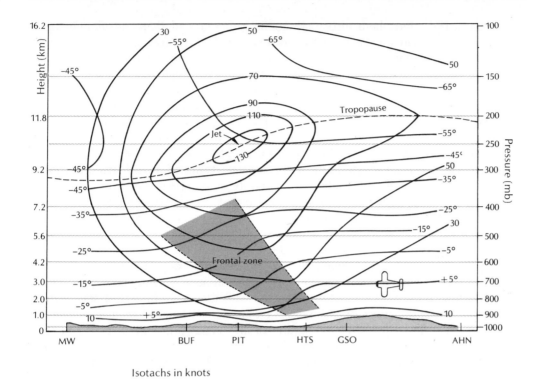

Isotachs in knots

FIGURE 7.10 *Vertical cross section through a cold front at 7 A.M. EST on October 16, 1973.*

front, the temperature would decrease much more slowly, reaching −15°C at the end of your flight.

In figure 7.10, the strongest winds aloft occur around 250 millibars over Pittsburgh and are associated with the sharp horizontal temperatures in the lower and middle troposphere. These temperature differences mark a frontal zone separating cold air to the north from warm air to the south. This frontal zone intersects the ground around Greensboro, North Carolina (GSO), and slopes upward to the north over the cold air. Notice that in general the westerly wind speed changes most rapidly with height where the horizontal temperature differences are large, such as over Pittsburgh. In contrast, where the horizontal temperature gradients are weak, such as over Athens or Ottawa, the wind speed changes slowly with height.

It is also important to note in figure 7.10 that the westerly wind speed increases with height when the atmosphere is colder to the north than to the south, which is the usual state of affairs in the troposphere. Above the tropopause (denoted by the heavy dashed line in figure 7.10), however, the temperature difference between the north and the south is reversed, with cold air lying to the south in the stratosphere. In this region, the westerlies decrease with increasing height. Therefore, the maximum wind speeds usually occur at the tropopause over strong frontal zones. Here, at about 200 millibars (12 kilometers), the jet stream reaches its maximum intensity.

The importance of jet streams to aviation interests is obvious. Winds of over 100 knots have a great effect on the fuel consumption and navigation of aircraft. The atmosphere also tends to be rather unsettled in the vicinity of jet streams. Turbulence is favored in regions where the wind speed or direction varies greatly with height. Also, because jet streams always occur with fronts, the clouds and precipitation associated with the front can make flying hazardous.

For the meteorologist, the importance of jet streams lies in their association with major precipitation-producing storm systems, or cyclones. Intensification of surface cyclones is most likely when a strong jet exists aloft. Cyclone development along the jet stream and ahead of an upper-level trough is often indicated by the appearance upstream of localized regions of higher wind speeds— jet streaks. Once a cyclone forms at the surface, its direction and speed of motion are determined by the orientation and intensity of the upperlevel jet.

Although we have discussed only the upper-tropospheric jet associated with the polar front, jet streams also exist in the lower troposphere, usually around 1 or 2 kilometers. These jets are usually weaker than the jets associated with fronts, perhaps averaging only 40 knots or so. They are often associated with frictional forces near the ground and vary in intensity from day to night. Over the United States, they usually occur in the warm southerly flow ahead of cold fronts. These low-level jets are closely connected with the development of severe thunderstorms and tornadoes over the Great Plains, as we shall see in chapter 10.

In terms of our discussion of air masses, the outbreak of a mass of cold polar air from Canada into the United States, with its front moving southeastward means that the polar front jet stream must shift southeastward with it. Unhappily, we don't yet know why either thing happens more frequently and more virulently some winters and not others, but the picture we wish to create is of a winding jet stream, here bending southward and there northward, always more or less parallel to the edges of the cold air masses. Because it is colder toward the poles *on average,* the jet stream winds blow from the west *on average,* and weather systems tend to move eastward. But now we can see why the jet stream will meander northward in one part of the world, carrying tropical moisture, clouds, and rain, and southward in another part, carrying cold, dry air from the poles. All of this happens near the periphery of great masses of cold air that distort the jet streams into their wandering paths. It is worthwhile to review figure 7.8 to see an example of this distortion when a pocket of cold air is located over the central United States.

7.4.3 JET STREAKS

Fronts and jet streams are physically linked by the horizontal temperature gradient, but it should not be thought that their relationship to each other is passive. It was pointed out in the previous section that small regions of especially strong winds in the jet stream, *jet streaks,* significantly affect the formation of new wave cyclones, such as that in figure 7.6. The waves may grow to be the great comma-shaped cloud patterns of a wild winter storm, but the precursor is often a little jet streak.

Figure 7.11 is an upper air chart at the 500-mb level, which depicts the streamlines (actually, the height contours) along which the wind is blowing

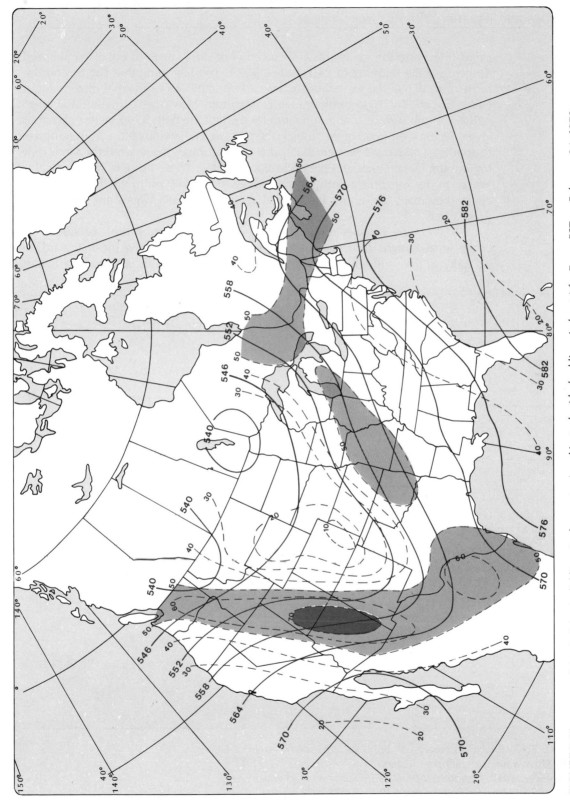

FIGURE 7.11 500-millibar heights (solid lines in decameters) and isotachs (dashed lines in knots) for 7 A.M. EST on February 24, 1979.

(solid lines) and the values of wind speed (*isotachs*—lines of equal wind speed, dashed) for the norming of 24 February 1979. The first thing that catches our eye is the distortion of the westerlies southward over the Rockies in what meteorologists call a *trough*. The base of the trough is in eastern New Mexico, where the height of the 500-mb surface is at a minimum for the latitude belt, 5500 meters above sea level. To the west (rear) of the trough, the northwesterly winds blow from northwest to southeast above the western states. East of the trough, southwesterly winds blow toward the Great Lakes and the mid-Atlantic states. These troughs are often called *waves* in the westerlies, partly because they look wavy on maps, but mainly because they really are waves in the fluid air, just as ocean waves are waves in the fluid water.

The weather is usually more cloudy and wet on the east side of the upper level troughs for three reasons. The air is usually more moist toward the equator, so the southerly winds are more moist. The air is usually warmer toward

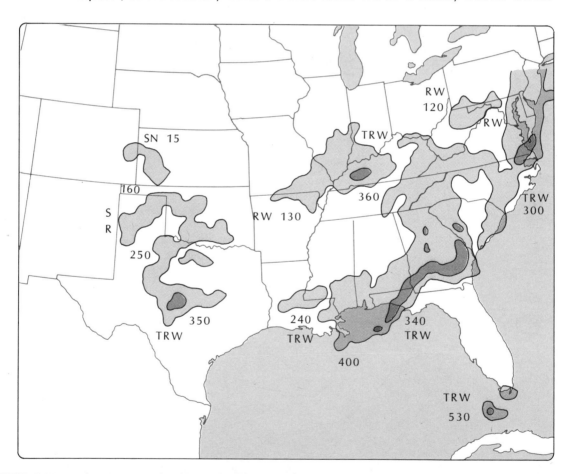

FIGURE 7.12 *Radar summary for 6:35 A.M. EST on February 24, 1979. Shaded areas depict precipitation. SW = snow shower, RW = rain shower, TRW = thundershower. Numbers are cloud tops in feet, with last two zeros omitted, so that 530 = 53,000 feet.*

the equator, so the southerly winds are often flows of warm, light air being forced upward over colder, denser air as it moves poleward; when air rises, it expands and cools to saturation. Finally, and most importantly, the air ahead of (downwind of) troughs tends to rise. Figure 7.12 is a composite radar chart for the same time (approximately) as figure 7.11. Radar charts depict precipitation patterns, and this one shows widespread precipitation east of the upper level trough where the air is rising. It isn't raining or snowing everywhere, but it is easy to see why meteorologists are concerned with waves, and why weather forecasters look for approaching troughs in the westerlies when they are thinking about precipitation.

Further inspection of figure 7.11 shows that there is a narrow belt of high speed air curling around the trough, with speeds from 40 to 80 knots (75 to 150 km/h). Here is our meandering jet stream. Even though there are no temperatures on the map, we now know that the air must be cold over the northern Plains and the east slopes of the Rockies and rather warm in California and Florida. Not only that, we know that the temperature gradient must be under the jet stream.

The next feature of the 500-mb chart that we notice is the relatively small region of high-velocity air moving southward in the rear of the trough. An 80-knot maximum is located over Arizona. This is an example of a jet streak (a wind maximum along the jet stream). It was mentioned earlier that jet streaks can be indications of developing storms. The present case is an excellent example, for when the jet streak appears in the rear of a trough, the conditions for formation of a new cyclone within a day or so are quite good. The mechanism is a redistribution of air from the left front side of the moving jet streak to its right side so as to increase the pressure difference necessary to accommodate such a large wind. Thus, the pressures fall to the left* of the jet, and a cyclone forms, usually near the bottom of the trough. In the example, a new cyclone formed in northeast Texas on the afternoon of 24 February, 6 to 8 hours after the time of the map.

We can see that jet streams and their disturbances, the streaks, are by no means decorations on the air mass/front packages, but they are actively linked to them by the cyclone. They contribute to cyclone formation, the cyclone grows into a huge comma cloud, altering the winds and moving the fronts. In turn, the jet streams move, and the atmosphere makes its fascinating, subtle and interactive accommodations to the new situation. But no laws are broken—only our concentration, sometimes, by the bewildering variety of outcomes that natural law permits.

CYCLONES AND ANTICYCLONES—BIG EDDIES IN THE WESTERLIES   *7.5*

On the tropical side of latitude 30 degrees, wind flows are quite regular (though not steady), with northeast trades near the surface in the Northern Hemisphere, and southeasterly winds quite persistent in the Southern. However, the winds in

---

*With the wind of the jet stream at your back.

middle and high latitudes are quite variable, as the eddies, which are the cyclones and anticyclones, form and move about. The cyclones rotate in the same sense as the earth, counterclockwise as viewed from above the Northern Hemisphere and clockwise as viewed from above the Southern Hemisphere. Anticyclones, as their name implies, rotate in the opposite sense, that is, clockwise in the Northern Hemisphere. Both cyclones and anticyclones have typical lifetimes on the order of a few days, or at most a couple of weeks. Their horizontal dimensions are from a few hundred to a few thousand kilometers. These vortices rotate about their centers at the same time that they move horizontally above the face of the earth. The situation is similar to what one sees when a paddle is drawn through a stream: the paddle causes whirls in the water, which then move away horizontally in response to the flow of the stream, all the while rotating about their own centers. This analogy can be carried one step further by noting that the direction of movement of the eddies in the stream is usually guided by the larger-scale current of the stream itself, that is to say, downstream. Similarly, in the atmosphere, the cyclones and anticyclones rotate about their own centers but also tend to be carried downstream by the large-scale flows in which they are embedded. As a result, cyclones and anticyclones are observed to move toward the east on the average, because the broad-scale flows aloft are usually from the west. However, we hasten to remind you that individual cyclones and anticyclones may travel very differently. They may meander, remain stationary for a while, travel westward, execute loops, speed up, slow down, and so forth. Besides their general eastward motion, anticyclones usually have a component of motion toward the equator. On the other hand, there is a strong tendency for cyclones to move with a component toward the pole, so that a typical one moves northeastward over the Northern Hemisphere.

There is much more of interest than wind changes in all these cyclones and anticyclones drifting about. Cyclonic rotations occur about points of minimum pressure, and anticyclonic rotations occur about pressure maxima. A cyclone is just another name for a low-pressure system, affectionately called a *low* by meteorologists. Similarly, the anticyclone is a *high*. It is the pressure difference between the highs and lows that causes the winds.

Later we will think again about how the wind and pressure differences are related. For the present, we are simply stating that the pressure distribution not only causes the wind and controls its sense of rotation, but also plays a role in determining the relative amount of cloudiness, as we have seen in the pictures. What is more, cyclones and anticyclones are the offspring of the temperature difference between the equator and the poles, and their formation and resulting north-south winds are the atmosphere's mechanisms to mitigate that difference by transporting heat poleward.

Although there is a certain amount of inevitability to cyclone formation, it is not obvious that all cyclones should have a common structure. But it is a remarkable fact that virtually all cyclones outside the tropics have many similar characteristics; they even bear notable resemblance to tropical storms. This is not to say that all cyclones cause the same weather, for some cyclones are stronger than others. In addition, available moisture supplies, hills and mountains, lakes and seas, the time of year, prevailing temperatures, and countless small-scale effects intervene to present the bewildering variety of actual weather that is both the

186

glory and the curse of weather forecasting. But these complications may also thrill astute observers who can learn to assess regional and local topographic effects, sense or measure the moisture and the local winds, watch the clouds, and thus anticipate the changing weather, often as well as the professional meteorologist. These are the reasons that many farmers and fishers are such good short-range weather forecasters; better, indeed, than meteorologists who believe that repeating weather patterns cause repeating weather. They are good observers, and accurate observation is at the heart of all good science.

But for all the local "tuning," the weather does march to the sound of a much larger drum, and when the weather is bad, that drum is the cyclone.* Knowledge of cyclone structure makes it possible to go a step beyond the observation of local effects, extending our vision beyond the horizon. With satellites, radar, and television, the task is getting somewhat easier for both the public and meteorologists; however, we can apply the ideas discussed in the following pages without the use of any modern communications. Cyclones are cyclones, however we observe them, and the local weather they produce can be understood and anticipated. But this task is not easy, and we have to watch carefully. Even so, we will be disappointed at times, for there are many aspects of the weather that remain mysteries.

THE NORWEGIAN CYCLONE MODEL 7.6

Throughout the twentieth century, the Scandinavian countries have produced a number of remarkably gifted meteorologists, all in the tradition of the founder of a noted school at Bergen, Norway—Vilhelm Bjerknes. Shortly after World War I, Norwegian meteorologists proposed a model of a typical cyclone, which has achieved wide acceptance. While it is true that we have learned to be skeptical of its overenthusiastic use and to apply it flexibly, it is also true that the Norwegian cyclone model is the best single concept for the interested layperson or the vacationing meteorologist to know. Satellite pictures, radar, and weather maps help us go beyond the typical, or idealized, cyclone; but it is a rare satellite picture that does not depict some part of a cloud structure that reminds us of the Norwegians' great insight. And they did it without satellites!

The Norwegian wave cyclone model links the cloud pattern, precipitation (duration, type, and intensity), wind, action of the barometer, temperature, visibility, and air quality to the distribution of sea-level pressure,** fronts, and wind direction and speed aloft. The basis of relating all these aspects of the weather is the idea that pressure differences at a given elevation (say, sea level) cause the air

---

*Snow, rain, thunderstorms, etc., are "bad" weather to the average person, but, of course, are "good" weather for the meteorologist, who would otherwise be without a job or a hobby.

**You can set a barometer to sea-level pressure simply by calling a neighbor, the nearby weather station, or anyone who knows the sea-level pressure in your locality at the moment. Set your barometer to the correct value. It doesn't matter if your observation point is not at the same level as theirs. Your correction is then different, but you don't care, or need to know, by how much.

to move laterally; that is, horizontal pressure differences produce wind. The winds then carry moisture, lift and cause clouds or sink and evaporate them, move the fronts, change the temperature, and so forth. This is why meteorologists are so preoccupied with highs and lows: lows are regions of low sea-level pressure, or cyclones, and high-pressure cells are anticyclones. Pressure differences cause the winds, and indirectly, the weather. Let us see how the wind and pressure are related to each other when we look at flows much larger than those across a beach or down the side of a mountain.

7.6.1 WINDS AND ISOBARS

Figure 7.13 is a surface weather map for 7 P.M. E.S.T. on February 24, 1979, twelve hours after the upper air map of figure 7.11. A key to the plotted reports is given in figure 3.4. Also drawn on the map are the isobars (lines of equal pressure) of the sea-level pressure. We can see that a cyclone really did form that afternoon, as our interpretation of the upper air chart suggested. At this evening map time, the center of the lowest pressure is in Mississippi, where the sea-level barometric pressures approach 1006 millibars (1013.2 is the mean sea-level pressure), a bit below normal. The value of the sea-level pressure is not so important as the fact that it varies about the map. Outward from the cyclone center, the stations are reporting higher pressures, and we have drawn the isobars* to show the pattern of horizontal variation in pressure, as well as to locate the cyclone and anticyclone centers, for it is the horizontal pressure difference (gradient) that causes the winds. Notice that it is a good bit colder in the north winds behind the cyclone than ahead of it, as the cyclone is starting to transfer heat northward by exchanging cold air for warm.

Of special interest to us now is the observation that the winds are tending to blow counterclockwise *around* the cyclone, rather than straight *at* the low from higher pressure. This is the effect of the earth's rotation, discussed in chapter 5. In fact, we get a better sense of the flow around the cyclone center by imagining that the wind is blowing parallel to the isobars rather than across the isobars (from high to low). But there is also a clear tendency for the wind to cut across the isobars somewhat, toward low pressure. (Recall the rotary wind controversy described in chapter 1.) Here we see that there was some truth in both positions with respect to the surface winds.

Far to the north, an anticyclone is stretched from Canada to the Rockies. Observe that the winds over the northern United States are blowing from the northeast, now clockwise out of that great arctic air mass center, but again with some tendency for flow across the isobars from higher pressure toward lower.

Thus, we observe that the large-scale winds around cyclones and anticyclones do not blow simply from high to low pressure the way the sea breeze does at noon, or canyon winds do at night, or the way water flows when we open a faucet. Their movement is complicated by a marked deflection toward the right

---

*Nowadays, we usually program computers to draw the maps.

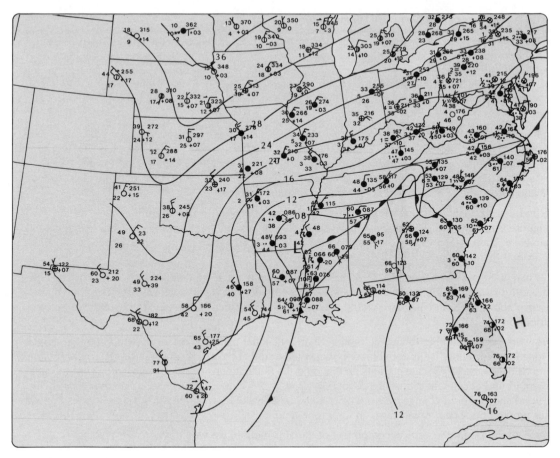

FIGURE 7.13 *Surface weather map for 7 P.M. EST on February 24, 1979.*

of lines connecting the highs and lows. We have seen (section 5.3.3) that the earth's rotation produces this deflection toward the right of the direction of motion (in the Northern Hemisphere).

Further, notice that the winds are stronger in Oklahoma, Arkansas, and part of Texas, where the isobars crowd together, than they are over Florida and Georgia. Crowded isobars mean a large horizontal pressure gradient—large horizontal pressure differences over relatively small distances—compared to a weak pressure gradient where the isobars are widely spaced. In the latter areas, you could travel a long way and find the pressure changes relatively little and the winds light. Table 7.1 summarizes what we see on this map and points out a few things that we do not.

Many of the observations summarized in table 7.1 can be explained by one of the most useful models in meteorology. This model relates the horizontal wind to the horizontal pressure distribution at a given level and latitude; it is called the *geostrophic wind*.

189

TABLE 7.1   *Relations of observed winds to the distribution of pressure*

1. We note that the wind blows mainly along the isobars, with low pressure to the left and high pressure to the right of the wind. If a similar chart was prepared for any region in the Southern Hemisphere, we would find the reverse—namely, that the wind blows mainly along the isobars with low pressure to the right of the wind. Because rotation of the earth's surface in the Southern Hemisphere is opposite that in the Northern Hemisphere, we suspect that the different behavior of the winds is due to the rotation of the earth.

2. We find that the wind is not altogether parallel to the isobars but tends to stream toward the side where the pressure is low. However, if we prepared charts for any level above 1 or 2 kilometers, we would find no systematic drift toward the lower pressure. We are thus led to believe that the drift toward lower pressure is caused by friction along the earth's surface, and that its effect is not noticeable above about 1 to 2 kilometers above the ground. This is true in both hemispheres.

3. We find that the wind is strong where the isobars are crowded, and weak where they are wide apart. If we ignore the drift toward lower pressure, we gain the impression that the wind blows in isobaric channels in such a manner that the speed stands in inverse proportion to the width of the channel. This is true in both hemispheres.

4. If we extended our chart from pole to pole, we would find that the relation between isobar orientation and wind direction is rather firm in high and middle latitudes but weakens as we approach the equator. Between about 10°N and 10°S, we would have considerable difficulty in relating the winds to the pressure distribution.

5. If we were in a position to follow the motion of an individual parcel of air and measure the rate at which its speed changes (that is, its acceleration), we would find that the accelerations are very small. In fact, if we considered the large-scale currents and ignored the fluctuations associated with brief gusts and lulls, we would find accelerations that are about 0.0002 m/s^2 (2.6 km/h^2). (If this rate continued for an hour, the wind speed would change by 2.6 km/h.) In the large wind systems, the air is a slow starter, but when it has worked up speed, it will carry on for a long time.

6. If we could measure the vertical component of the air's motion, we would find that it is large in thunderstorms, tornadoes, etc., and also in the very small eddies which we call turbulence. However, if we considered the large-scale currents of the atmosphere, we would find that the average vertical motion is small and that the wind is overwhelmingly horizontal.

7.6.2 GEOSTROPHIC FLOW

In chapter 5 we introduced the idea of balanced flow (figure 5.11).

In this idealization, the horizontal wind results from a balance between the horizontal pressure gradient force, directed to the left of the air, and the Coriolis force, directed to the right. In this *geostrophic balance*, the wind blows parallel to the isobars with low pressure on the left. In figure 7.13, however, we see that the geostrophic wind model doesn't really hold at the earth's surface, because it requires that the wind blow exactly parallel to the isobars. Indeed, we hasten to admit that the real wind rarely behaves as a true geostrophic wind should, but it is often a close enough approximation that we can neglect all other forces and consider what happens when the pressure gradient force is exactly balanced by the Coriolis force. We should not expect the approximation to be especially accurate near the ground where friction is important, nor for strongly curved flows where centripetal forces (forces directed toward the center of a circle) cannot be neglected. But where the flow is straight and more than a kilometer above the ground (where friction is weak), the following conditions of geostrophic motion apply (refer to figure 7.14):

1   The pressure gradient force and Coriolis force are equal in strength and opposite in direction.

2   The wind is strong (Coriolis force proportional to wind speed) when the pressure gradient is large, that is, when equal-interval isobars are crowded together. Similarly, the wind is weak when the pressure gradient is small (small Coriolis force balancing a small pressure gradient force).

3   When there is no pressure gradient, there is no wind and no Coriolis force. This is true in pressure maxima and minima, the centers of highs and lows, respectively. It is more noticeable in highs, which tend to have much larger centers.

4   Isobars are by definition perpendicular to the pressure gradient force. The geostrophic wind is also perpendicular to the pressure gradient force and, thus, blows parallel to the isobars. This is because the pressure gradient force and Coriolis force, in order to balance each other, must be along the same line, and the Coriolis force is always at right angles to the wind flow. Check the validity of each of these statements in figure 7.14.

5   Toward the equator, Coriolis forces are weak and are balanced by weak pressure gradient forces; therefore, the isobar spacing for balanced winds of given speed increases toward the equator. Stated another way, for a given isobar spacing, the geostrophic winds are stronger near the equator.

6   Inasmuch as the Coriolis force acts to the right of the wind flow in the Northern Hemisphere, the pressure gradient force must act to its left. The pressure gradient force acts from high pressure toward low; therefore, an observer looking downwind in the Northern Hemisphere finds low pressure on the left and high pressure on the right. This rule was formulated in 1857 by a Dutch meteorologist, Buys Ballot, into Buys Ballot's law: In the Northern Hemisphere with the wind at your back, high pressure is on the right and low pressure on the left; in the Southern Hemisphere with the wind at your back, high pressure is on the left and low on the right.

The last part of Buys Ballot's law is true because the Coriolis force acts to the left in the Southern Hemisphere. However, in either hemisphere, application of Buys Ballot's law is sometimes chancey at ground level where frictional effects or local perturbations to the winds may obscure the geostrophic wind direction. Notwithstanding, the rule is often of value, even at the surface.

FIGURE 7.14 *Geostrophic wind (V) resulting from a balance between the horizontal pressure gradient force (P) and the Coriolis force (C). Solid lines are isobars, labeled in millibars with leading 10 dropped. Example (a) represents high latitudes, strong pressure gradient force and moderate wind. Example (b) represents the same latitude as (a), but with ½ the pressure gradient force and ½ the geostrophic wind. Example (c) represents the same pressure gradient as in (a), but at a lower latitude so that the geostrophic wind speed is greater.*

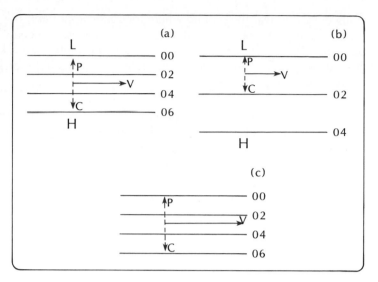

7.6.3 HOW FRICTION MODIFIES GEOSTROPHIC FLOW

Friction's effects on wind speed are fairly obvious: they tend to slow the wind. Quite different are friction's effects on the direction. Picture what happens when friction is introduced to previously geostrophic flow: the brakes are applied, and the speed diminishes. As a result, the Coriolis deflection, proportional to the wind speed, is reduced. Accordingly, the pressure gradient force dominates, pushing the air to the left, and we see a component of flow from high pressure to low. And, indeed, we do see it in figure 7.13; in fact, we can see that the flow cuts across the isobars toward low pressure at an approximate angle of 30 to 40 degrees. We then infer that surface friction turns the wind in toward low pressure by this amount, while it also slows the wind. Both of these effects increase with the roughness of the terrain, so we cannot be rigid about the exact angle of inflow. This can especially be seen in some of the wind reports in the Appalachian regions of Tennessee and the Carolinas.

7.6.4 EXTENSION OF THE GEOSTROPHIC WIND TO CIRCULAR FLOW AROUND CYCLONES AND ANTICYCLONES

When we consider the wind above the friction layer, from typically a few hundred meters upward, there is very little flow across the isobars, and geostrophic conditions are much more likely. Accordingly, we rely on geostrophic wind ideas to interpret flow patterns through much of the atmosphere. The beauty of the geostrophic wind is that it is simple, reasonably faithful to nature, and involves the pressure field, which is easy to measure accurately. Hence, if we know the spacing of the isobars and the latitude, we can evaluate the geostrophic wind speed and direction and be confident that they approximate the real winds without measuring the wind at all. Hence, it is possible to plot a finite number of pressure observations on a map, obtain the pressure gradients, and then calculate the wind velocity at an infinite set of points on the map. This is the genius of good analysis—it extends the utility of limited resources. Current practices increasingly involves going the other way—observing the winds and determining pressures that are consistent with them. But the basic idea is the same: we use dynamic relationships between the wind and pressure to extend the information provided by a limited number of observations.

We now want to extend the relationship between pressure and wind to the pressure patterns commonly encountered on weather maps. In particular, circular cells of high and low pressure, elongated pressure maxima called *ridges*, and the elongated pressure minima called *troughs* usually display flow parallel to the isobars even though centripetal effects must be present where the isobars curve. Careful examination will show that the geostrophic wind speeds are erroneous by 50 percent or more with curved flows; however, the geostrophic wind direction is still reasonably accurate. Because the geostrophic-direction rule states that high pressure is on the right of the flow in the Northern Hemisphere, the wind will blow clockwise (as seen from above) around anticyclones. Similarly, the geostrophic wind direction and the actual wind direction are such that low pressure is to the left of the flow; thus, the wind blows counterclockwise about cyclones in the Northern Hemisphere. However, in the Southern Hemisphere, the wind blows counterclockwise about highs and clockwise about lows. Low-pressure centers are still called *cyclones* there, so it is not correct to say that the wind always blows counterclockwise about cyclones.

Although the actual wind direction does not depart significantly from the geostrophic direction around circular highs and lows, the speed does. For the same pressure gradient, the flow is faster around highs than around lows. This may seem surprising, because our experience tells us that strong winds are most likely to be associated with storms (low-pressure systems) than with fair weather (anticyclones). The paradox is explained by noting that the pressure gradients associated with lows are normally much greater than those associated with highs.

Let us now see what effect curvature has on the relationship between the isobar spacing and wind speeds for flow around highs and lows (figure 7.15). We first note that to make the air turn in a circle, a net force (called the *centripetal*—"center-seeking"—*force* on the air must be directed inward toward the center of the circle. Otherwise, the air would continue to flow in a straight line. The only two forces that can produce this net centripetal force are the pressure gradient force, which acts from high to low pressure, and the Coriolis force, which acts to the right of the wind direction. By assumption, the magnitude of the wind speed (and therefore the Coriolis force) is the same in both cases shown in figure 7.15. However, the pressure gradient associated with the low is directed inward (toward the center of rotation), whereas the pressure gradient force around the high is directed outward. Because the sum of the Coriolis and pressure gradient forces must be directed inward, a greater pressure gradient force is required in the case of flow around the low because it must overcome the outward-directed Coriolis force.

193

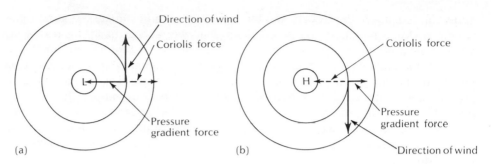

FIGURE 7.15 *Unbalanced flow around circular highs and lows: (a) cyclonic flow and (b) anticyclonic flow around lows and highs when the wind speed is the same. To make the air turn in a circle, the air must have a net force directed toward the center of the circle.*

Thus, we have shown that for a given wind speed, the horizontal pressure gradient force for cyclones must be greater than that for anticyclones. Note also that, in contrast to straight geostrophic flow, the pressure gradient and Coriolis forces are unequal in magnitude in curved flow. The resultant force is simply the centripetal force that is required to make the air turn.

If we now refer back to figure 7.6 and to the composite satellite pictures for May 1–5, 1973 (figure 7.1–7.5), we can appreciate why the wind is swirling in a counterclockwise direction about those cloudy cyclones. The pressure is low there, and the pressure gradient force is trying to accelerate air toward the centers to fill them. But the Coriolis force intervenes and deflects the air paths to the right, so the air ends up blowing around the center, not perfectly symmetrically, but still in the direction of counterclockwise rotation. Furthermore, we can see that what inflow there is occurs in the lowest kilometer or so, as a result of friction.

Although we now see why the wind blows as it does about high- and low-pressure centers, ridges of high pressure, and troughs of low pressure, we still have to deal with the reason we can see the cyclones so clearly in the pictures. Low-pressure systems are cloudy because upward motion of air is favored near cyclones, and high pressure centers are usually less cloudy because downward motion is favored near anticyclones. Let us turn to those aspects of the cyclone.

7.6.5   VERTICAL MOTIONS

The association between the pressure and wind fields is so strong that, even when the wind departs rather strongly from geostrophic conditions, we find useful relationships between the winds and ridges, troughs, anticyclones, or cyclones. What is more, we can construct the pressure distribution from temperature data. Thus, three principal atmospheric variables—pressure, temperature, and wind—are intimately related to each other. But all that would be nothing if we could not relate winds, cyclones, and temperature fronts to the upward motions that produce clouds and precipitation, for in the last analysis, the upward and downward motions of air govern the weather. Unfortunately, these

194

vertical motions are often weak, erratic, prone to operate on small scales, and variable with height, and therefore are notoriously difficult to measure and forecast accurately.

When we deal with vertical motions, it is especially important to distinguish between scales. On the one hand, we have small-scale motions, such as turbulent eddies, individual clouds, and thunderstorms. In these cases, regions of rising and sinking motion are relatively close to each other, perhaps 100 meters to a few kilometers apart. Vertical motions in such small-scale systems may be quite large, from 1 m/s in ordinary turbulence near the ground on a clear day (figure 2.9) to many tens of meters per second in severe thunderstorms. Such vertical motions can be measured directly from instruments such as bidirectional vanes near the ground or on airplanes.

In contrast are large-scale vertical motions, which represent averages over areas at least the size of a typical American state. Such synoptic-scale vertical motions are quite small, typically on the order of a few centimeters per second. Even though these motions are small, they cover large areas and last a long time; thus, they can lead to large-scale cloud formation and precipitation, or to large-scale drying of the air. These small vertical motions cannot be measured directly. They can only be inferred from patterns of wind and temperature by use of the various equations in chapter 5. Let us now see how wind, temperature, and pressure cause upward motion and clouds.

The causes of upward and downward motion are associated with different processes, depending on the scale of the vertical motion. On the smallest scale, the vertical motions may be caused by horizontal winds that vary rapidly in the vertical (vertical wind shear) or by buoyancy. The former is most likely to occur close to the ground and can generate the small turbulent eddies near the earth's surface. Buoyancy is responsible for dry and moist convection that occur when volumes of air are lighter than their surroundings. Convection can also occur through the sinking of heavy air (recall the mamma clouds of section 6.4.2).

Buoyancy becomes especially significant when the lapse rate is steep (temperature decreases rapidly with height). It can then produce very strong upward and downward motions of from 1 to 20 or 30 m/s (40 to 60 mi/h). The horizontal scale of buoyancy is then always quite small (a few kilometers or so), and any clouds invariably will be of the cumulus variety. Buoyancy is often represented by the shorthand statement "warm air rises," but it really means "light air rises." Water vapor weighs less than nitrogen or oxygen, which comprise 99 percent of the air, so the optimum condition for buoyant exchange in the vertical occurs when the lapse rate is steep and the low-level air is humid.

On larger scales, four different processes lift the air. In two of these processes, the upward motion is related to the upglide of light air over something that is more dense, either the ground or cold, dense air. Air rising over sloping terrain is called *orographic lifting;* when it rises over cooler air, it is called *overrunning.* Overrunning is really buoyancy operating on a large horizontal scale. It is nearly always associated with warm advection aloft, which means that warmer air is replacing cooler air. Either of these two upgliding processes is likely to generate weak vertical motions on the order of 5 cm/s (0.1 mi/h) over a relatively large area, although lifting by terrain can be much stronger and on a smaller scale with strong winds in the vicinity of steep hills. Nevertheless, these weak motions frequently persist long enough to produce significant amounts of precipitation.

Conversely, either downslope motion or cold, dense air advancing toward warm air tend to produce sinking motions. Air flowing down mountain slopes leads to familiar rain-shadow (precipitation minimum) effects on the lee of mountains. Advancing cold air, called *cold advection,* is just the opposite of overrunning: the advancing cold air sinks, warms somewhat as it is compressed, and dries out relative to saturation. Thus, it usually turns colder after a storm. The cold advection is one of the factors that causes the storm to end, for the turn to colder produces sinking motions.

One has to be careful to consider the possibility that the temperature advection aloft may be stronger, weaker, or even of opposite sign to that near the ground. In one common case, there can be surface cold advection and sinking to the north of a cyclone, but warm advection and upglide aloft. Then the low atmosphere may remain cloud free or have only rag-tag cloudiness, but a deep layer aloft may be overrunning and generating clouds and precipitation, which may fall through the cold air to the ground or may evaporate on the way down. Sometimes a very threatening, cold day can exhibit this process without producing any rain or snow at the surface. Snow lovers have been known to complain about such days! Figure 7.13 shows that on 24 February 1979, however, rain, freezing rain, and snow were able to reach the ground from Kentucky to Oklahoma, despite surface cold advection. We infer that there is overrunning aloft, and reference to the southwest winds aloft in figure 7.11 confirms this.

The third, and most difficult to understand, cause of upward motion is associated with imbalances in the atmosphere. When heating, strong upper winds, or any other cause transports air aloft out of a region (divergence), vertical motions arise to compensate for the changed pressure. These vertical currents then induce horizontal convergence (the flowing together of air) at some other level, as shown in figure 7.16. The compensating horizontal motions always occur at different levels of the atmosphere, and the vertical circulations that couple them are important weather-producers, even though they may be very weak (5 cm/s). That is because these inflow-outflow patterns not only produce vertical circulations; to the extent that they are unbalanced, they produce pressure changes as well. Thus, net divergence can reduce the surface pressure and form lows or destroy highs, whereas net convergence destroys lows and forms highs. The changing highs and lows then can modify the geostrophic winds and feed back through overrunning or orographic effects to other vertical circulations. In summary, dynamic vertical motions arise when disturbances aloft cause pressure changes through their associated divergence patterns.

A fourth process that can produce large-scale vertical motions is friction at the earth's surface. In section 7.6.3 we saw how friction causes the low-level air to deflect across the isobars toward low pressure. This causes a general convergence of air in the low levels toward the center of lows, even when they are dying and there is no mechanism aloft to remove the air as fast as it flows in. Thus, old cyclones may remain cloudy and rainy at least in spots, for a few final days. Also, regions of cyclonically curved surface isobars may very well be sites of enhanced shower activity, in middle or high latitudes or in the tropics, as a result of

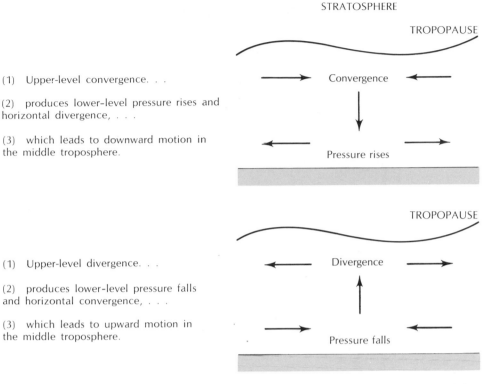

(1) Upper-level convergence. . .

(2) produces lower-level pressure rises and horizontal divergence, . . .

(3) which leads to downward motion in the middle troposphere.

(1) Upper-level divergence. . .

(2) produces lower-level pressure falls and horizontal convergence, . . .

(3) which leads to upward motion in the middle troposphere.

FIGURE 7.16 *Relationships among horizontal divergence, horizontal convergence, and vertical motion.*

frictional convergence. Conversely, anticyclones have frictional divergence near the surface, and usually display sinking motion and rainfree skies as a result. A change from cyclonic to anticyclonic isobars—especially in the tropics or in the warm seasons when the atmosphere is unstable but has no strong disturbances aloft—is sufficient to change the weather. Compare the region of anticyclonic isobars in southern Florida in figure 7.13 with the region of cyclonic isobars over northern Florida. The northern stations have rain, while the southern stations do not, despite the distance from the upper-level disturbance. The showers are probably attributable to surface frictional convergence, while the good weather further south is in a region of surface frictional divergence and sinking. Watch the isobars: sometimes they are the key!

7.6.6 FORMATION OF CYCLONES AND ANTICYCLONES

The key to the formation of cyclones and anticyclones lies in disturbances in the upper westerlies. Figure 7.17 shows a vertical cross section of the relationships among upper-level divergence–convergence patterns, surface highs and lows, and the vertical-motion pattern associated with a cold air mass and its front. The cyclone is usually found at the surface along the front and

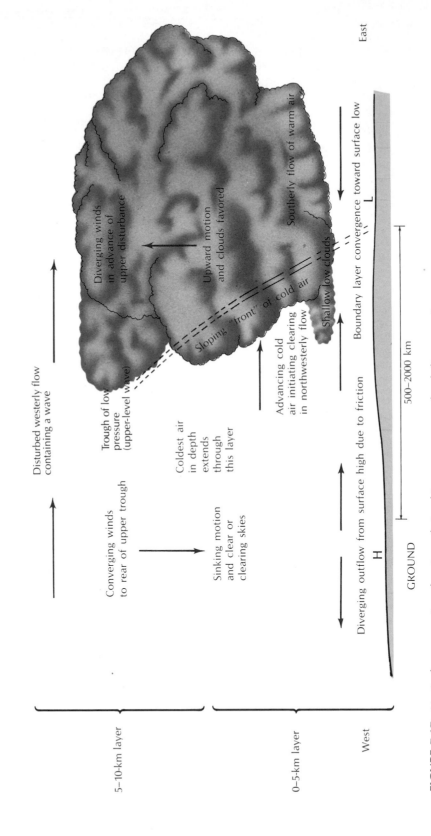

FIGURE 7.17 *Vertical cross section showing relationships among upper-level divergence/ convergence patterns, surface highs and lows, and vertical motions in a frontal system.*

then slopes upward toward cold air. Thus, lows are at temperature boundaries near the surface and typically have a warm side and a cold side (figure 7.13), but aloft, lows are cold. As a result, the highs found in these cold air masses tend to die out at higher elevations. (All highs don't vanish upward; only cold ones do. Warm highs in a tropical airmass, such as the Azores/Bermuda high or the Pacific high, extend to great depth. It all hinges on cold air masses being heavy and hugging the ground.)

If the divergence of the winds aloft are stronger than that of the converging winds near the surface (*net column divergence*), the low intensifies, and the central pressure falls. Then, the inflow gradually strengthens and ultimately overcomes the upper divergence, killing the cyclone. For this example, we can see that the upper wave disturbance is the driving force behind the formation and intensification of surface cyclones. In fact, because the intensification of cyclones must be characterized by net divergence in an atmospheric column to reduce the surface pressure, and because low pressure at the surface results in frictional inflow or convergence, divergence must occur aloft. Therefore, storms are created as well as steered by the upper-level winds. We have seen that jet stream disturbances (jet streaks) are one mechanism for the upper winds to provide the necessary divergence. Now we can see that they are part of a larger class of upper air wave disturbances which precede the surface cyclone formation and whose task it is to provide upper level wind divergence.

In the early stages of the cyclone, called the *open wave stage*, the most prominent features are falling barometers, rapidly deteriorating weather conditions, and quite often, marked cumulus convection. The front "buckles" at the point of the incipient cyclone, with the sinking cold air just to the rear of the low advancing as a cold front. Just ahead of the low, the cold air stops advancing, or at least slows down. This produces the kinked wave shape that we see on the front (figure 7.13). If the cyclone forms on a stationary front, the cold air starting to move is an indicator of cyclone formation. For most cyclones, the cold front behind the storm advances readily; ahead of the low, the cold air may retreat as a warm front or it may not move much until the cyclone passes; then it advances as a cold front, for now it is at the rear of the storm.

The other distinguishing feature of the early stages of the cyclone is the formation of the comma head of active cloud pushing into the cold air. Figure 7.18 shows a satellite picture taken at 2200 GMT (1600 CST) on 24 February 1979, just a few hours into the life of the surface low and just two hours prior to the map of figure 7.13. The comma head is already very prominent from eastern Oklahoma to northern Mississippi, sloping up over the cold air. Figure 7.13 shows moderate rain and moderate snow in eastern Oklahoma and Arkansas, rain in southern Missouri and western Kentucky, and thunderstorms under the comma head in western Tennessee and northern Mississippi. Pressures are falling ahead of it and rising sharply behind it. The comma tail shows up well through Mississippi, Louisiana, and the Gulf of Mexico. As is often the case, it has several cloud bands in it, with two groups being especially prominent. Thus, it is usually difficult to tell, from the pictures, exactly where the cold front is.

The lumpy appearance in Mississippi (and also in southern Arkansas) at the northern ends of the major bands is caused by deep thunderstorms catching

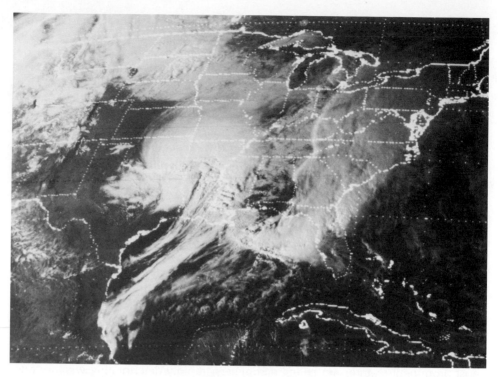

FIGURE 7.18   *Visible satellite photo at 5 P.M. EST on February 24, 1979.*

the late afternoon sun. Four hours later, severe thunderstorms struck northern Mississippi and Alabama, an observation that should not surprise us now. An intensifying cyclone swinging a comma tail through warm, moist unstable air (note temperatures and dewpoints in the warm sector), with strong upper level divergence and low level convergence can build vigorous thunderstorms in a hurry. This brand new cyclone is telling us that it is in a hurry to grow up!

As the cyclone intensifies, the cold front continues to wrap around the locus of the surface center, as the center itself moves along steered by the upper level winds. Ahead of the well-developed cyclone, the cold air often retreats ahead of a warm front. Figure 7.19 is a schematic diagram of the principal features of a well-developed cyclone. It depicts the northeastward path of the storm together with the different wind that is experienced on each side of the path. It also shows the cold and warm fronts, and an extension of the cold front (called an *occlusion*) where the cold front has overtaken the warm air. The occlusion point is the northernmost point that gets in to the warm air, called the *warm sector*; its path is also shown. The pattern of precipitation and the comma-shaped cloudiness pattern are represented, together with some reminders of the causes of the clouds (overrunning, frictional convergence where the isobars are cyclonic, and regions of cumulus convection associated with steep lapse rates).

Overrunning is favored where warm air flows toward cold air. This usually occurs ahead of (or northeast of) the counterclockwise rotating circulation

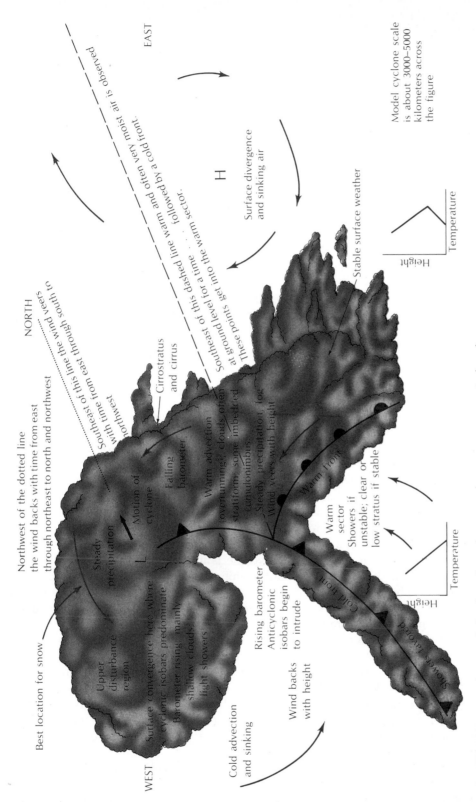

FIGURE 7.19 *Distribution of clouds and weather around a mature extratropical cyclone.*

201

FIGURE 7.20 *Probability of precipitation in various regions with respect to the extratropical cyclone.*

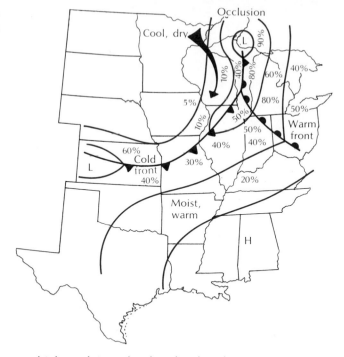

in the Northern Hemisphere, which explains why the cloud-and-precipitation pattern associated with a typical cyclone is not symmetric about the low center. Most cyclones become "right-handed," with more clouds and precipitation on their east sides, where the wind is from the south. Figure 7.20 shows the typical precipitation probabilities in various regions of the model cyclone. Because most cyclones travel from southwest to northeast at any given place, rain mainly falls when the cyclone is approaching, that is, when we are on its northeast side. Thus, rain occurs when the barometer is falling as a result of a cyclone's motion toward us. Once the center of the low arrives, at the time of the lowest barometric pressure, the rain is usually nearly over. Of course, if the cyclone is intensifying, the barometer falls will be stronger, and so will the upward motion and rain. If the cyclone is old and dissipating, the barometer will hint at this weakened state by not falling so fast.

Once the cyclone is well developed, the effect of the rotation on the cloud pattern becomes more marked. The cyclone that we are considering, which formed on the afternoon of 24 February, was already mature by the following (Sunday) morning at 7 A.M. EST. Figure 7.21 is the weather map for Sunday morning, showing the center near Nashville, the cold front sweeping toward the warm sector over Georgia and Florida, and a warm stationary front ahead of the low in the Carolinas. As is often the case, this cyclone is having trouble swinging the surface fronts much to the north because of the blocking effect of the Appalachian Mountains, but warm air is overrunning aloft as rain and freezing rain now extend northward to Cincinnati and Indianapolis.

The warm sector is still susceptible to cumulus convection, with a line of thunderstorms that we call a *squall line* extending southward through the Gulf of Mexico. However, nighttime cooling in the warm sector has easily cooled

FIGURE 7.21 *Surface weather map for 7 A.M. EST on February 25, 1979.*

the surface air to its high dewpoints, and much fog is also reported. When we hear of fog, we usually think of stable conditions with light winds and little vertical mixing. That is certainly true in much of the warm sector, but it is only true of a shallow layer of air close to the ground. Most of the troposphere above that shallow layer has the steep lapse rate of a maritime tropical air mass and is quite ready to produce thunderstorms when it is lifted, as the observations show.

Of special interest when a cyclone spins up is the tendency for dry air to rotate around south of the storm and produce a marked dry slot or *surge region* of rain-free, even cloud-free, air. This clear tongue often extends right to the vicinity of the surface low, so the barometer may be very low while the sun is shining. The surge region for the 24–5 February storm is very clear in figure 7.22, an 11 A.M. EST satellite picture made four hours after the map of figure 7.21. By 11 A.M., the cold front had moved east to eastern Georgia and northern Florida, then off the coast to the vicinity of the Yucatan Peninsula. Look how rapidly it clears behind the cold front! The sinking of the advancing cold air (*cold advection*) is wiping out all clouds, all the way north to the Tennessee border, almost in the low pressure center. Compare this to Ley's remarkable model (plate 2b). If he had

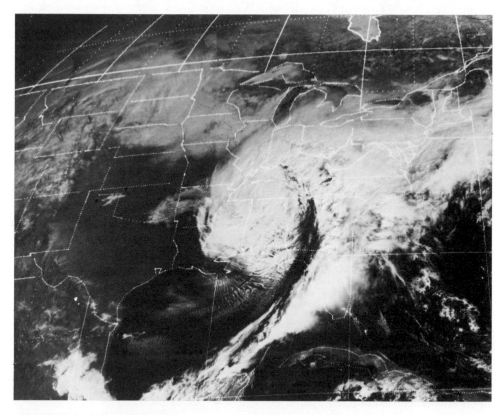

FIGURE 7.22   *Visible satellite photograph at 11 A.M. EST on February 25, 1979.*

204

elongated the comma tail, he would have almost completely anticipated what we see in satellite pictures a century later!

Clouds return to the surge region in the northwest, but they are mainly shallow ones over Alabama, Mississippi, and Tennessee. Only sprinkles occur there; sometimes the entire surge region will have these low clouds, but that doesn't alter the change to much less precipitation that usually occurs behind cold fronts.

Still further northwest, from southeast Missouri to Indiana, a band of deeper clouds curls back along the front section of the comma head. Here there is overrunning and divergence aloft—the dry surge has yet to cut it off—with heavy rain and snow to the left of the storm track.

We can see from all this that the comma head can have considerable structure as the storm ages. On the second day of the cyclone's life, the deepest precipitating clouds are in thunderstorms along the cold front, in isolated thunderstorms over the ocean in the warm sector, and over a wide area north of the cyclone. Later, the cyclone will swing dry and cold air completely around it. It will no longer slope toward cold air, because it will be in the cold air. Then it will degenerate to bands of clouds and showers, and finally, it will die, as a new cyclone takes over where the air is warm and the waves along the jet stream are ready to start their exciting life cycle.

7.6.7 USE OF THE CYCLONE MODEL IN INTERPRETING LOCAL WEATHER

The principal advantage of the cyclone model is that it is a good tool for understanding the weather we observe at any one point. In other words, the observer who knows the model can usually diagnose the weather situation without weather maps, satellite pictures, or anything else. A barometer is some help, but two good eyes are all you really need. First, you can determine the winds both at the surface and aloft (watch the clouds that are straight overhead). Applying Buys Ballot's law at both levels, you can determine where the pressure systems lie, and by applying the steering principle using the winds aloft, where they will go. For example, if the wind at the surface is from the southeast, and clear skies have recently been replaced by cirrus or cirrostratus, you can surmise that a cyclone is somewhere to the west or northwest. By watching the cirrus cloud motion, you can then predict whether the cyclone is approaching or passing off to the north. If there is little cloud motion aloft, you can guess that the cyclones and anticyclones are moving very slowly, implying that any local changes in weather can only come about as a result of sea breezes, valley breezes, or the like. This principle of *steering* is worth some emphasis. If there were a swarm of bees or a cloud of dust suspended through a great depth of atmosphere, you would not be shocked to find it moving in the direction of the winds aloft, which occupy a deep layer and usually blow from a consistent direction that may be different from that of the low-level winds. So, too, with clouds and cyclones; they tend to be steered downwind by the winds aloft. All these signs are available for the looking.

7.6.8 WARM FRONTAL PRECIPITATION

There is much more in the cyclone model. Notice that in regions of overrunning (figure 7.23), the warm air over cold air stabilizes the atmosphere with respect to buoyancy forces, suppressing cumulus clouds in favor of deep and extensive sheets of stratus clouds. The schematic temperature soundings in figure 7.23 show the stable conditions that are favored (but not guaranteed) in the region ahead of the warm front. Notice that the upper-level front is associated with the top of an inversion in the soundings. Thus, to the extent that the atmosphere ever produces steady rain over flat terrain, regions of overrunning on the cold side of warm fronts or stationary fronts are the places to find it. Notice that we do not insist that the frontal boundary at the surface, that is, the edge of the cold air mass, be moving. If it is moving in the sense of cold air receding back toward its source regions—usually to the north—the front is called a *warm front*. If it is not moving at all, we call it a *stationary front*. In any case, the behavior of the surface front does not really matter; as long as there is significant overrunning aloft, stratiform (layered) clouds and steady precipitation will occur on the cold side of the front. What really matters for overrunning is warm flow relative to the front—that is, warm flow toward the front when the front's motion has been subtracted from the winds.

The actual frontal type depends on the surface winds just on the cold side of the front, with northerly winds indicating a cold front, southerly a warm front. The main overrunning will be hundreds of kilometers back over the cold air. Significant overrunning is usually associated with warm, or at least stationary, fronts, because cold advection, to persist, would have to infect a large enough volume to cause sinking and drying in the cold air. Then the structure would be a *cold front,* with cold air advancing and rapidly clearing skies on the cold side of the front, which is the situation to the rear of typical cyclones.

Note also in figure 7.23 that the steadiness of the overrunning precipitation depends intimately on the initial stability of the overrunning warm air. If the warm air is stable, layered clouds such as altostratus (gray, layered clouds occurring in the middle troposphere) will prevail in the warm air, and precipitation will be steady. If the warm air is unstable, however, thunderstorms may erupt as the warm air is lifted. But underneath, in the cold air, it will always be stable, so stratus-type clouds will almost certainly predominate when viewed from below.

One process of low cloud formation that is of particular interest for pilots frequently occurs in an overrunning situation. The clouds that are created by this process are marked *P* in figure 7.23. Initially, the low-level air in the cold air mass is dry, because cold air masses originate in the arid and often ice-covered portions of the globe. The overrunning rain, however, humidifies this air as the first drops come down and evaporate. Because the surface air usually has been pre-humidified to some extent by surface vapor sources, there frequently follows the rapid formation of low stratus clouds when rain starts falling from higher clouds. The unwary amateur pilot who needs to maintain visual contact with the ground may be flying at 2 km (7000 ft) with an overcast far above, say at 4 km (13,000 ft).

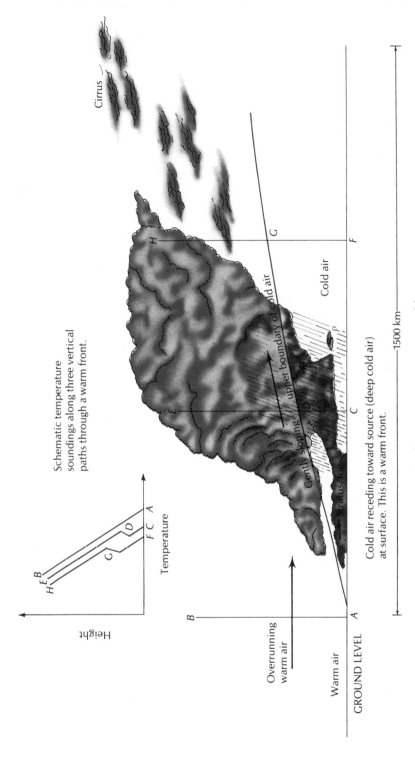

Schematic temperature soundings along three vertical paths through a warm front.

Cirrus

upper boundary of cold air

Gently sloping

Cold air

P

Cold air receding toward source (deep cold air) at surface. This is a warm front.

1500 km

Overrunning warm air

Warm air

GROUND LEVEL

Height

Temperature

H E B

G

F C

D

A

B

A

FIGURE 7.23 Clouds and precipitation associated with warm air overrunning cold air.

But suddenly an undercast appears down near 1 km (3000 ft) or so, and with it, real problems. The tipoff, of course, is spatters of rain on the windshield.

7.6.9 WINDS AROUND THE CYCLONE

In addition, the cyclone model can be used to forecast temperature changes by watching the variations with time of the wind. We use the term *veering* when the wind changes in such a way that our wind arrow turns in the clockwise direction when looking down on it. Conversely, if the wind arrow turns counterclockwise when viewed from above, we say that the wind is *backing* (see figure 7.24). Notice that at all points south and east of the dotted line (the storm track) on the schematic cyclone model (figure 7.19), the wind veers with respect to time through the south, and the weather generally improves. This is the history of the warm (south-wind) side of the storm track. On the other hand, points to the north and west of the track experience wind backing with time. These locations are on the cold side of the track, and thus much more likely to observe snow, sleet, or freezing rain during the cold seasons. This model explains the following proverbs, valid for the Northern Hemisphere only:

A veering wind, fair weather;
A backing wind, foul weather.

If wind follows sun's course, expect fair weather.

So if you want to forecast temperature (and especially snow) when a cyclone is headed in your general direction, watch whether the wind veers or backs with time; you'll become the local groundhog!

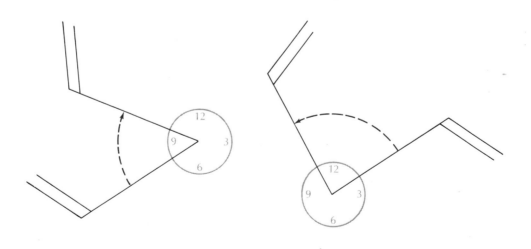

Veering wind   Backing wind

FIGURE 7.24 *Veering and backing winds.*

208

7.6.10 TEMPERATURES AND PRECIPITATION AROUND THE CYCLONE

Points south and east of the dashed line in figure 7.19 will have a period of high temperatures because they will get into the warm sector for a time.* Then they will be approached from the west by the cold front, which is typically associated with a band of showers or (during warm seasons) thundershowers. The small scale of these showers means that not everyone in the path of the cold front will get one. However, the cold frontal region and the zone a couple of hundred kilometers wide just ahead of it are the favored regions (figure 7.25). The schematic sounding labeled *G-H* in figure 7.25 shows that in these regions which have been in the warm sector, and especially at the times of the day and year when the sun is strong, the lapse rate is quite steep, and buoyancy production of deep cumulus and cumulonimbus clouds is quite likely. Of course, the intensity of the convection depends on the details of moisture availability, strength of the low-level convergence, frontal intensity, and steepness of the lapse rate. Some of our most devastating thunderstorms, and even tornadoes, occur along squall lines (lines of thunderstorms) that are just ahead of cold fronts.

In contrast to the unstable low-level soundings ahead of the cold front, the air is generally more stable behind the front (note soundings *G-H* and *A-B-C* in figure 7.25. Therefore, the clouds, if any, tend to have little vertical development. As in the case of the warm front, the cold front aloft appears as an inversion in the soundings.

In discussing the composite satellite pictures, we called the cyclones comma clouds. Now we can see that the main features of these cloud patterns resulted from the extensions of cloudiness along fronts. We have not, however, focused on the causes of fronts. No one would challenge the idea that temperature variations should exist between low and high latitudes and along coastlines, mountains, etc. When the flows are such that the horizontal variation of temperature gets crowded into a small space, fronts are produced. Recognizing that low-level winds that are convergent can pack the variation into a small space, we should not be surprised to find cloudiness maxima along fronts, for low-level convergence results in upward motion and, hence, clouds. Further, we should thus expect fronts to be asociated with lows rather than highs, and indeed, fronts always occur in troughs of low pressure where the low-level air is convergent.

7.6.11 THE OCCLUDED FRONT

So far, we have discussed the typical structures of cold and warm fronts, but have not yet discussed what happens when the surface cold front overtakes the surface warm front, a phenomenon which occurs in the later stages of a cyclone's life. Let us first consider the case when the overtaking cold air is colder than the cool air ahead of (on the cold side of) the warm front (figure 7.26a). The coldest air pushes under the cool air, just as it does with the

---

*Cold fronts often "pinch off" (see section 7.6.11) warm sectors at the surface because they travel more rapidly than warm fronts. Thus, this line could shift southward with time.

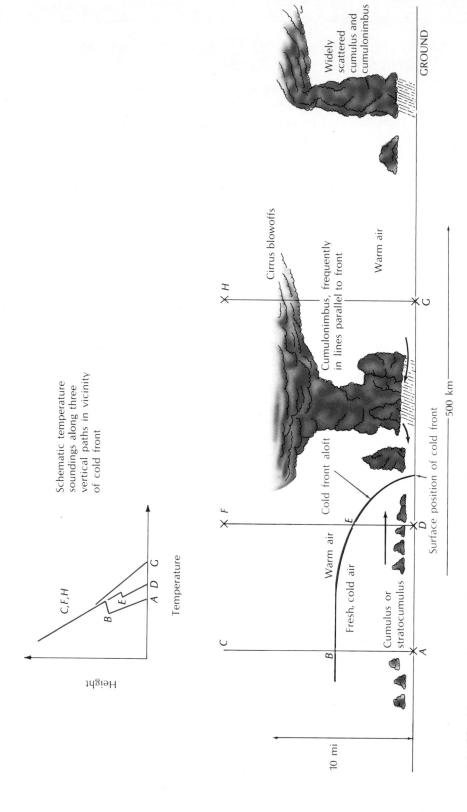

FIGURE 7.25 *Vertical cross section showing clouds and precipitation in warm sector ahead of cold front and vertical temperature profiles in warm sector and behind cold front.*

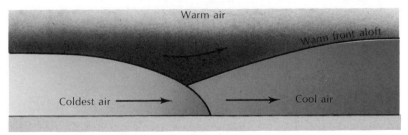

(a) Cold-type occluded front

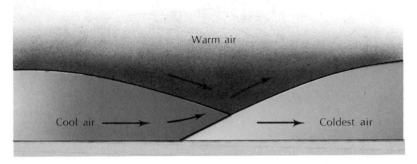

(b) Warm-type occluded front

FIGURE 7.26 *Vertical cross sections through cold-type and warm-type occluded fronts.*

warm air, but the warm air mass is still present aloft even though the warm front disappears from the surface. The boundary between the coldest air and the cool air is called a *cold-type occluded front*. Because the advancing air is colder than the air being displaced, the sequence of weather events that accompanies a cold-type occlusion is similar to that which accompanies a cold front. Indeed, the distinction between these most common occlusions and cold fronts is quite small. A cold-type occluded front is depicted in figure 7.19, the model of the extratropical cyclone. The occluded portion of the front extends from the low-pressure center to the point where the cold and warm fronts intersect.

A second, less common, type of occlusion occurs when the retreating cold air ahead of the warm front is colder than the overtaking cool air mass behind the cold front (figure 7.26b). In this case, the advancing cool air rises over the colder air mass, producing weather that is similar to the weather associated with warm fronts. The *warm-type occlusion* occurs occasionally in winter west of the Rockies when westerly flow behind Pacific cold fronts is warmer than the cold air over the continent.

A third type of occlusion occurs somewhat spontaneously when two currents of air with nearly the same temperatures and dew points converge. The front can be observed with its shift in wind direction, cloud bands, and precipitation, but little temperature change occurs, except that related to cloudiness changes and precipitation.

7.6.12 WARM AND COLD ADVECTION INDICATED BY VEERING AND BACKING WINDS

One last concept for use with the cyclone model may be a bit difficult to apply, but it is worth the effort. Because the temperature and pressure aloft (not at the surface) are closely correlated, we can usually tell where cold air and warm air are and how the temperature is likely to change by observing the winds aloft. You can do this in your backyard (see figure 7.27). Just watch the middle or high clouds, preferably those straight overhead. Then remember Buys Ballot's law and "cold—low, warm—high." Now, pretend that you are standing on a very high tower with the upper-level wind at your back. High pressure and warm air are on your right, and low pressure and cold air are on your left. Check whether the low-level wind is bringing in that cold air, the warm air, or neither,

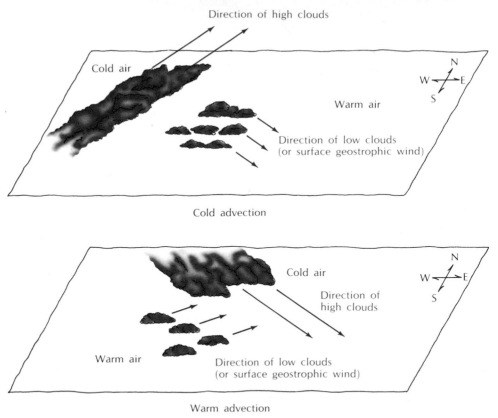

FIGURE 7.27 *Relationship of change in wind direction with height to horizontal advection of temperature.*

and you are ready to make a temperature forecast that will be highly reliable until the next front comes along to change things. If the winds are bringing in air of different temperatures, you can diagnose the situation more precisely by using the following Northern Hemisphere rule, which is just a generalization of the ideas noted above: If the geostrophic wind direction veers upward through the atmosphere, it is getting warmer (warm advection); if it backs, the atmosphere is getting colder (cold advection). Notice that you have to determine the geostrophic wind direction for at least two different levels. The direction at the higher level can be obtained by the motion of the middle or high clouds. If lower-level clouds are also visible, these cloud motions may be used to estimate the lower-level geostrophic wind direction. However, if low clouds are not present, the surface winds may be "corrected" for the effects of friction to yield an estimate of the geostrophic wind. Because friction causes the surface wind to deflect about 30 degrees to the left of the geostrophic wind, we need to make a 30-degree correction to the actual wind to obtain the surface geostrophic wind direction. To do this, face into the surface wind direction, then turn 30 degrees to your right to allow for frictional inflow across the isobars. Now you are facing into the low-level geostrophic wind, and you are ready to determine whether the geostrophic wind is backing with height and the atmosphere getting colder, or veering with height and the atmosphere getting warmer. Of course, if the geostrophic winds at the two levels are from the same direction, no large-scale temperature change is taking place.

Questions

1. Using the principal temperature classifications (tropical, polar, arctic) with the moisture classifications (maritime and continental), list the six principal air masses of the world, and give one geographic region which would be a good source region for each air mass.

2. Explain why the cirrus outflow at the top of a precipitating cloud system has low absolute humidity, while it plainly had high relative humidity.

3. In terms of air masses and fronts, explain why the jet stream is often located over the Gulf States of the United States in winter but seldom in summer.

4. Sketch a comma cloud pattern on a blank map of the United States. Let the northern edge of the comma head be at the U.S.–Canada border and the southern end of the comma tail extend to the southern boundary of the U.S. (either the Gulf of Mexico or Mexico). Label the comma lead, tail, surge region, region of most likely precipitation, probable position of surface low pressure center (if there is one—not all commas are associated with surface lows), and the axis of the associated upper level trough.

5. Assume that there is a surface low pressure center (cyclone) at the position selected in question 4 above. Place the following fronts on the map in positions that are consistent with the comma cloud and with the Norwegian cyclone model for an intensifying storm: cold front, warm or stationary front, and occlusion (optional). Label the warm sector, and the region most likely to observe snow or cold rain.

6 An observer in Paris knows that Britain is toward the northwest, Germany toward the northeast, Italy southeast, and Spain southwest. The person notices that low clouds, located above ground about one kilometer (about 3000 feet—above the influence of friction) are moving from southwest to northeast, from Spain to Germany. Supply the names of the countries for the following. (A small sketch map may help).
   a. The sea-level pressure is highest in _____.
   b. The pressure gradient force is acting to push the air toward _____.
   c. The isobars run from _____ toward _____.
   d. The Coriolis force (which balances the pressure gradient force) is acting toward _____.
   e. The sea-level pressure is lowest over _____.

7 If the observer in question 6 could see higher clouds crossing Paris from west to east, is warmer or colder air flowing toward Paris?

8 Suppose a continental polar front is moving southward across the Great Plains of the United States at a speed of 30 km/h. Suppose that the warm air above the (sloping) front is at rest (no horizontal wind aloft). What direction would the warm air move, up or down? Would there be clouds above the front? What is this process called? A sketch is helpful.

Problems

9 If the front in question 8 had a slope of one km upward for each 100 km horizontally (which is typical), what would be the magnitude of this upward motion? (Answer: 8.3 cm/sec)

10 Suppose the conditions were the same as in question 8, except that the air aloft was moving southward at 15 km/h. What would be the (only) difference?

11 Suppose the conditions were the same as in question 8, except that the air aloft was moving southward at 50 km/h. What would be different from the result of question 10 with these upper winds? Can you describe a way to determine whether a front will have overrunning clouds and/or precipitation?

12 When cold arctic air flows across the Great Lakes of the United States in late fall or early winter (when the water is still relatively warm), it flows downhill toward an aerodynamically smooth warm water* body and accelerates out across the Lakes. Often the air is clear on the upwind side of the lake(s), but heavy snow is observed on the downwind side. List all of the physical processes and causes of clouds and precipitation which contribute to the heavy snow.

---

*Also Hudson Bay, the North Sea, the Sea of Japan, and numerous other places.

EIGHT

Watching the Weather

rotation of winds about cyclones/anticyclones • upslope clouds/
precipitation • scud clouds • daily temperature range • stratocumulus
clouds • false clearing • cumulus and cumulonimbus clouds •
evaporative cooling • relative humidity • frictional convergence •
severe thunderstorm conditions • dew point temperature •
altocumulus turrets • unstable air • frontal clouds • steering
principle • cyclonic isobars • cold advection • Norwegian cyclone
model • surface divergence and sinking • Bermuda High

All the pictures, diagrams, maps, and descriptions of Chapter 7 would mean little or nothing if we did not put the ideas behind them to everyday use. To be sure, we can learn from weather maps; they are included in this book for that purpose. But if we apply the ideas and concepts that emerge from an understanding of weather maps by using our eyes and brains, we will not need weather maps, except the one in our mind's eye.

So, let us set out in our imaginations on a cross-country trip. We will drive across the Great Plains, from the Rockies to the Appalachians, using Interstate 70 most of the way, from Denver to Pittsburgh. We choose for our trip the period in early May, 1973, that we have already "watched" from the satellites. We saw in figures 7.1–7.3 that between May 1 and May 3, a slow-moving cloud system spread across the United States into eastern Canada, finally abandoning the Plains states on May 3.

8.1 THE WEATHER ACROSS THE UNITED STATES ON APRIL 30, 1973

Figure 8.1 shows the weather map for the morning of April 30, 1973. Because this synopsis sets the scene for the events that follow, it is important to understand what is pictured. At each reporting station, the weather elements are plotted according to the scheme described in chapter 3. For example, the temperature is plotted to the upper left of the station; it is 46°F (8°C) at Goodland, Kansas, but 66°F (14°C) at Wichita, just to the east. The heavy line lying through Kansas denotes a front, a fact that the temperature contrast between Goodland and Wichita fairly screams at us. Indeed, everywhere to the north of that frontal line, the temperatures tend to be cool or cold; Churchill, Manitoba, is coldest of all at 7°F (-14°C). South of the front, temperatures are warmer; most stations report temperatures above 50°F (10°C) with Corpus Christi and Brownsville, Texas, at 73°F (23°C). A few spots are cooler, the result of local effects such as elevation or shallow pockets of cool night-time air settling in regions of light wind and relatively clear skies.

The sea-level pressure in millibars is plotted at the upper right-hand side (for convenience, the leading 9 or 10 and the decimal point are not plotted; thus, the number 981 means 998.1 millibars; the number 120 means 1012.0 millibars, etc.) The large-scale pressure pattern is depicted by the sea-level isobars. Note that the high, or anticyclone, which lies to the north of the front is cold;

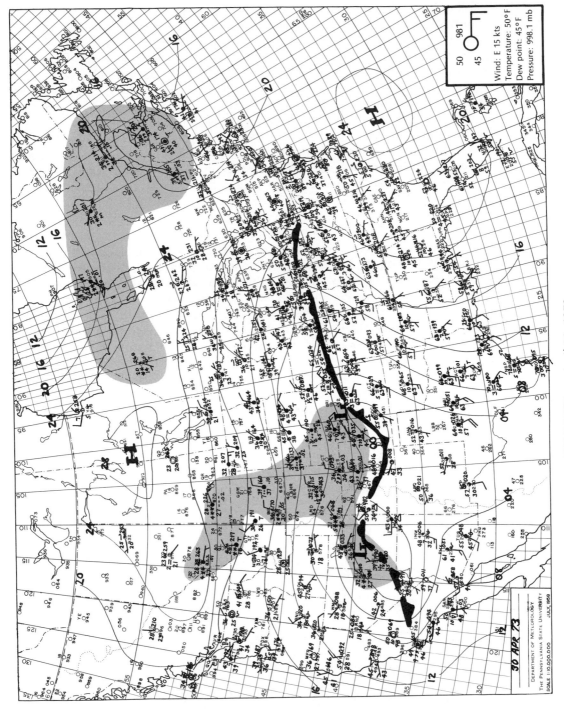

FIGURE 8.1 *Surface weather map at 7 A.M. EST (1200Z) on April 30, 1973.*

Wind: E 15 kts
Temperature: 50°F
Dew point: 45°F
Pressure: 998.1 mb

30 APR 73

DEPARTMENT OF METEOROLOGY
THE PENNSYLVANIA STATE UNIVERSITY
SCALE 1:10,000,000   JULY, 1958

another high, off the East Coast, is warmer. Between the highs, low pressure is observed along the front, as we have come to expect. Finally, we note that the lowest pressures of all are found in the southern Rockies. Recalling what we have learned about the wind circulations, it is reassuring to see that the surface wind reports, plotted as before, confirm the tendency for counterclockwise rotation about the lows and clockwise rotation about the highs. There is also a notable tendency for inflow and surface convergence near the lows and the low-pressure trough along the front, and a concomitant outflow from the highs. We are not confused by odd wind reports such as those from Boise, Idaho, or Prince George, British Columbia, which do not fit the large-scale pattern; we have seen that local effects can dominate large-scale effects, especially in mountainous terrain. At present, however, we are interested in the broad picture painted by all the reports.

We can now see the reason for the abundant cloudiness over the Rockies and northern Plains that is visible on the satellite photographs during the first days of May (figures 7.1–7.3). On the last day of April, the atmosphere was generating clouds there, with surface convergence pushing the air upward, forming clouds. What is more, the shading on the map and the symbols at stations (∇ = shower, $\bullet\bullet$ = rain, $\ast\ast$ = snow) show us that these clouds are producing precipitation at many locations. Also notice that the precipitation is occurring mainly (but not exclusively) on the cold side of the front. It makes sense to attribute at least some of this precipitation to overrunning, for we have seen that overrunning of cold air by warm air always puts the surface precipitation into the cold air (recall figure 7.23). And where the air is overrunning the Great Plains slope east of the Rockies, there is even more widespread precipitation. Easterly winds blowing clockwise around the cold high to the north and counterclockwise around the cyclonic system in the southern Rockies are climbing upstairs as they flow westward. Look at the weather—a cold, rainy morning in Denver, and it gets worse farther north along the east of the Continental Divide, where some snow is falling. What little precipitation that falls along the east slopes of the Rockies usually comes with upslope east winds. For us travelers, the day does not look propitious for embarking on a long journey, but let us start out anyway.

8.2 THE FIRST MORNING (APRIL 30)—LOW CLOUDS AND RAIN IN COLORADO

We head eastward along I-70, planning to spend the night of April 30 just outside Kansas City. All morning, and into the early afternoon, from Denver to Hays, Kansas, low clouds, rain, drizzle, and fog make the driving miserable. Notice Denver's $2^{1}/_{2}$-mile visibility on the 7A.M. EST (1200Z) chart (figure 8.1). The visibility improves only slightly across eastern Colorado, where warm air, gliding upward and westward over higher and higher ground, as well as over the cold air west of the Kansas front, is saturated almost at ground level. Clouds in such situations are a widespread stratus type with ragged, low bases. Even lower, patchy scud clouds race southwestward on the low-level winds from the northeast. It is a raw morning for late April in this cool northeasterly flow, and the temperature is still only 49°F

(9°C) when we stop for lunch at Oakley, just east of Goodland, Kansas. Goodland was 46°F (8°C) at 6 A.M., and as is normal for a cloudy, drizzly day with north-easterly wind, the temperature has risen very little during the morning hours. When this day is over, Goodland will have experienced a range of only 9°F (13°C) between the warmest and the coldest readings.

We can study the clouds a bit now while we are stopped. The steady rain, which was falling over Colorado's higher ground, has diminished. Low stratus clouds still cover the sky, but now they have the peculiar rolled shape that is common in cold, low-level flows behind cold fronts. Meteorologists call these *stratocumulus clouds*. The rolled shape does give some cumuliform appearance, but these are really stratiform clouds, formed mainly by the mixing of cool, moist air near the ground upward to a level where the air is cold enough to produce saturation.

THE FIRST AFTERNOON—DRIVING THROUGH THE FRONT IN KANSAS 8.3

After a welcome hot lunch, we approach Hays, in west central Kansas, where a remarkable transformation occurs. We see the skies brightening to the east, and the stratocumulus layer begins to break up. Now it is easy to see some higher, broken clouds, mainly globules of altocumulus, with some cirrus; our spirits are also lifted by patches of blue, which widen to the east.

We are approaching the narrow band of clearing, sometimes called *false clearing,* that often lies just behind cold fronts and is caused by sinking air there. Figure 7.22 shows an example of this band, which is the dark streak behind the front where the satellite can "see" down to the ground. We call it *false clearing* because with a cyclone's normal motion toward the east or northeast, the clear band, being only 80–160 km wide, is at most over any one point only a few hours. Today, we are driving eastward at 80 km/h and overtaking this slow-moving (15km/h) cold front, and those bright skies ahead tell us that we are getting close to the front.

As we drive eastward across Kansas, we notice that the temperature is rising rapidly; we know at least two reasons why this is so. First, we are escaping the cold air by driving east. As we continue, we will cross the front into the warm air and escape the cold air altogether; even now, we have already driven close to the edge, and it is not as cold here. Second, daytime solar heating of the ground, which was suppressed in western Kansas by the deep, reflecting clouds, is strong in this brighter part of the state, so afternoon temperatures are higher.

We left the steady rain in Colorado, but now there are some showers scattered about, and we can see the bulging cumulus and cumulonimbus clouds which are strung out along the front. Here is where the surface winds that comprise two great currents of air—one warm, the other cold—are converging along a great arc. We saw the boundary of these two currents (the front) on the morning map and in the satellite picture. Now, we see it from the ground—warm and humid surface air along the very edge of the cold flow being lifted by the converging surface winds. It is obviously unstable air; a look at the towering cumulus clouds

221

tells us that. And suddenly we catch up to and disappear under some hard showers and thundershowers.

The rain beats on the windshield; we are not making 80 km/h now. It is dark enough to see a lightning flash or two under these huge cumulonimbus clouds that are 11 or 13 kilometers deep. The raindrops pelting the windshield have grown to monstrous sizes; it is raining at the rate of 2 centimeters per hour. A partial evaporation of these raindrops has chilled the air to 64°F (18°C). We almost wish that we had stopped, but we reason that if we caught up with this thunderstorm, we ought to be able to overtake it and break through it. And suddenly we do, for it soon brightens to the east, and we drive out of the rain and under the cirrus clouds that are stretched out ahead of the thunderstorm cell. It is warm again, and rather humid, with a south wind. The dark clouds obliterate the western horizon behind us. We are in the warm air east of the front and moving toward a marvelous expanse of blue skies dotted with patchy cumulus clouds.

The weather that is observed in warm flows such as this one can be quite varied. Often there is abundant sunshine, for warm air can sustain much water vapor and, therefore, frequently has a low *relative humidity,* which is the fraction of the maximum possible amount of water vapor actually present. However, other factors can cause even this warm air to become saturated. If the earth's surface is cold, the low-level air can be cooled to its dew point, causing fog and stratus clouds. Warm, humid winds in winter often produce this depressing situation on the southern Plains. Also, if the there is a cyclone close by, surface frictional convergence into the low center can lift the warm air to saturation. If the warm air is unstable, bands of showers can be produced. Since warm air can be rich in moisture, even when unsaturated, these warm-air thunderstorms can deliver copious rains, which are blessings for American framers and for all in the world who use the huge crops grown on these plains.

But just occasionally, when the winds aloft are very strong, the thunderstorms become too vigorous, generating hail, destructive winds, or even tornadoes. The signs to watch for reflect the factors that produce vigorous convection— warm, humid surface air; altocumulus clouds that grow vertically, indicating unstable conditions aloft; strong surface winds that help produce strong convergence; a low barometer (more convergence); and cirrostratus clouds blowing off the tops of thunderstorms which are located upwind. These *anvil* clouds are often present with ordinary thunderstorms, but if the anvils indicate strong jet-stream winds aloft when the other factors listed are present, some of the thunderstorms are likely to be severe. On the Plains, where the ground is smooth, large rotating thunderstorms can organize their spin into the dreaded tornado.

Thus, we can see why tornadoes are favored on the Plains in springtime. They must occur in cyclonic circulations (low barometer and surface convergence), when the low-level wind is southerly (warm and moist), and when the jet stream aloft is strong and from the southwest (wind veering with height, indicating warm advection and rising air). So, we are not surprised to learn that five out of eight tornadoes in the United States move northeastward, and we remember now to watch the direction of these streaking anvils. Maybe we should have stayed in the cold air, where tornadoes are very rare.

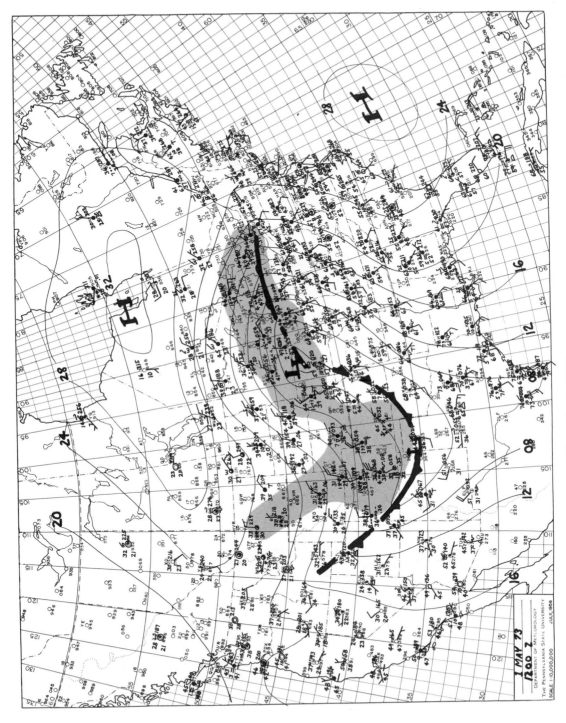

FIGURE 8.2 Surface weather map at 7 A.M. EST on May 1, 1973.

8.4 THE FIRST NIGHT—A THUNDERSTORM IN KANSAS CITY

We reflect uneasily on these matters on the muggy night of April 30 as lightning flickers and thunder rumbles over Kansas City. There are a large number of night-time thunderstorms over the Midwest on warm nights, when it is difficult to see the sky. They often form when strong southerly winds carry heat and moisture rapidly northward, and when the southerly winds are forced aloft by a front or even a cold outflow from an earlier thunderstorm (a mini—cold front), excessive rains can occur. Such rains have deluged Kansas City in recent years, as well as Rochester, Minnesota, and Rapid City, South Dakota, causing loss of life and heavy property damage.

Recently, new tools for detecting severe thunderstorms and flash floods have been developed. Doppler radar, which can detect cloud water velocities and even tornado rotations, infrared satellite imagery to detect cold-topped, deep clouds at night, and new insight into the structure of thunderstorms are making it possible to issue *watches* and *warnings** of such events with better precision. Fortunately, tonight we do not hear the roar of a tornado or the rush of a flash flood. There is considerable moderate thunderstorm activity, however, and on the morning weather map for 7 A.M. EST of May 1 (figure 8.2), we can see a thunderstorm (symbol $\dot{\bar{R}}$) at Cape Girardeau, Missouri. Thunder during the past six hours is reported at Des Moines, Iowa; Moline, Illinois; Wichita, Kansas; and Springfield, Missouri; as well as at Kansas City. Cape Girardeau received about 2 centimeters of rain in its thunderstorm, and we can infer that plenty of rain fell this night between the reporting stations. The winds aloft charted on the 500-mb map (figure 8.3) show that there is a strong jet stream overhead. We should not be surprised to hear of tornadoes today.

8.5 THE SECOND DAY—RACING THE COLD FRONT ACROSS THE MISSISSIPPI VALLEY

As we prepare to depart on the morning of the second day (May 1), the cold front is lurking to the northwest and threatens to overtake us again. The winds across eastern Nebraska have turned into the northwest as the first of the Rocky Mountain lows has raced northeastward to Wisconsin. Those winds will swing the front across Kansas City today, but we will drive eastward and stay ahead of it. This 57°F (14°C) air in the morning is certainly more comfortable than yesterday's chilly weather; looking back at Denver and Cheyenne, Wyoming, we find snow in May!

This morning's weather at Kansas City gives us many clues to what the weather map looks like. It is warm for the date, and humid, as indicated by the 54°F (12°C) *dew point* (the temperature at which saturation and appearance of visible waterdroplets occur). We observe patches of fog in the low places along the highway. A south wind and the thunder of the previous night confirm that there

---

*A *watch* indicates that conditions are favorable for development of severe weather; a *warning* indicates that severe weather has been sighted.

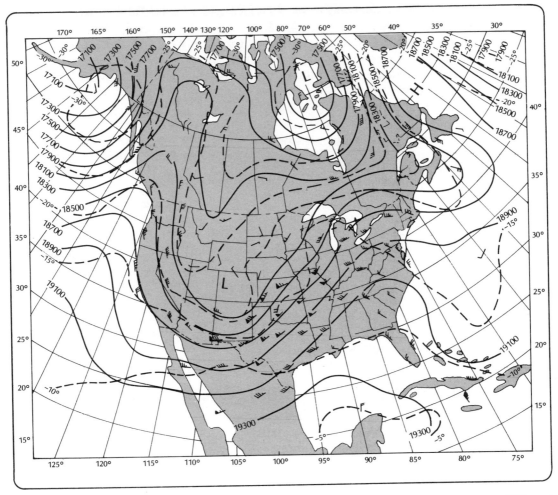

FIGURE 8.3 *Upper-air (500-millibar map) at 7 A.M. EST on May 1, 1973.*

is high pressure to the east, low pressure to the west, and a warm, unstable air mass south and east of us. But there is more. Altocumulus clouds in the morning sky are moving northeastward (see the winds-aloft chart in figure 8.3). The wind at the surface is southerly, so the wind is veering with height. When we correct the surface winds by 30 degrees or so to account for the effects of surface friction, the veering is reduced, but in any case, the wind is certainly not backing. Therefore, the atmosphere may be getting somewhat warmer. With abundant blue sky, the May sun will drive the surface temperature up quite rapidly today. The little turrets on the altocumulus clouds also tell us that the middle troposphere (about 3 km, 10,000 ft) is unstable, so residents of Kansas City may expect more thunderstorms to form as the surface heating progresses.

We can observe one other thing with our eyes. The rapid northeastward movement of the altocumulus clouds tells us there is a strong jet stream, and strong jet streams are found over fronts. The front is close, but the southwest winds aloft are nearly parallel to it, so it is moving very slowly. In fact, we can see on the May 1 weather maps (figures 8.2 and 8.3) that these southwest winds aloft are carrying another cyclone toward Kansas City. It is the second member of this cyclone family, following a track southeast of its parent. This morning, it is over the Texas Panhandle, but Missourians will sense its approach today as barometers, thundery rains, and hopes for the spring planting fall together. Falling weather!* Once this second cyclone goes by, temperatures will also fall as northerly winds follow the passage of the cold front.

As we drive on eastward across the Mississippi Valley toward St. Louis, enjoying this warm sector, we know that a cold front is nipping at our heels. However, at our average speed of 75 km/h, we gradually pull ahead of it, since its speed is only about 30 km/h. We also know that we may see some scattered showers or thundershowers today in this unstable air. However, the chances are not nearly as high as they would have been had we stayed in Kansas City, close to the cyclone path.

By driving toward higher pressure, we reduce the odds of our meeting bad weather, and so we have a mostly sunny, warm day with afternoon temperatures in the 70s (°F) and numerous swelling cumulus clouds reminding us of the high humidity. Late in the day, a few of the towering cumulus grow to cumulonimbus clouds, but they are widely scattered, and we do not drive under any of them. But the telltale cirrus anvils visible on the afternoon horizon remind us of the persistent cold front and its associated squall line following us.

8.6   THE SECOND NIGHT—MORE THUNDERSTORMS IN INDIANA

During the twilight of our second day, we stop in western Indiana. We have built up a good lead on the front, but fronts do not sleep or tarry for meals, and after we stop, the front begins closing the gap between us. Although the scattered daytime cumulus clouds are dying with the sunset, the western sky reveals fresh bands of altocumulus merging with thickening altostratus, indicating the approach of the front. The night will be cloudy, and because this warm air is rather unstable, we expect showers and thunderstorms to be embedded in the layers of altocumulus and altostratus. And sure enough, later that night, we are partially awakened by rumbles of thunder and flashes of lightning as thundershowers visit Indiana.

8.7   THE THIRD DAY—OVERTAKEN BY THE FRONT IN PITTSBURGH

By Wednesday morning, May 2, the surface map (figure 8.4) shows us that we are practically in the same position relative to the front as we were on the previous

---

*This expressive term is reportedly of Pennsylvania Dutch origin. Modern barometer watchers have learned that the pressure tends to rise every morning and fall every afternoon, so they usually concentrate on large or persistent trends in the barometer reading rather than on the instantaneous change. In the present example, barometers in Missouri would fall all day, but especially in the afternoon.

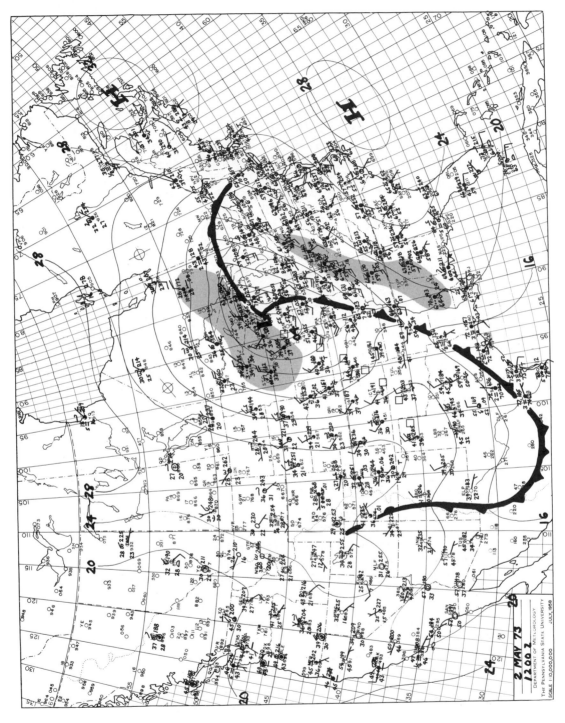

FIGURE 8.4 Surface weather map at 7 A.M. EST on May 2, 1973.

morning. Yesterday's drive kept us ahead of the front, but now the second cyclone has moved northeastward to the western shores of Lake Michigan, and the front has continued its relentless march eastward. Note that the cyclone track (indicated by squares in figure 8.4) shows that the center of this second cyclone passed right over Kansas City yesterday after we left. The Kansas City airport received well over 3 centimeters of rain.

On the morning of May 2, the 7 A.M. Indianapolis weather report shows overcast skies and a temperature 5F° (3C°) warmer than Kansas City was yesterday at the same time. We are still in the warm sector with a southerly wind at the surface. However, through breaks in the low clouds, we can see a more westerly movement in the upper-level clouds, and, as expected, figure 8.5 shows that the flow aloft is definitely more westerly than it was over Kansas City yesterday. (Compare the two upper-air charts in figures 8.3 and 8.5.) The difference is that the low-pressure center in the upper atmosphere has moved from western Colorado to western Wisconsin. Thus, the steering winds aloft have shifted, and now, instead of being parallel to a slow-moving front, the flow aloft is blowing perpendicular to the front, sweeping it eastward. This front may catch us today.

We can see from the importance of the steering principle why weather forecasters want to be able to anticipate changes in the steering winds. But we do not need to know all of the details of the upper-air chart. The strong flow from the west aloft, as indicated by the rapidly moving upper-level clouds, tells us that our pursuing cold front should approach quite rapidly now. And the Norwegian cyclone model tells us that colder, drier air will flow behind it from a northerly quarter. We know all that, standing by our wet car in western Indiana on a warm, humid morning in early May, by using just our eyes and remembering the cyclone model.

Looking again at the surface map for the morning of May 2 (figure 8.4), we can see that there is no third member of this cyclone family. The surface winds west of the front over Illinois, Iowa, and Missouri have shifted to west and northwest. If there were another low along the front, near Little Rock, Arkansas, for example, these winds would be different, probably from the northeast, like the winds we had in Colorado and western Kansas on the first day. And if that were the case, the front would move slowly, just as it did on Monday. But in the absence of a third low, the resulting strong west-to-northwest winds across Iowa and Illinois are driving the cold air and its leading edge, the front, eastward quite rapidly. So, we draw the same conclusion from weather maps that we did with our eyes. This front is moving eastward more rapidly today than it was yesterday.

Tonight, the front will overtake us and give us a taste of colder air. It is sticky enough on this warm Indiana morning to make the thought welcome. Today's drive of 650 kilometers will take us to Pittsburgh, and since we can drive an automobile somewhat faster than the front moves, we remain in the warm, moist air. As the cyclonically curved isobars on the map show, this moist air is converging near the ground and rising, so we do not see much of the sun. Many low clouds and occasional showers may mar the driving.

We watch for signs of the front that evening in Pittsburgh, but the thickening low clouds defeat us. After dark, however, hard showers set in, and the

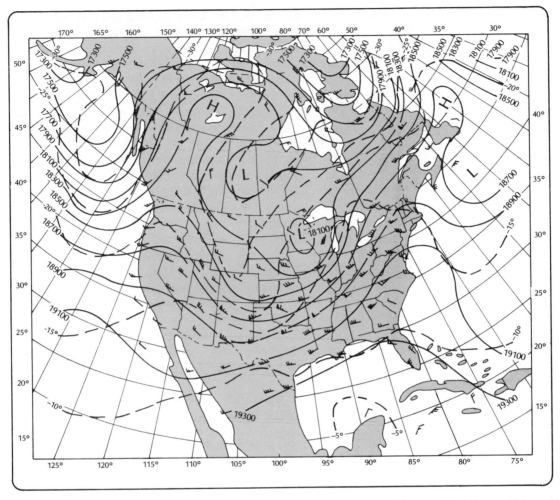

FIGURE 8.5 *Upper-air (500-millibar map) at 7 A.M. EST on May 2, 1973.*

wind shifts from the south around to the west. Now, if we had a barometer, we would observe its lowest reading, followed by a rather sudden upsurge as the cold air rolls in. And we feel the difference in this air after the hard showers end—west winds, cold and drier in the absolute sense even though it is still cloudy. The dew point tumbles from 60°F to 50°F (16°C to 10°C) in just an hour or so.

THE FOURTH DAY—COLD ADVECTION AND LOW CLOUDS OVER WESTERN PENNSYLVANIA 8.8

We take our own morning observation at Pittsburgh at 7 A.M. EST (1200Z), May 3, and what a difference from our previous morning! It was 62°F (17°C) in Indianapolis at the same latitude yesterday morning, and today in Pittsburgh it is 52°F

229

(11°C). The air feels much less humid, and the wind is more gusty and from the west rather than the south. The low clouds have broken, but the clearing affects only the low clouds, and we are able to observe the winds aloft by watching the movement of the high clouds. It is very hard to see the motion of the amorphous layer of altostratus clouds over us, but by looking straight up and watching carefully, we see that they are moving from the southwest. Well, that clinches it! The wind is backing with height, for the surface geostrophic wind must be from west-northwest (about 30 degrees clockwise from the surface wind). The atmosphere over Pittsburgh is cooler and is cooling further with the cold advection. The May sun may shine enough today to drive up the temperature a good deal, but it will very likely be cooler than yesterday, and tonight should be a lot cooler.*

Here in Pittsburgh, on the morning of May 3, we bring our road trip to an end, but our observations are not quite finished. We have followed a cyclone family across the United States. Now, let us wrap up the sequence of events with a last synoptic view—that is, an analysis of a number of observations taken at the same time. Figure 8.6 shows the May 3, 7 A.M. EST (1200Z) map for the United States for comparison with the other morning surface charts. We see that all of the features that we observed this morning in Pittsburgh are consistent with the map. The cold front has passed east of Pittsburgh, and a westerly flow of cooler and drier air has established itself behind the front over the Ohio Valley. Indianapolis is 17F° (9C°) cooler today than at this time yesterday. On toward the east, the warm, humid air of the warm sector, with 60°F (15°C) morning temperatures and dew points nearly as high, extends as far northward as Burlington, Vermont, and possibly Montreal, Quebec. Farther north, over New England, the Maritimes, and much of Quebec, cold air at the surface shows that the warm, moist air is overrunning colder air; a surface warm front marks the boundary where the warm air leaves the surface. Portland and Houlton, Maine, Chatham, Ontario, and Yarmouth, Nova Scotia, are in the 40s (5−9°C), and Seven Islands, Quebec, has rain falling into 34°F (1°C) air. Sleet is reported at Kapuskasing, Ontario, and it is easy to guess that snow is reaching the ground just north of Seven Islands. This cold side of the storm track is just where the Norwegian model would have us expect snow or sleet. Indeed, we can now identify the great band of clouds that we saw in the composite satellite pictures of figures 7.1−7.3, which spread northeastward from the central United States all the way to Labrador between May 1 and May 3. Those clouds mark the northeastward movement of these two cyclones under the southwest flow aloft, and, in particular, the overrunning of the cold air in Canada by warm, moist air from the south.

But now, let us look at the satellite pictures at 10 A.M. EST. Figure 8.7 is a photograph taken in the visible wavelengths of the clouds over the United States and Atlantic Ocean. We see a view of a classic middle-latitude cyclone. Notice that the extensive cloudiness over the eastern United States and Canada gives the appearance of counterclockwise (cyclonic) rotation about a point near

---

*The data at Pittsburgh: Yesterday's maximum, May 2, 67°F (19°C); May 3, 62°F (16°C). Minimum on the first night of our visit, 51°F (11°C); the following night, 38°F (3°C) (with snow flurries!).

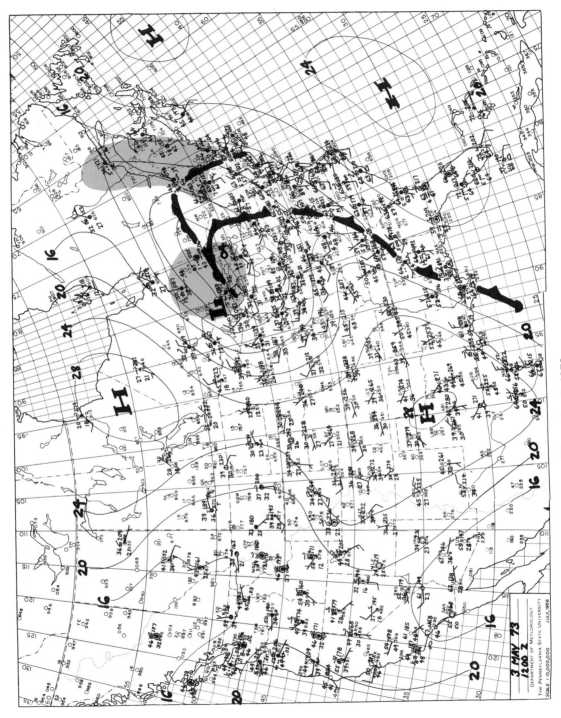

FIGURE 8.6 *Surface weather map at 7 A.M. EST on May 3, 1973.*

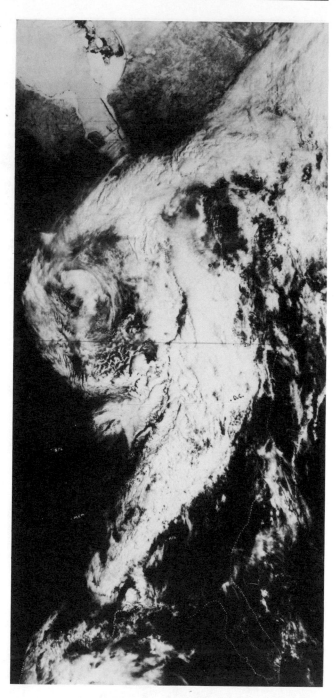

FIGURE 8.7 *Visible satellite photograph at 10 A.M. EST on May 3, 1973.*

the Great Lakes; to its rear (west), we can almost sense the air flowing in from the west and northwest. We notice that the skies clear behind the front, which is expected, both because cold advection produces sinking and because the following high-pressure center produces surface divergence and sinking air. What is more, we can detect more extensive bands of clouds along the fronts, where there is

FIGURE 8.8 *Infrared satellite photograph at 10 A.M. EST on May 3, 1973.*

overrunning, and where there is strong surface convergence near the cold front. Off to the southeast, the air is nearly clear again, with just a few small cumulus and cirrus clouds and a couple of scattered showers over the Atlantic marring an otherwise clear sky. We can even infer the presence of tiny fair-weather daytime cumulus over Florida and the southeast coast. The individual clouds are too small

233

to see, but their combined effect whitens the land. Notice how few there are over the oceans, which are not heating up as rapidly as the land. And notice that to the rear of the cyclone, the air is so dry that there are not even any fair-weather cumulus clouds.

The imaginative observer even may be able to see the wind patterns in the clouds as the warm air swirls around the western end of the fairweather Bermuda High.* That air is flowing in toward our second cyclone; by tomorrow, it will be feeding the next member of the family over the Gulf of St. Lawrence. But let us look more carefully at the clash of air currents along the cold front. To do this, we look at another picture taken at the same time (10 A.M. EST, May 3) from the same satellite, but for this picture, the sensor is tuned to "see" infrared radiation, which is proportional to the temperature of the objects viewed rather than to their color. In this picture (figure 8.8), cold objects are white, and warm ones dark. We notice how warm the surface of the earth is, especially the warm ocean. The Florida peninsula, which is about the hottest land area, shows up dimly.

But now, we can tell where the deep clouds, with cold tops, are found. Along the cold front is certainly one place. In fact, we can use the infrared picture to locate the front. Comparing the visible and infrared, we can see that the deep clouds occur along the leading edge of the front. In contrast, behind the front, the clouds are shallow because their tops are warm. (Compare the appearance of this cyclone with the pictures of the cyclone on May 20 in figures 3.6 and 3.10).

Finally, let us consider the surface weather at Pittsburgh later in the day. Looking outside, we find that low stratocumulus clouds have rolled in since morning, and it looks dark and threatening. We are momentarily baffled; cold air behind a cold front is supposed to sink and produce clearing weather. It is not raining, but the air feels damp and cool, and we wonder what mechanism is producing all these dark, low clouds. Two factors could produce low clouds, and low clouds only. One would be convergence across cyclonically curved surface isobars, the other, air flowing upward over the lower Allegheny Mountains. The surface map for 7 A.M. EST (figure 8.6) shows that the isobars behind the cold front are curved in a cyclonic sense over the Great Lakes and about as far south as Indianapolis, which might explain the cloudiness there. But that is not the situation at Pittsburgh, where the isobars are actually curved anticyclonically. We conclude that these low clouds at Pittsburgh must be very shallow, produced by west winds blowing upward along the western slopes of the Alleghenies. We also conclude that no more than a few sprinkles can fall from these shallow clouds, and the ball game at Three Rivers Stadium will not have to be called off.

We can verify the preceding reasoning on the infrared picture (figure 8.8). Notice how warm the cloud tops are over Pittsburgh and everywhere to the west and southwest. Warm tops mean low tops, and low tops mean little or no rain.

We call a friend in Roanoke, Virginia. "Has the front reached you yet?" "No, it's been raining off and on all morning. It's very humid, with clouds at many levels, some of which are towering like big castles through the low clouds."

---

*The Bermuda High is the semipermanent anticyclone centered near Bermuda.

We could make a forecast for Roanoke, for we know that the front is near. The westerly winds over the Ohio Valley make the relief from the humidity and an end to the showers virtual certainties for Roanoke.

A call to Memphis verifies what we expect. The weather over Tennessee is clear, bright, dry, and beautiful. It is almost too cool (39°F or 4°C) back near the high center at Kansas City, where the dry, cool air lost much heat last night.

We call a friend in Georgia, near Augusta, where it is a lovely day, with the temperature already 75°F (24°C) and rising fast. She reports a few cumulus clouds and some high, thin ones. "We know about that; we've seen them on the satellite pictures. Your wind must be from the south; it's probably a bit sticky there. Well, watch the western and the northwestern sky this evening; Georgia will have another northern visitor tonight."

We make a quick call to Augusta, Maine. Just as we thought—at this Augusta, it is another story; rainy, gloomy, foggy, and cold—45°F (7°C). The warm front is south of here, and it will be something to watch its approach today as the cyclone runs up the St. Lawrence Valley. Augusta may get into the warm sector, which is over at Burlington, Vermont; if it does, our friends "down east" are going to witness a dramatic rise in temperature before the cold front arrives to pinch off the warm air. As it happens, the temperature at Boston shoots up close to 80°F (27°C), but the cold front comes through Maine before the warm air reaches the surface.

A last call to New York City brings some excitement. The air is warm and humid, and there have already been some showers. The barometers are falling sharply, and although the sky has cleared partially, thunderheads already are appearing in the western sky. Sharply falling morning barometers; warm, humid air; and building cumulus clouds foreshadow a great day for convection. And as we tell of the front rolling toward New York, we know that later today the convection will be invigorated by the frontal convergence. It looks like a good day for a squall line to spring up in the Poconos or the Adirondacks and sweep across the Hudson River. Over such rough terrain, tornadoes are rare, but they are possible; at least our friends in New York should be on the lookout for strong winds, lightning, and possible hail today. Tomorrow they can relax.

From sleet at Kapuskasing to hail at New York; from snow in Quebec to steady rains at Roanoke; from high, thin cirrus in the southwestern sky over Newfoundland to the last edge of the shallow boundary-layer stratocumulus just west of Indianapolis; from the cool, dry, clear air over Missouri to the warm, moist, clear air over the Atlantic; from all these facets of the cyclone circulation emerge some sense and order when we consider them all together. And if we know the cyclone model and use our eyes, we can often guess what those reports would be, whether we are hiking in the forest or driving along the highway. The details may vary from cyclone to cyclone and from season to season, but the keen eye is rarely unrewarded.

Questions

1 Make a graph of the barometric readings at Kansas City, Missouri, for the period 30 April 0600 CST to 3 May 0600 CST. Use an ordinate scale from 1000 to 1030 mb. Interpolate the time from the surface maps and from the text. Use the graph to estimate the time when the cold front passed.

2 Why were there showers and thunderstorms from cumulonimbus clouds where it was warm and sunny (or had been the previous daytime), but not (at least, not as many) where the stratus, stratocumulus, and scud clouds were observed?

3 Compare the visible and infrared satellite pictures of figures 8.7 and 8.8. Notice that on pictures the bright, white clouds in the visible between Washington, D. C. (DC on pictures) and Pittsburgh (PIT) are also bright white in the infrared; however, the infrared picture is dull gray west of Pittsburgh (to the left). Are the cloud tops west of Pittsburgh closer to the ground (note that the ground is black in both pictures) than those east of Pittsburgh? What does your answer mean in terms of possible rainfall?

4 Describe the clouds that an airline pilot descending into Pittsburgh from the west would see on the morning of May 3.

Problem

5 On the basis of the satellite pictures, the surface maps, the Norwegian cyclone model, and the text, estimate a map of rainfall amounts for the 24-hour period ending at 0700 EST on 3 May. Make your estimates by drawing contours enclosing regions receiving more than 1 mm (0.04 in) of rainfall. Then, inside that region, draw another contour to enclose regions where you think it is possible that they received more than 1 cm (0.40 in).

NINE

Hurricanes

hurricane • cyclone • tornado • tangential velocity • angular
momentum • streamlines • eye • computer model • spiral bands

The sun sets weeping in the lowly west,
Witnessing storms to come, woe, and unrest.
[Shakespeare, *King Richard*
the Second, 2.4.21–22]

9.1 CYCLONES, HURRICANES, THUNDERSTORMS, AND TORNADOES

The relatively infrequent severe weather phenomena that most threaten us also hold a fascination that ordinary events do not provide. Few people fail to be stirred, excited, or frightened by the disquieting rumbling of a thunderstorm on a warm summer night, the ominous pounding of the surf as a hurricane approaches, or the terrifying roar of a nearby tornado. These events have rightly been the subjects of a large fraction of meteorological research. Like the news coverage of extraordinary "bad" events, the misbehaviors in meteorology attract most of the attention.

In the previous chapter, we saw how the common, middle-latitude cyclones affect our weather during the year. Many people confuse tornadoes, hurricanes, and cyclones, as evidenced by this quotation from *The Wizard of Oz*, by L. Frank Baum: "The north and south winds met where the house stood, and made it the exact center of the cyclone." The tornado which swept Dorothy's house from its Kansas foundation and deposited her in the Land of Oz was, in fact, a cyclone; however, the vast majority of cyclones are not tornadoes. A cyclone is merely any circulation around a low-pressure center, regardless of size or intensity. Thus, tornadoes and hurricanes, as well as extratropical cyclones, are all different kinds of cyclones.

The large extratropical cyclones are usually beneficial, providing necessary precipitation, exchanging cold polar air for warm tropical air, and cleansing the air at the same time. Hurricanes and tornadoes are smaller, and frequently more vicious, cousins of the extratropical cyclone, although hurricanes have certain beneficial as well as destructive aspects. Before discussing the origin and structure of these storms, we must first carefully distinguish between these three distinct phenomena.

The hurricane is an intense cyclone that forms over tropical oceans. Although smaller in size than the middle-latitude cyclone, it is much larger than either a tornado or thunderstorm, averaging about 800 kilometers (500 miles) in diameter. An unusual nighttime photograph (figure 9.1a) taken by moonlight, of

FIGURE 9.1(a) *Moonlight view of Hurricane David (1979).*

Hurricane David (1979) gives an idea of the size of a mature hurricane in comparison with familiar distances on a map. A photograph of Hurricane Becky (1974) approaching a cold front over the Atlantic, with an extratropical cyclone located north of the cold front, provides a comparison of the size of hurricanes with extratropical circulations (figure 9.1b).

The most prominent characteristic of the hurricane is a doughnut-shaped ring of strong winds exceeding 120 km/h (75 mi/h) surrounding an area of extremely low pressure at the center of the storm. (You can get an idea of the strength of a wind speed of 120 km/h by leaning out the window of a car going at this speed.) It is impossible to stand in such a wind; it is also difficult to breathe.

241

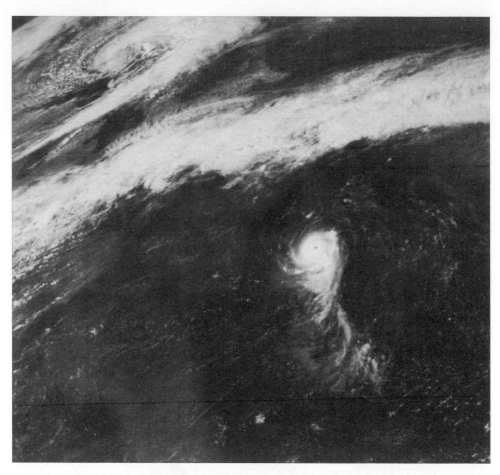

FIGURE 9.1(b) *Satellite photograph of Hurricane Becky at 1 P.M. EST on August 29, 1974.*

These winds may drive the ocean into waves exceeding 6 m (20 ft) in height. The hole of the doughnut, called the *eye*, is a relatively calm region of clear skies (figure 9.1a, b and plate 8). A hurricane may live for three weeks or more as it drifts along in its odyssey over tropical waters.

Tornadoes are similar to hurricanes in that strong winds blow counterclockwise around a center of very low pressure.* In fact, the strongest surface winds on earth occur with tornadoes. Although the maximum tornadic velocities that have occurred are somewhat controversial (direct measurements being impossible), there is indirect evidence that winds have exceeded 480 km/h (300 mi/h) in extreme cases.

The diameter of the tornado is much smaller than that of the typical hurricane (figure 9.2), averaging about ¼ kilometer. Thus, it is quite feasible to

---

*A few cases of winds blowing clockwise rather than counterclockwise have been observed, but these are rare.

avoid an approaching tornado by driving, or even running, at right angles to its path of motion.

Size is not the only important difference between tornadoes and hurricanes. While hurricanes originate over warm water in a uniform tropical air mass, tornadoes occur most frequently over land, and are produced when cooler, drier air streaks over warm, moist air, which is less dense. Tornadoes occur most frequently in the spring, when the strongest contrasts between cold and warm air are present. They are also usually associated with severe thunderstorms.

Tornadoes and thunderstorms frequently are found embedded within the much larger hurricane vortex. Such hydralike storms are not impossible nightmares, but terrifying realities. For example, Hurricane Beulah (1967) spawned at least 115 tornadoes as it moved ashore in Texas, killing five persons and causing nearly $2 million in damage.

Thunderstorms are so common and familiar to everyone that they scarcely need differentiating from tornadoes and hurricanes. Unlike the horizontal winds around the hurricane or tornado, the air circulation of the thunderstorm is primarily up and down. Updrafts and downdrafts may exceed 80 km/h in strong storms as warm, light air rises and cold, dense air sinks. Winds near the ground in the vicinity of thunderstorms are variable, gusty, and difficult to predict. They do not follow the counterclockwise vortical pattern that typifies cyclones.

By definition, the thunderstorm must produce lightning, which is the cause of thunder. It surprises most people that lightning kills about 200 people a year in the United States, a higher toll than that for hurricanes. The energy from a

FIGURE 9.2 *Tornado.*

243

single lightning stroke can light a 60-watt bulb for several months. The total power from lightning in the 2000 thunderstorms in progress at all times over the earth is about 500 million kilowatts, which is about equal to the total U.S. power generating capacity in 1978.

9.2 THE LIFE CYCLE OF HURRICANES

In contrast to extratropical cyclones, which derive their energy from horizontal temperature gradients between different air masses, the hurricane winds are generated and maintained by the condensation of water vapor within a uniform, tropical air mass. As we saw earlier, the condensation of one gram of water vapor liberates about 580 calories of heat. The rate of condensation heating in one typical hurricane is 10^{11} kilowatts (kW); in a day, the hurricane produces 24×10^{11} kilowatt-hours (kWh). By comparison, the entire U.S. use of electric energy in 1978 was 23×10^{11} kilowatt-hours. Anyone who has visited the tropics in summertime can appreciate the enormous amount of latent heat stored in the form of water vapor. A cold can of beer is warmed to air temperature within minutes by the condensation on the can. Day after day, mixing ratios exceed 20 grams of water vapor per kilogram of air, an amount that is twice the annual average in Washington, D.C. Indeed, with the abundance of water vapor available for condensation, it is somewhat remarkable that hurricanes are relatively infrequent, in spite of the numerous thunderstorms and convective showers in the tropics.

On the other hand, with the frequent occurrence of thunderstorms and the associated release of the latent heat of condensation, we might wonder why hurricanes form at all. Let us consider how thunderstorms are organized to form the larger hurricane vortex.

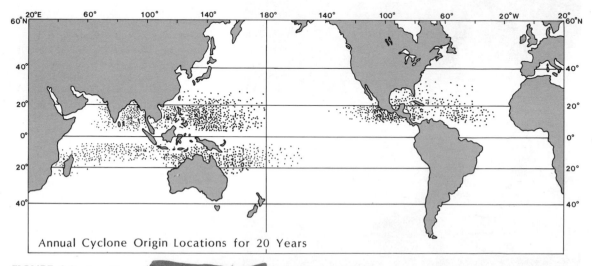

Annual Cyclone Origin Locations for 20 Years

FIGURE 9.3 *Location of initial genesis points of tropical cyclones over a 20-year period.*

244

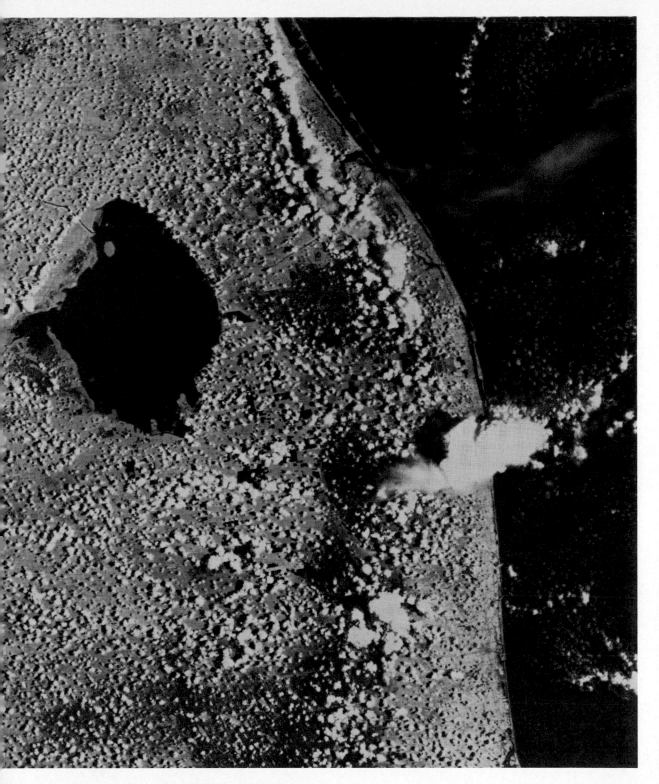

Plate 9 Landsat photograph of cumulus clouds over Florida, 10 A.M. EST, August 18, 1972. The line of towering cumulus clouds along the east coast marks the "sea-breeze front" (courtesy of NASA).

Plate 10a (above)   Mature thunderstorm showing well-developed anvil (courtesy of Ronald Holle).

Plate 10b (below)   Photograph from space showing anvil (courtesy of NASA).

9.2.1 COOPERATION MAKES A HURRICANE

To understand the genesis of hurricanes, we must consider a cooperation between two weather systems of distinctly different horizontal scales, the hurricane vortex and the much smaller cumulonimbus, or thunderstorm, clouds. We know that if air with the temperature and moisture structure of the tropics is lifted, condensation will liberate enormous amounts of heat. But theory and observations show that the release of this heat energy will occur in narrow cumulonimbus towers, not in large hurricanes. Why, then, do not all tropical disturbances consist of groups of thunderstorms, with very little organization of horizontal winds on the larger scale? The explanation is that, under the most favorable conditions, the circulations associated with the individual clouds and the hurricane vortex cooperate, with each circulation acting to reinforce the other. If the favorable conditions persist for a sufficiently long time, this mutual cooperation can result in the awesome hurricane.

9.2.2 CLIMATOLOGICAL ASPECTS OF TROPICAL CYCLONE FORMATION

Tropical cyclones form in all of the tropical oceans except the South Atlantic and the eastern South Pacific (figure 9.3). About two thirds of all cyclones occur in the Northern Hemisphere. Tropical storms do not form within 4° to 5° of the Equator, and only a few (approximately 13 percent) develop poleward of 22°N. The majority (65 percent) form in the zone between 10° and 20°

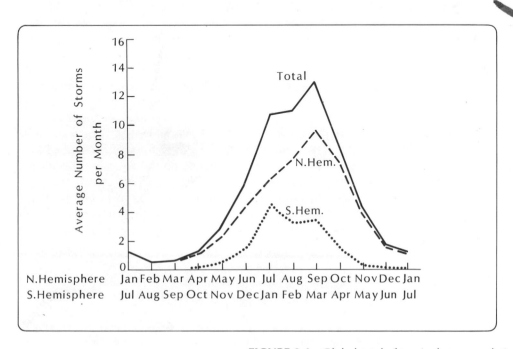

FIGURE 9.4 *Global total of tropical storms relative to solar year.*

245

FIGURE 9.5 *Upper-level flow (solid lines) and favorable and unfavorable regions for tropical cyclone formation.*

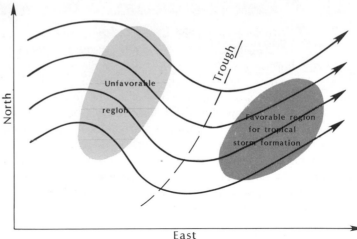

of the Equator. The absence of storm formation close to the Equator suggests the importance of the Earth's rotation about the local vertical axis in causing disturbances to begin spinning. The distribution of tropical cyclones by months (figure 9.4) shows that most storms occur in late summer and early autumn, when water temperatures are highest.

Climatological studies of tropical cyclonesis have revealed several large-scale aspects of the tropical circulation that are favorable for storm formation (table 9.1). High sea-surface temperatures mean that a rich supply of water vapor and heat is available to fuel the tropical storm. High relative humidites in the middle troposphere are favorable for intense thunderstorms which release large amounts of latent heat. If the middle troposphere is dry, evaporation of cloud and rain water can produce cooling rather than warming and inhibit the growth of thunderstorms and the tropical storm.

Environments with strong vertical shear of the wind are unfavorable for tropical storm formation because the heat and moisture supplied to the middle and upper levels of the disturbance by the cumulus convection is carried away by the winds, thus *ventilating the storm.* Therefore, environments in which the wind is light and uniform in the vertical are more favorable for tropical cyclogenesis.

Since tropical storm formation depends on the release of latent heat in cumulus clouds, circulation patterns that favor the growth of clouds are favorable for storm development. We saw earlier that precipitation in extratropical systems occurs mainly east of an upper-level trough where the air is being lifted. Upper-level waves also exist in the tropics, providing a lifting of air east of the

TABLE 9.1 *Factors favorable for tropical storm formation*

- High sea-surface temperature (greater than 27°C).
- High relative humidity in middle troposphere.
- Weak vertical shear (change of wind direction or speed) with height
- Location of low-level disturbance east of an upper-level trough.
- Larger than normal vorticity (spin) of low-level flow.

trough and a sinking west of the troughs. Therefore, tropical storm formation is favored when low-level disturbances are located under the southwest flow aloft associated with an upper-level trough (figure 9.5).

Finally, hurricane development is favored in regions where the spin, or the vorticity, of the low-level air is above normal. In a sense, this air already has a head start in attaining the rapid spin associated with mature hurricanes.

9.2.3 AN EXAMPLE OF HURRICANE GENESIS

To illustrate the formation of a major hurricane through the cooperation of cumulus clouds and the large-scale vortex, we consider the warm, humid air mass in the Caribbean in August, 1969, which would eventually spawn the notorious hurricane Camille, one of the most intense hurricanes on record. For weeks, the high summer sun had warmed the ocean and the air and enriched the humidity by evaporation. A weak disturbance in the normally steady easterly trade winds had been drifting westward across the Atlantic for the prior week. Evidence of this disturbance on August 14 is found in the mass of clouds located in the vicinity of latitude 20°N, longitude 80°W in figure 9.6.

The origin of all these disturbances is not fully understood, although some originate over central North Africa. Like the wave disturbances in the middle-latitude westerlies, a portion of these disturbances is associated with weak converging air near the surface. As the air flows together, it ascends and cools by expansion. In the moist environment, even a small amount of lifting and cooling is sufficient to trigger intense but short-lived thunderstorms, which produce copious

FIGURE 9.6 *Satellite photograph of an early stage of Hurricane Camille in the Caribbean.*

amounts of rainfall before dissipating. Many such disturbances cross the Atlantic and Caribbean each summer, but most never develop beyond this immature stage.

But the environment of this disturbance was unusually favorable: the water was warmer than normal, and the humidity of the air unusually high. The disturbance itself was moving rather slowly, allowing the numerous cumulonimbus clouds to warm and gradually moisten their environment at higher levels. A day later, the effect of the many individual thunderstorms was a mean warming of the larger-scale environment by 1C°, and the cooperation between the incipient vortex and the embedded clouds began to increase.

The high-level horizontal divergence from the warming of the environment, even by only 1C°, produces a pressure fall at the surface. Warm air weighs less than cooler air. The slowly converging horizontal winds near the surface respond to this slight drop of pressure by accelerating inward, driven by the increased horizontal pressure difference. On the morning of August 14, the growing storm had developed a circulation center, the pressure had fallen to 991 millibars, and winds had increased to 80 km/h.

But the increased outflow aloft away from the center of falling pressure produces increased lifting of air, so that the thunderstorms become more numerous and intense. The feedback cycle is now established. The inflowing air fuels more intense thunderstorm convection, which gradually warms and moistens the environment. The warmer air in the disturbance weighs less, and so the surface pressure continues to fall. The farther the pressure falls, the greater the inflow and the stronger the convection. The limit to this process will occur when the environment is completely saturated by cumulonimbus clouds. Further condensation heating would not result in additional warming, because the heat released would exactly compensate for the cooling due to the upward expansion of the rising air.

9.2.4 THE DISTURBANCE SPINS FASTER

Although we have explained the fall of surface pressure and the inward acceleration that fuels the thunderstorms, we have not yet mentioned why the winds blow around the storm faster near its center. The production of this *tangential* component to the flow arises from the simple *law of conservation of angular momentum*. This law states that *the product of tangential (rotational)*

FIGURE 9.7 *Schematic diagram of conservation of angular momentum.*

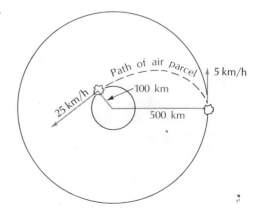

velocity of a parcel of air about an imaginary vertical axis and the distance of the parcel from this axis of rotation is constant, in the absence of friction or other torques. This law is fundamental to understanding the rotational nature of many atmospheric motions; it is illustrated in figure 9.7. If a rotating parcel of air moves inward toward the center of rotation, the product of the distance and velocity must remain constant. As the distance decreases, the rotational velocity must increase, so air that originates at a distance of 500 kilometers from the center with a tangential velocity of 5 km/h, upon reaching a point 100 kilometers from the center, would have a velocity of 25 km/h in the absence of friction. Should the same parcel reach a radius of 10 kilometers, it would be moving at 250 km/h. Friction reduces these values somewhat. Nevertheless, the basic principle of conservation of angular momentum—that air converging toward the center of a low tends to rotate faster—explains many wind circulations.

9.2.5 HURRICANE-FORCE WINDS ARE ACHIEVED

As the young tropical cyclone intensifies, the feedback between the growing vertical circulation and the cumulonimbus clouds increases. As the low-level air moves inward from large distances, the law of conservation of angular momentum demands a more rapid spinning. The intensification process, once started, may proceed rapidly; in this case, hurricane-force winds (exceeding 120 km/h) were reached on the afternoon of August 15 as Camille approached western Cuba. The pressure in the eye had fallen to 964 millibars by this time.

As Camille passed over the western tip of Cuba during the night of August 15−16, the intensity of the storm weakened slightly. However, upon reentering the Gulf of Mexico, where water temperatures were about 30°C (86°F), the storm rapidly intensified, with the minimum pressure dropping to around 910 millibars during the day of August 16. Figure 9.8 shows a *Nimbus* satellite photograph of the mature hurricane Camille, and figure 9.9 shows the surface streamline analysis for 7 A.M. EST, August 16. (*Streamlines* are lines that are parallel to the direction of the wind at every point and therefore indicate the flow of air at a given time.) In figure 9.9, there are no surface observations near the center of the storm where the winds are greatest. However, the observations around the storm show clearly the general counterclockwise circulation around the storm center, located at about 24°N and 85°W. In addition, there is an appreciable component of inflow, especially in the southeast quadrant. The fastest winds occur where the inflowing air penetrates closest to the center, thus explaining the doughnut of strong winds surrounding the storm center.

But, the inflowing air cannot reach all the way to the center. As the radius becomes very small (less that 30 kilometers or so), the conservation of angular momentum demands higher speeds (and therefore kinetic energy, which is proportional to the square of the wind speed) than can be produced by the available amount of condensation heating. In other words, there is a limit to the total energy, which prohibits velocities higher than about 320 km/h (200 mi/h). Therefore, instead of penetrating all the way to the center, the inward-moving air turns upward and outward, as indicated in the vertical cross section of figure 9.10. In

249

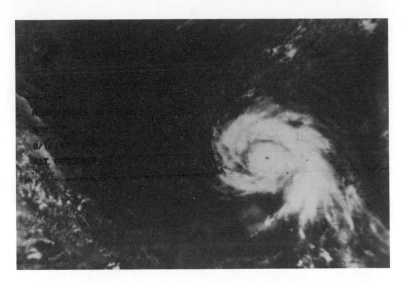

FIGURE 9.8 *Nimbus satellite photograph of mature Hurricane Camille, 12:10 P.M. EST on August 16, 1969.*

this spinning ring of rapid upward motion, condensation and rainfall are intense; a ship unlucky enough to remain under this practically continuous ring of thunderstorms would collect about 250 centimeters (100 inches) of rain per day.

Finally, in the cold upper atmosphere, the now-dry (in absolute humidity or mixing ratio) air turns outward and moves away from the storm center. As the distance from the center increases, the law of conservation of angular momentum operates in reverse, and the cyclonic rotational speed subsides, eventually becoming anticyclonic at a distance of 300 kilometers or so from the center. At greater distances, the air merges with the large-scale circulation of the tropics.

9.2.6 DEATH OF THE HURRICANE

If the hurricane remains in a favorable environment, i.e., over warm waters and away from hostile cold, dry air masses, the storm may continue indefinitely. A particularly long-lived storm, Hurricane Ginger (1971), traveled halfway across the Atlantic twice over a period of a month (figure 9.11). In fact, hurricane paths may be extremely erratic, as the sample tracks in figure 9.11 indicate. We can easily appreciate the forecaster's problem when dealing with such behavior.

Eventually, however, most hurricanes drift northward and are captured by the middle-latitude westerlies. When this occurs, the storm may quickly die, or it may acquire some characteristics of the middle-latitude cyclone as contrasting air masses are brought into the circulation. Hurricane Hazel (1954) is an example of the latter type. After moving inland on the Virginia coast, it moved nearly due north into Canada, increasing in diameter and producing copious amounts of rainfall. The storm retained some of its tropical character however, as

250

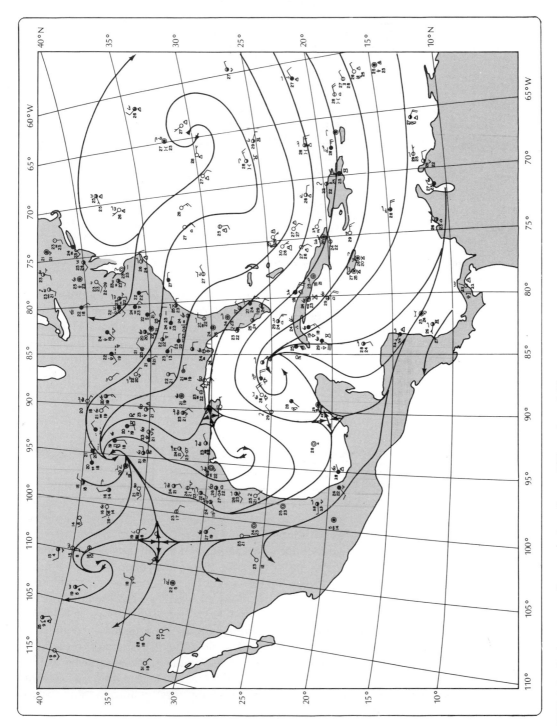

FIGURE 9.9 Surface streamline analysis at 7 A.M. EST on August 16, 1969, showing Hurricane Camille in the Gulf of Mexico.

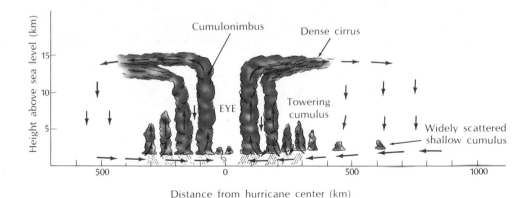

FIGURE 9.10 *Vertical cross section through a hurricane. A vertical slice through the center of a hurricane shows typical cloud distribution and direction of flow.*

residents of Toronto were treated to the rare spectacle of the hurricane eye as it passed overhead.

Hurricane Camille, whose birth and growth we have followed, met the fate of many hurricanes that move north. After reaching a maximum intensity of 320 km/h on August 17, it moved ashore in Mississippi, killing 135 people, causing losses to over 63,000 families, and leaving a $1000 million damage in Mississippi alone. As with the majority of hurricanes, most of the damage was caused not by wind, but by water in the form of extreme tides, called the storm surge. In Pass Christian, Mississippi, the storm surge was 7.5 meters (24.6 feet), higher than any previous tide on record.

After moving ashore, the sudden cutting off of evaporation from the sea starves the hurricane of its vital water supply, and a rapid reduction of the maximum wind occurs. The increased friction associated with the roughness of the land also contributes to the wind reduction. Killer winds in Camille dropped from around 290 km/h to 95 km/h only six hours after landfall. Moving northward into Mississippi and Tennessee, Camille's rainfall proved beneficial, breaking a severe drought.

Even though the maximum sustained hurricane winds are reduced by 50 percent or more in just a few hours, the rainfall and flooding risk may continue for days as the remnants of the storm continue to move overland. For example, after drifting slowly northward over Tennessee, Camille swung eastward into Kentucky, West Virginia, and Virginia. As the moist air was lifted by flowing over the Appalachians, rainfall increased to record amounts. Unofficial reports in Virginia claimed 64 centimeters (25 inches) of rain in the mountainous areas, enough to cause massive mud slides and flooding. The flooding in Virginia's mountains claimed 112 lives, nearly as many as were lost when the hurricane moved ashore on the Gulf Coast three days earlier. Finally, on August 20, Camille moved back off the Atlantic Coast, and its old circulation was absorbed in a frontal zone separating polar air to the north and tropical air to the south (figure 9.12).

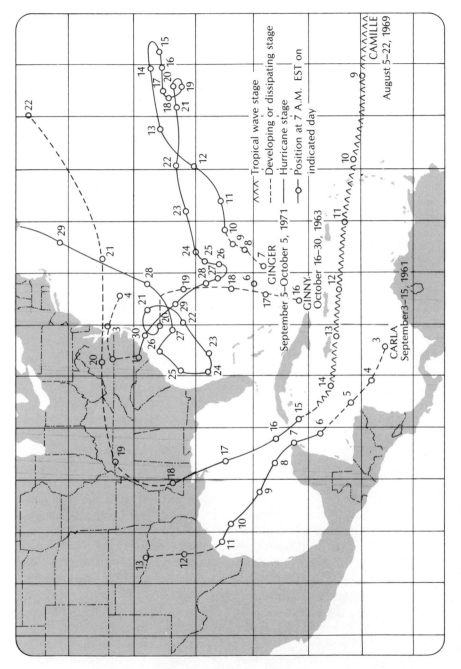

FIGURE 9.11 *Examples of hurricane tracks.*

253

9.3 A COMPUTER MODEL OF THE HURRICANE

Having explained in considerable detail the life history and structure of one intense storm, it is instructive to consider how a hurricane might be modeled using a high-speed computer. Besides the hope of increasing our understanding of the hurricane, there are several practical motivations behind such an effort. The first practical use of a realistic computer model would be the prediction of the motion and intensity of real storms. The advantages in terms of warning and preparation from an accurate prediction are obvious. It is also true that an inprovement in hurricane forecasting would obviate many unnecessary and expensive preparations. For example, because of the uncertainty in the hurricane's landfall position, a large number of people either evacuate or make preparations which later turn out to have been unnecessary because of a slight deviation of the storm's path. The second practical motivation for modeling the hurricane on a computer is to produce a realistic model that may be used as a guinea pig for testing hurricane modification

FIGURE 9.12 *Nimbus–3 satellite photograph showing Hurricane Camille dying in polar front off Virginia coast on August 20, 1969.*

254

hypotheses. It is relatively inexpensive, and far safer, to test modification theories on a computer model than on real storms.

9.3.1 A NUMERICAL MODEL OF THE ATMOSPHERE

Although we will discuss how a computer model of a hurricane might work, the concept of modeling the atmosphere on computers is quite general. Our daily forecasts are based to a large extent on a computer model of the atmosphere that is run on a daily basis at the National Meteorological Center, Camp Springs, Maryland. This model, as well as many others, is very similar in principle to the hurricane model discussed in this section.

What is the basic idea behind computer models of the atmosphere? The computer certainly does not produce a physical model of the hurricane, with miniature clouds and scaled-down winds. Instead, the computer allows us to solve complicated mathematical equations that describe the physical laws of the atmosphere. For example, the force of gravity is a physical effect that can be described completely by one equation, and a mathematical model of a stone dropped from a cliff is relatively easy to construct. In a similar way, other effects, such as the expansion of air as it is lifted, or the warming of air as it is heated, can also be expressed by mathematical equations.

9.3.2 EQUATIONS OF THE ATMOSPHERE

The mathematical forms of the equations that describe the motion, temperature, pressure, and humidity of the atmosphere are treated in many tests. Although necessary for *quantitative* calculations, they are not necessary for a *qualitative* understanding of the physical effects they describe. Here, we schematically write each equation and briefly explain the meaning of each term.

The equation that predicts the change of wind over short time intervals is the *equation of motion,* which may be written:

Change of velocity = advection of nearby + acceleration due to
at a point in   air of different   pressure gradient
space   velocity   forces (toward
                                              low pressure

          + Coriolis force   + friction (slows   **(9.1)**
            (due to earth's   down air)
            rotation, deflects
            moving air
            current to right)

Equation (9.1) is an expanded form of Newton's second law [equation 5.4, section 5.3] and really consists of two equations (one for the south-north component of velocity and the other for the west-east component. Advection represents the movement of air with higher or lower velocities to the point at which the change is being computed. The pressure gradient force represents the acceleration of the

255

air caused by horizontal differences in atmospheric pressure. A nonmeteorological example of the pressure gradient force is the acceleration of toothpaste when the pressure is increased at the end of the tube. Like the motion of the toothpaste, air accelerates from high to low pressure whenever pressure differences exist. If there were no other forces, air would flow directly from high to low pressure.

As discussed in section 5.3.3, the Coriolis force represents the effect of the earth's rotation on moving air and, in the Northern Hemisphere, appears as an acceleration to the right (facing downstream) of the flow. It is primarily the Coriolis force that prevents air from flowing directly from high to low pressure. Instead, a balance between these two forces results in a wind that blows parallel to the isobars, with low pressure on the left and high pressure on the right of the moving air current.

The equation that predicts the rate of change of temperature with time is the *thermodynamic equation*, which may be expressed as

$$
\begin{array}{l}
\underline{\text{Change of temperature}} \\
\text{at a point in space}
\end{array}
= \;
\begin{array}{l}
\text{advection of nearby} \\
\text{air of different} \\
\text{temperature}
\end{array}
\; + \;
\begin{array}{l}
\text{cooling or warming} \\
\text{caused by expansion} \\
\text{or compression of air} \quad \textbf{(9.2)}
\end{array}
$$

$$
+ \; \text{change due to} \;
\begin{array}{ll}
\text{(a)} & \text{radiation} \\
\text{(b)} & \text{condensation} \\
 & \quad\;\; \text{or evaporation} \\
\text{(c)} & \text{conduction}
\end{array}
$$

Equation (9.2) is an expanded form of the first law of thermodynamics [equation (5.2), section 5.2.3]. The processes that cause the temperature to change at a point are easier to understand than the processes that cause the velocity to change. Advection of temperature is the transport of warmer or colder air to the point. The rapid fall of temperature as a cold front passes is primarily caused by advection. We may also easily describe the cooling of air by expansion and the warming by compression in terms of familiar examples. As compressed gas rushes out of an aerosol can, it expands rapidly and cools by 10 degrees or more. On the other hand, the temperature and pressure of air in a tire rise as more air is pumped in. Finally, the addition or removal of heat energy (expressed in calories) may change the temperature at a point. The important heating and cooling processes in the atmosphere are radiation, condensation, and conduction. Short-wave radiation absorbed from the sun warms the air, whereas long-wave radiation emitted by the atmosphere usually cools the air. Radiative temperature changes above the ground are normally small, 1 or 2 C° per day. Condensation of water vapor or evaporation of liquid water can release or absorb enormous amounts of heat, as discussed earlier. Changes in temperature of 100 C° per day would result in heavy rain situations if the compensating cooling effects of advection and expansion were not present. Of course, we never observe such extreme changes, so these compensating effects must nearly offset the release of latent heat.

Two other equations describe the conservation of mass and water (vapor + liquid + solid). For *mass,* we have the continuity equation [equation (5.9), section 5.5]. Because the change of mass in an atmospheric column is usually very small, the continuity equation provides a simple relationship between vertical and horizontal velocities, as discussed in section 5.5 (see figure 5.14). The equation describing the *water vapor budget* is

$$
\begin{array}{c}
\underline{\text{Change of water vapor}} \\
\text{in a volume}
\end{array}
=
\begin{array}{c}
\text{amount of vapor} \\
\text{flowing into} \\
\text{volume}
\end{array}
-
\begin{array}{c}
\text{amount of vapor} \\
\text{flowing out of} \\
\text{volume}
\end{array}
\tag{9.3}
$$

$$
+ \text{ evaporation} - \text{condensation} - \text{sublimation}
$$

The last two equations needed to describe completely the state of atmosphere are the *hydrostatic equation*, which simply relates the pressure at any level to the total mass of air above that level [equation (5.7), section 5.4], and the *equation of state,* which relates pressure, temperature, and density at a point [equation (5.1), section 5.2.2].

We thus have seven equations and seven unknowns—vertical velocity, two components of horizontal velocity, pressure, temperature, density, and water vapor. We are now ready to solve these equations for the seven variables, which will then provide a detailed description of the atmosphere. Note that such principles as the conservation of angular momentum do not appear explicitly but are implied within the preceding equations.

9.3.3 DATA FOR THE MODEL

Because the solutions to the predictive equations for wind, temperature, pressure, and water vapor depend on the three-dimensional structure of the atmosphere at a given moment, we must first make analyses of these variables at many points on a three-dimensional weather map. Figure 9.13

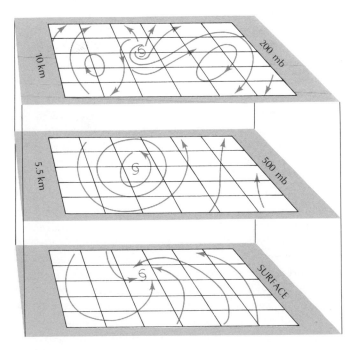

FIGURE 9.13 *Schematic three-dimensional streamline analysis of a mature hurricane. Note the low-level inflow toward the center of the storm (represented by the symbol Ϛ) and the upper-level outflow away from the center.*

257

shows a schematic three-dimensional streamline analysis of a tropical cyclone. Such an analysis is based upon measurements from weather balloons, aircraft, satellites, and surface observing systems.

Starting with a three-dimensional analysis of wind, temperature, pressure, and moisture, the prediction equations are solved at each point on the map for a short time later (for example, 10 minutes), thereby creating a weather map of the future. Then, this future analysis is used as a starting point for a new forecast 10 minutes farther in the future. The process is repeated many times, until the desired forecast time is reached. The complexity of the equations, the large number of points needed to resolve properly the detailed structure of the atmosphere, and the large number of successive short forecasts needed to arrive at a 24-hour forecast necessitate extremely fast computers. For example, a 24-hour forecast of the weather over the Northern Hemisphere made by the operational model at the National Meteorological Center requires about 300 million calculations!

9.3.4 THE COMPUTER FORECASTS A HURRICANE

The earlier part of this chapter described the formation of Hurricane Camille. To illustrate a numerical model of an atmospheric phenomenon, let us follow the development of a hypothetical hurricane from a weak disturbance as predicted by a computer model. Figure 9.14 illustrates the structure of a circulation associated with a weak low-pressure center. Now, let us follow the computer through several forecast cycles.

At the initial instant of time, the equation of motion, which accounts for advection, accelerations due to horizontal pressure differences, the rotation of the earth, and friction, predicts a small net acceleration of the wind inward toward the center of low pressure. This acceleration is the result of a slight predominance of the pressure gradient force over the other forces. One result of the increased

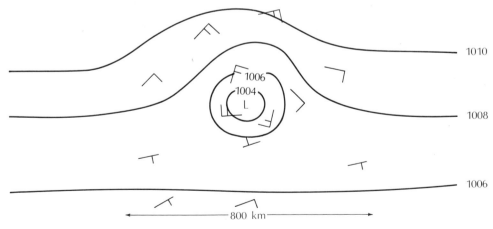

FIGURE 9.14 *Circulation around a weak low-pressure center in tropics. Each full barb represents 10 knots. Isobars are labeled in millibars.*

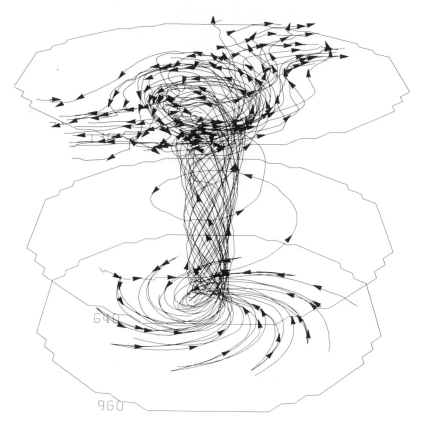

FIGURE 9.15 *Three-dimensional trajectories in model hurricane.*

inflow toward the storm center is an increase of the rotational speed, a consequence of the conservation of angular momentum discussed earlier. Thus, the equation of motion predicts a slight increase in both the inward and tangential parts of the wind. The horizontally converging air must eventually rise, so another consequence of the increased horizontal inflow is a slight increase in upward motion, which affects the thermodynamic equation in two ways. The first effect is an increased cooling due to the expansion of the air as it is forced to rise. However, because the cooler air is able to hold less water vapor, condensation occurs quickly. The latent heat of condensation then acts to warm the ascending air. Although the net result of these two opposing effects is not immediately obvious, the latter predominates under the right conditions, so the mean temperature of the inner region rises.

The water vapor equation predicts a loss of moisture by condensation but at the same time predicts an increase from inward horizontal advection and from evaporation from the sea, which enables the convection and condensation to continue.

As the inner region of the developing storm warms slightly and expands upward, an outflow of mass away from the center at high levels occurs. The

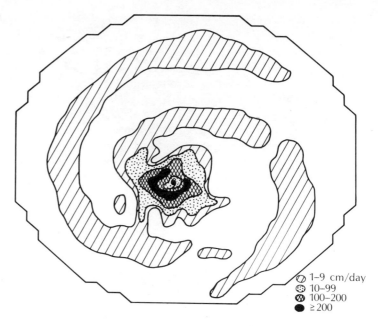

FIGURE 9.16 *Spiral bands of rainfall computed from a model hurricane.*

1–9 cm/day
10–99
100–200
≥ 200

upper-level outflow exceeds slightly the low-level inflow, and the surface pressure drops a small amount.

All of these changes predicted in the 10-minute forecast made by machine are small in magnitude; for example, the wind might increase by 0.1 km/h, the mean temperature might increase by 0.05 C° and the surface pressure might drop by 0.1 millibar. But, then, we have a new, slightly stronger storm, and the new values of wind, temperature, and moisture may be used to generate another 10-minute forecast.

A repetition of the sequence described in the preceding paragraph over many 10-minute intervals gradually describes the intensification of the model storm: A slight predominance of the inward-directed pressure gradient force continues to accelerate the air inward. The increased inflow also produces stronger vertical velocities and increased condensation. Water vapor fuel for the convection is supplied at faster rates as the wind speed increases. The thermodynamic equation predicts higher temperatures as a result of the condensation, and the warmer temperatures lead to still further decreases in surface pressure.

The intensification process does not continue indefinitely, however, because the imbalance of forces gradually changes. For example, as the winds increase, so does the effect of friction, which tends to decrease the wind speed. The increased upward motion, besides increasing the convection, also increases the cooling due to expansion. Both the frictional and expansion cooling effects tend to weaken the circulation. At some point, these two weakening effects may cancel, or even slightly exceed, the forces of intensification. At this point, the storm reaches a steady state in which the generation and the dissipation effects are balanced. Such a mature state is shown in figure 9.15, which illustrates the three-dimensional trajectories of air parcels in a model hurricane. A comparison with figures 9.10 and 9.13 shows that the storm model accurately reproduces many aspects of the mature hurricane. In particular, the model vortex consists of strong cyclonic low-level inflow, upward motion near the center, and outflow of air in the upper atmosphere. A view of the rainfall rates predicted by the model (figure 9.16) shows intense rainfall near the center, and even some spiral bands which resemble the spiral bands of rainfall that are visible in radar and satellite pictures of real storms (see plate 8 and figures 9.1 and 9.17).

Several physical effects may contribute to the model hurricane's downfall. Movement of the storm over land or colder water may reduce evaporation from the surface so that the storm starves from a lack of water vapor, or cool, dry air from middle latitudes may be drawn into the circulation, destroying the core of warm, light air. In spite of its enormous amounts of energy, the hurricane is quite fragile and may weaken quickly under unfavorable conditions. All of these effects may be represented in a mathematical model of the hurricane.

9.3.5   A WORD OF CAUTION

Because the preceding example of a computer model appears quite realistic, it is easy to be overimpressed by the capability of computer models in general to predict the weather. As everyone knows, however, weather forecasts are still far from perfect, even with the greatly increased use of computers, because computer models are only as good as the physical models upon which

FIGURE 9.17 *Radar picture of the rainbands of Hurricane Betsy, 1965.*

they are based. Because the number of points in a model is limited by the finite speed of computers, it is not possible to resolve simultaneously all scales of motion. Thus, the hurricane model cannot predict both small cumulus clouds and the larger-scale hurricane vortex. However, the net effect of the cumulus clouds must be accounted for in some simple and approximate way. Such approximation produces errors. The uncertainties in initial data fields also lead to errors in the forecast. Thus, although many models have demonstrated a skill in predicting behavior of the atmosphere, they are not completely dependable.

Questions

1   Compare and contrast extratropical cyclones, hurricanes, and tornadoes.

2   List five factors that favor hurricane development.

3   How do cumulus clouds and a larger-scale circulation cooperate to produce the hurricane?

4   What is the major source of energy for the hurricane?

5   If a parcel of air has a tangential velocity of 1 km/h at a radius of 100 km, what will its tangential velocity be if it moves to a radius of 1 km and conserves its angular momentum? (Answer: 100 km/h)

6   Why do hurricanes weaken rapidly after they move over land?

7   The temperature change due to the release of latent heat alone in a hurricane is several hundred C° per day. Why are these huge changes never observed?

8   Summarize how a computer model of the atmosphere works.

Problems

9   In which sense does the vortex rotate in your bathtub drain? Is the earth's rotation responsible?

10   Compute the kinetic energy per unit mass, $\frac{V^2}{2}$ (in J/kg $= m^2/s^2$), for each of the following vortices:

$$\text{Tornado: radius} = 0.2 \text{ km}, V = 250 \text{ km/h}$$
$$\text{Hurricane: radius} = 50 \text{ km}, V = 150 \text{ km/h}$$
$$\text{Extratropical cyclone: radius} = 500 \text{ km}, V = 50 \text{ km/h}$$

Estimate the total volume and mass (m) in each vortex system and then compute the total kinetic energy $m \frac{V^2}{2}$ in each system. Assume a mean depth of 8 km and a mean density of 0.8 kg m^{-3}. (Answers: Tornado—2411 J/kg; 1.94×10^{12} J. Hurricane—1736 J/kg; 8.73×10^{16} J. Extratropical cyclone—193 J/kg; 9.70×10^{17} J.)

11 How many grams of water per day are condensed in a hurricane that produces 24 × 10^{11} kWh of energy in one day? If this rain were distributed uniformly over a circle of radius 200 km, how deep would the water be? (1 watt = 1 Joule; 1 cal = 4.187 Joules) Assume a latent heat of condensation of 597 cal/g. (Answers: 3.45 × 10^{15} g; 2.7 cm)

TEN

Thunderstorms and Tornadoes

condensation level • entrainment • updrafts and downdrafts •
pileus • gust front • Doppler radar • hail • flash • stepped leader •
dart leader • return stroke • F-scale

Severe thunderstorms produce the most awesome and destructive atmospheric phenomena—tornadoes, lightning, hail, and flash floods. And yet, they provide essential rainfall to much of the agricultural bread-baskets of the world. This chapter looks at thunderstorms and their associated severe weather phenomena.

10.1 THUNDERSTORMS

It is a warm, humid, but clear July morning in Miami, Florida. Soon after sunrise, small puffy cumulus clouds appear over the south Florida peninsula. By noon, purple-black thunderheads drench the Everglades with cool rainfall. It is a sunny, pleasant morning in the Rockies. By early afternoon, the mountain peaks are shrouded in clouds, and thunderstorms begin to drift eastward over the plains, persisting long after sunset. In the Midwest, after four days of a sultry heat wave, rumbles of thunder herald the approach of a squall line from the west. Shortly after the line of thunderstorms moves through the area, a cold front with brisk northwesterly winds brings in drier air and a welcome drop in temperature.

10.1.1 LIFTING PRODUCES THUNDERSTORMS

The preceding three examples of thunderstorms (illustrated schematically in figure 10.1) have one thing in common—a rapid lifting of moist, warm, low-level air to the high troposphere. The lifting in the first two examples was originally caused by differential (uneven) heating on a small scale.

Differential heating on a global scale produces the general circulation, and horizontal differences in heating on the small scale produce local circulations. As the sun warms the Florida land faster than the surrounding water, a mean lifting of the air over the land occurs. Within this lifted air, smaller-scale buoyant parcels of air are created that are warmer than their environment, and therefore rise (figure 10.1a). In the Rockies, the sun heats the air in contact with the mountain peaks more than air at the same level over the plains, again producing buoyancy and lifting (figure 10.1b). In the Midwest example, the lifting is associated with horizontal convergence of low-level moist air in the vicinity of a cold front (figure

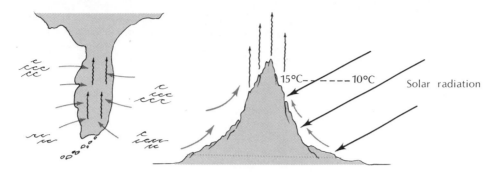

(a) Daytime heating over
Florida produces lifting.

(b) Air over mountain peaks warms
faster than air at same level over plains.

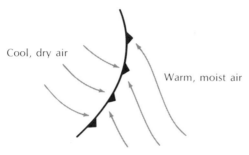

(c) Convergence of low level
air in vicinity of cold front.

FIGURE 10.1 *Three lifting mechanisms that can produce thunder-storms.*

10.1c). Under such conditions, thunderstorms tend to align themselves in squall lines (squall lines will be discussed later in this chapter). It is important to note that here the processes which produce the lifting occur on a larger scale than the thunderstorm cells themselves. As discussed in the hurricane section, large-scale lifting of a warm, moist air mass produces small-scale convection; i.e., local "hot spots" develop on the ground and generate bubbles of warmer, lighter air, which rise through their denser environment and produce the actual thunderstorm cells.

10.1.2 SURVIVAL OF THE FITTEST

Most hot bubbles of air never survive to become visible clouds or to develop into a mature thunderstorm. As with newborn insects, nature produces far more infant thermals than ever reach the adult stage. Normally the rate of expansion cooling in the rising parcel (10C° km) exceeds the rate of decrease of environmental temperature, so parcels soon become cooler and heavier than the air around them and return downward, as illustrated by parcel *A* in figure 10.2. However, a few large and unusually warm (or moist) bubbles may rise high enough to reach the condensation level, the level at which expansion cooling has

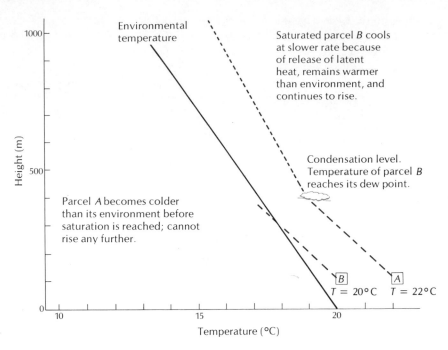

FIGURE 10.2 *Temperature variation with height of a parcel of air reaching condensation level.*

lowered the temperature of the cloud to its dew point. This situation is illustrated by parcel *B* in figure 10.2. With the onset of condensation, the chances for the growing cloud's continued rise are increased greatly because condensation adds heat to the parcel and helps negate the expansion-cooling effect. The net result of these two processes is a cooling rate in the parcel (neglecting *entrainment,* which is the mixing of the cloud with its environment) of about 6C°/km instead of 10C°/km. It is therefore more likely, although by no means certain, that the new cloud will remain slightly warmer than its environment and continue to rise.

A typical sequence of events leading to isolated thunderstorm cells on a hot summer day is illustrated in figure 10.3. The isolated cumulus stage, as shown in the *Landsat* satellite photograph of cumulus clouds over Florida (plate 9), consists of many small, white cumulus clouds with low tops. As we know from watching the hundreds of fleeting cumulus clouds form and die on a summer day, very few thunderstorms are actually produced from the many little clouds. The fledgling cloud faces a hostile environment of dry, cool air. Mixing of this air through the sides and top of the cloud evaporates waterdroplets and cools the parcel. Thus, most cumulus clouds decay a short time after they become visible. However, in their death by evaporation, the clouds enrich the humidity of the air, so future bubbles find a slightly less unfavorable environment and are able to last a little longer and grow a little taller. Time-lapse photographs show a series of rising bubbles or towers of cumulus clouds, each one growing bigger than its predecessor, until several clouds merge to form a cloud that is large enough to grow into the mature thunderstorm stage. In this larger cloud, the updrafts near the center are protected from the cooling effects of entrainment by the outer portion of the cloud. During this merging stage, the cell grows from about one to several kilometers in diameter, and the height of the top of the cloud increases rapidly.

As cloud drops are carried above the freezing level (indicated by the 0°C isotherm in figure 10.3), freezing of some of the drops is initiatied by ice nuclei. Then the formation of precipitation begins to occur rapidly as ice crystals grow, splinter, and cause additional water drops to freeze. Large ice crystals become too heavy to be supported by the updraft and begin to fall. The drag of the precipitating ice and water creates a downdraft in one or more portions of the cloud. The mature stage of the thunderstorm begins roughly when the precipitation reaches the ground. In the mature stage, strong updrafts (as high as 100 km/h) and downdrafts coexist in different parts of the cloud (figure 10.3c).

The mature stage is the most intense period of the thunderstorm. Lightning is most frequent during this period, turbulence is most severe, and hail, if present, is most often found in this stage. The cloud reaches its greatest vertical development near the end of this stage, usually reaching about 10 kilometers and sometimes penetrating the tropopause to altitudes greater than 15 kilometers.

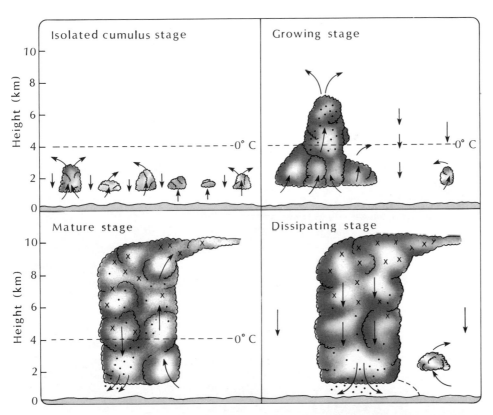

FIGURE 10.3 *Four stages of thunderstorm development. Water drops are indicated by •, ice particles by x. The freezing level is shown by the 0°C isotherm.*

Occasionally, when the updraft of air in the growing thunderstorm is very rapid and the air above the growing cumulus cloud top is moist, a very thin veil of clouds forms above the visible top of the cumulus cloud (figure 10.4). This thin web of clouds, which is frequently comprised of ice crystals, is called the *pileus,* or *cap cloud,* and is caused by the rapid expansion and cooling above the rising cumulonimbus or by air flowing over the cumulus top.

The mature thunderstorm is an impressive sight—a dark mass of clouds towering 8–16 kilometers over the surface and capped by an anvil of brilliant white cirrus clouds. This familiar anvil (plate 10) occurs when the rising air currents reach the tropopause and encounter the stable stratosphere. The stratosphere is so stable that the rising air parcels quickly become cooler than their environment and lose their buoyancy. The updrafts then spread out horizontally, fanning the ice crystals into the characteristic anvil pattern.

About fifteen minutes to a half hour after the mature stage, the isolated thunderstorm cell begins to die, and the large number of falling raindrops drag the air downward, producing downdrafts instead of updrafts. The rain also cools the surface underneath the storm by evaporation and may cut off its supply of warm, buoyant air. As the cool downdraft hits the ground and spreads out horizontally, it forms the *gust front,* a miniature cold front separating the cool downdraft air from the warmer environmental air. This front can propagate several kilometers away from the thunderstorm that generates it, producing cool, gusty winds as it passes. An example of a gust front is shown in plate 11 and figure 10.5.

Rainbows are most common during the dissipation stage of the thunderstorm (plate 12), when the sun can shine through the thinning clouds onto the

FIGURE 10.4 *Pileus, or cap clouds, forming over three growing cumulus clouds.*

FIGURE 10.5 *Gust front over Florida. Dark color is caused by dirt mixed up from surface.*

still-falling raindrops. As discussed in chapter 15, the rainbow is formed when parallel light rays from the sun are bent (refracted) upon entering the raindrop, reflect off the inner rear surface of the drop, and are bent again as they emerge from the forward side of the drop.

Severe thunderstorms, which often spawn tornadoes, are considerably more complex than the more or less isolated thunderstorms of moderate intensity described previously. When the atmosphere is very unstable and abundant moisture exists in the low levels, thunderstorms may organize themselves into mesoscale circulations with typical diameters of 20 to 40 kilometers. As the thunderstorm cells pump low-level air into the upper troposphere at velocities of 40 to 80 km/h, low-level air flows in to compensate for the vertical motion. Through the conservation of angular momentum, the entire thunderstorm system may begin to rotate, producing a mesoscale cyclone. Such a circulation is shown in figure 10.6, which shows the winds at a height of 0.3 km in a thunderstorm system in Oklahoma on June 8, 1974. The winds in figure 10.6 were diagnosed by dual-Doppler radar measurements. Doppler radars emit a radio signal which is reflected by precipitation particles back to a receiver. The speed of the particle toward or away from the source of radiation causes a change in the frequency of the radiation (the Doppler shift). Although one radar can give only the velocity component along a line from the emitter to the particle, the 3-dimensional flow can be reconstructed if there are two Doppler radars separated by some distance and observing the same particles.

The winds in figure 10.6 show strong rotation, with winds exceeding 144 km/h in places. The wind shift from southeast to northwest in the southern part of the storm (indicated by a heavy black line) is the gust front and marks the leading edge of cool air originating from a downdraft.

271

FIGURE 10.6 *Horizontal wind analysis of a severe thunderstorm near Harrah, Oklahoma. Mean flow has been subtracted. Contour lines denote relative radar reflectivity.*

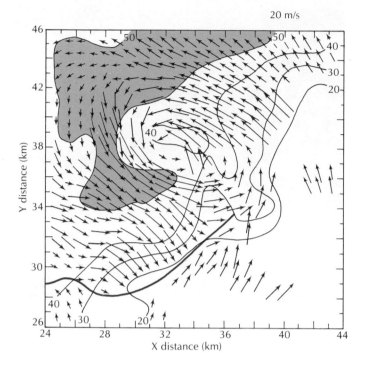

From Doppler radar analyses such as the one presented in figure 10.6, a reasonably complete 3-dimensional picture of the circulation associated with long-lived severe thunderstorms is emerging (figure 10.7). In figure 10.7 a storm is moving toward the east and is being continually supplied with warm, moist, low-level air around its leading edge. In the updraft fed by this inflow, condensation produces rain below the freezing level and ice at higher levels. To

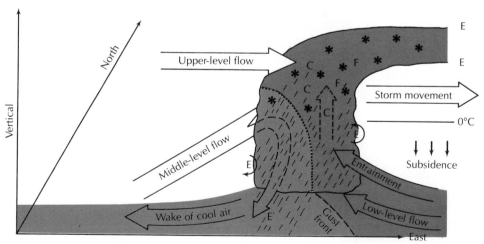

FIGURE 10.7 *Schematic diagram showing types of precipitation and flow in severe thunderstorms.*

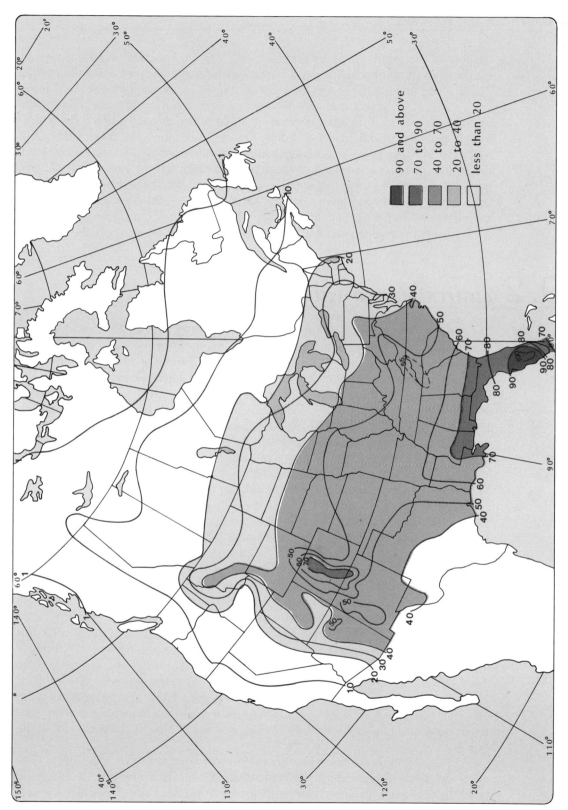

FIGURE 10.8 *Mean number of thunderstorms per year.*

the rear of the storm, some dry middle-level air is incorporated into the storm. As evaporation of rain cools this air, it becomes negatively buoyant and sinks. When the resulting downdraft reaches the ground, it spreads out and forms the gust front.

Thunderstorms generally occur within the moist, warm (maritime tropical) air masses that have become unstable either through surface heating or forced ascent over mountains or fronts. Figure 10.8 shows the mean number of thunderstorms per year. In general, thunderstorm frequency decreases away from the source of tropical air (Gulf of Mexico). Superimposed on the general pattern, however, is a strong local maximum associated with the Rocky Mountains of Colorado, and a lesser maximum over the Appalachians. A third strong maximum of thunderstorm occurs over central Florida, where heating of the peninsula produces a low-level convergence and thunderstorms occur almost every day of the summer.

10.2 *LIGHTNING AND THUNDER*

The term lightning identifies the intense visible electrical discharges produced by thunderstorms. Thunder is the sound which results from the rapid expansion of gases in the lightning channel. In tropical regions lightning and thunder may occur in warm cumulus clouds, i.e., those which are everywhere warmer than 0°C, and which may be no more than 3 km deep. Lightning activity is most commonly associated with cumulonimbus clouds which extend well above the 0°C isotherm. The most intense lightning displays are observed in severe thunderstorms which may also be tornadic. Lightning and thunder also occur in volcanic clouds, snowstorms, and sandstorms, where they have a distinctly different character.

It was established more than two centuries ago that clouds contain electricity. Benjamin Franklin (1753) found that the bases of thunderclouds usually contained negative charge. It is now known that a net positive charge resides in the upper regions of thunderclouds, with a negative charge in the lower regions. These oppositely charged regions constitute an electric dipole. As discussed in the preceding section, the thunderstorm develops out of a succession of cumulus towers of increasing vigor. During the early stages the small cumulus towers appear to be electrically inactive. The first indications of electrical activity occur during the vigorous vertical growth of larger towers, and are detected as a rapidly increasing electric potential gradient. After a few minutes the potential gradient in the cloud may have become high enough to cause dielectric breakdown of the air—this occurs at about one million volts per meter—and lightning ensues. The development of cloud electrification is closely associated with precipitation formation, so that lightning activity is normally accompanied by intense showers of rain and hail. This type of weather persists throughout the vigorous convective stage. The dissipation stage which follows is characterized by sporadic lightning activity and steady rain of diminishing intensity. A typical storm is electrically active for about one hour.

The problem of how thunderstorms become electrified still remains unsolved. Electrification, precipitation formation, and dynamics are all closely related thunderstorm features; however, the central problem lies in establishing precisely how they are related. One view is that electricity is generated by the inter-action of precipitation with cloud particles. According to this theory, rain or hail fall through a field of small particles (water droplets or ice crystals), and in colliding with them become negatively charged. The negative charge is transported by the precipation to lower regions of the cloud. The positive charge, which was acquired by the rebounding cloud particles, remains at higher levels in the cloud. In another theory, it is argued that intense electrical activity results from the action of strong updrafts and downdrafts. The atmosphere itself is considered as the source of elec-tric charge, because it is weakly ionized, i.e., a small fraction of molecules com-prising the atmosphere carries positive or negative charge. A small initial electrical imbalance draws ions into the cloud, where they are transported vertically to cause ever-increasing numbers of ions to be attracted to the cloud. Acting in this way the thundercloud may be regarded as a gigantic electrostatic machine. Thus, we have two very different mechanisms which purport to explain how the thunderstorm dipole could develop.

The physical character of cloud-to-ground lightning has been inves-tigated with streak cameras using a moving film, so that in the photographs the lightning features are spread out in time. A typical image is shown in figure 10.9. Such records show that a lightning flash consists of a number of very bright strokes, each of which is preceded by a weakly luminous leader. The first leader in a flash

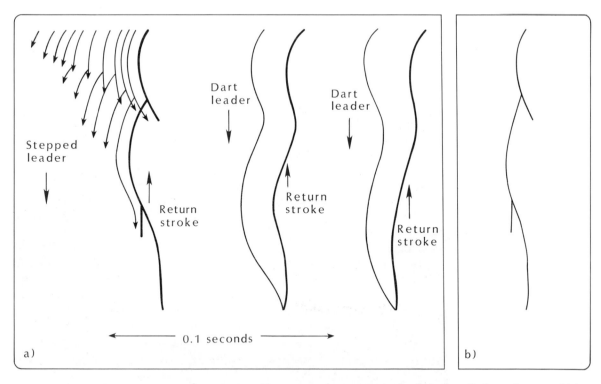

FIGURE 10.9 (a) Components of a lightning flash as they would be recorded by a streak camera. (b) Same flash as photographed by an ordinary camera.

275

is called the *stepped leader,* because it moves downward in intermittent steps. This brings down negative charges, creating a high potential gradient as the leader tip approaches the ground and intitating the highly luminous return stroke. The return stroke travels rapidly from the ground to the cloud and is soon extinguished. After a short interval the cycle is repeated. The leaders now move down a continous path, and are called *dart leaders.* Each is followed by a bright return stroke. Typically a lightning flash contains three or four return strokes, although twenty or more are sometimes observed.

Observers hear different sounds accompanying lightning depending on their proximity to the channel. If they are within a hundred meters the first sounds heard are a sharp click and a whip-like crack; if within a few hundred meters they may hear sound like the tearing of cloth or paper. These sounds are associated with the approach of the leader and the initiation of the return stroke. The most common sound, the rumble which we identify as thunder, is emitted from the column of hot gas which comprises the lightning channel. The channel is a region only a few centimeters in diameter which is momentarily heated to temperatures of the order of 30,000°K—5 times the temperature of the sun's surface. The hot gases expand explosively, creating a shock wave. After a short time this becomes a sound wave travelling at a speed of about 1200 km/h, which can be heard over a wide area as thunder. Thunder propagates at a much slower speed than light, which travels at 10^9 km/h, and the time interval between the flash and the sound can be used to estimate the range of the lightning flash at the rate of three seconds for each kilometer of range. One does not usually hear thunder over distances greater than about 15 kilometers. This is because the temperature distribution in the atmosphere causes sound waves to be bent upwards. If the range is

FIGURE 10.10 *Lightning at night.*

Plate 11 Gust front near Las Cruces, New Mexico (courtesy of Richard A. Anthes).

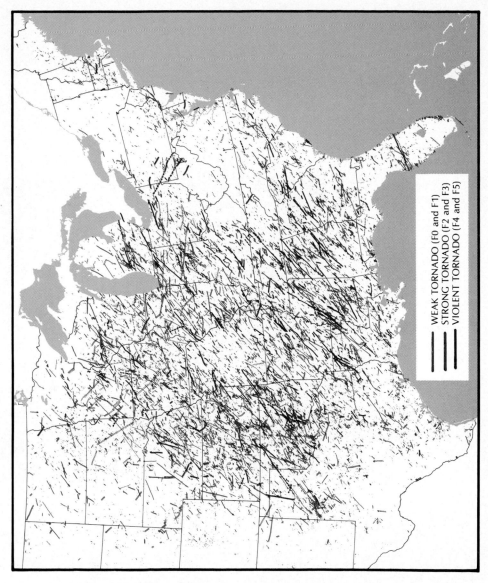

Plate 12   Tracks of tornadoes over the United States 1930–1974 (courtesy of T. Fujita).

too great, all the sound radiating from the lightning channel passes over one's head.

After witnessing a spectacular lightning display, one may wonder whether to look upon thunderstorm electricty as a valuable energy resource, and whether to try harnessing it for industrial or domestic use. We can estimate how much power is associated with cloud-to-ground lightning. A single lightning flash has the energy equivalent of about one gallon of fuel oil. The total power associated with worldwide cloud-to-ground lightning is about 10,000 megawatts. This is not a very significant amount when compared with global power demands—well in excess of 1,000,000 megawatts. Also, considering the enormous technical difficulty of harvesting it, other potential sources of energy, such as solar power, look much more attractive.

In the U.S. lightning is responsible for more deaths (200 to 300 per year) than any other weather phenomenon. It is also a major cause of property damage. Attempts to mitigate its effects are obviously desirable, but still in an exploratory stage. One technique consists of introducing large numbers of metallic fibers into the bases of thunderstorms in order to discharge the strongly electrified regions through a process other than lightning. Instead of a sudden lightning flash, the more or less continuous weak discharge might produce a glow similar to St. Elmo's fire, a glowing halo seen around wings, ship masts, and power lines during electrical storms.

HAIL 10.3

One of the most damaging byproducts of severe thunderstorms is *hail*, stones of ice which may under extreme conditions reach the ground the size of baseballs. Although hail causes few injuries, economic losses in the United States are about $300 million annually. Hailstones are formed when a small embryonic ice pellet remains for a sufficient time in a region of the thunderstorm in which there exists supercooled water (liquid water below freezing temperatures). Under such a favorable condition, the hailstone grows as the waterdroplets strike the ice pellet, freezing on contact. To grow to a size large enough to survive the final fall through the warm air to the ground, the hailstone must remain in the presence of abundant supercooled water for at least several minutes. Also, in order to hold the growing stone above the freezing level, very strong updrafts are required, because the typical hailstone's terminal velocity (velocity a hailstone would fall in still air) is 10–70 km/h. The updrafts in most thunderstorm cells are somewhat lower than these high velocities, which is one reason why hailstones that reach the ground are relatively rare. But in the most severe thunderstorms, updraft velocities of sufficient magnitude to keep the stones suspended above the freezing level for long periods of time are possible. Then, the hailstones grow until they finally become too heavy or until they are tossed into a part of the cloud where the updraft is weak or absent entirely. When the latter happens, the hailstones fall rapidly to the ground.

Occasionally, a hailstone may get carried up and down past the freezing level several times in successive up- and downdrafts before finally reaching the ground. The partial melting that occurs in the warm air, and differences in the cloud droplet size and concentration above the freezing level, produces a stone with laminations of ice of varying textures and appearances.

Because of the enormous crop damage caused by hailstones, considerable research has been directed toward reducing their size. Results have been promising enough that the Soviet Union has established an operational hail suppression program in the Caucasus Mountains. The Soviet program consists of firing cloud-seeding agents, such as silver iodide crystals, into growing cumulonimbus clouds. These seeding crystals encourage the formation of many small ice crystals rather than a few large ones. The hailstones that eventually form, although more numerous, are too small to survive the fall to the ground through the warm air and therefore arrive as beneficial raindrops rather than as destructive hailstones.

In spite of the success in hail suppression claimed by the Soviet Union, controversy over the possibility of significantly reducing hail continues. Recent experiments in the United States under the National Hail Research Experiment (1976) have produced somewhat ambiguous results. Apparently, seeding can either increase or decrease the amount of hail depending on the structure of the thunderstorm. Under certain conditions of temperature, amount of supercooled water present, and velocity of the updraft, the introduction of freezing nuclei may increase the number of small hailstones at the expense of the larger ones. Under other conditions, however, enough supercooled water may be present to increase the number of large hailstones as well.

As with any weather modification program, the main difficulty is that there is no way of knowing for sure what would have happened without the seeding. To circumvent this obstacle, statistical techniques must be used; that is, the behavior of a large number of unseeded (control) thunderstorms must be compared with a large number of seeded storms.

An additional problem with weather modification programs is political rather than meteorological. For example, cloud seeding may reduce hail (a beneficial result for almost everybody), but may also increase rain, perhaps producing local floods. Such a controversy arose in the Black Hills of South Dakota in June of 1972, when a number of freak thunderstorms produced over 30 centimeters of rain. Unfortunately, two of the clouds had been seeded in an experiment. Although the evidence suggests that very unusual local wind-flow patterns, and not the seeding, caused the floods, some people blamed the flood on the seeding. Thus, even if thunderstorm modification by cloud seeding became a proven fact, important political problems would remain concerning the control and application of this weather modification tool.

10.4 SQUALL LINES AND SEVERE THUNDERSTORMS

The types of thunderstorms considered so far in this chapter have been more or less isolated, created by local differential heating within a horizontally homogeneous air mass. In many parts of the United States, however, thunderstorms frequently occur in a highly organized pattern, with individual cells strung out in a line extending hundreds of kilometers. These *squall lines* are generated and maintained by especially favorable large-scale weather patterns and are often forerunners of cold fronts. Unlike the isolated storms, the squall-line thunderstorms move rapidly eastward and may persist for several hours. The most severe thunderstorms, including those that produce tornadoes, are usually embedded in a squall line.

The basic large-scale weather patterns that produce squall lines are fairly well known. Because some of the most severe squall lines in the world occur on the Great Plains, we will discuss the circulation pattern that is favorable to formation of squall lines in this region.

The typical weather situation conducive to squall line and severe thunderstorm formation occurs often in the spring months over Texas and Oklahoma. With a high-pressure center (anticyclone) located over the southeastern Atlantic states, and a cold front with its associated trough of low pressure approaching from the west, extremely warm, moist air is carried northward at low levels into the Great Plains by the clockwise circulation around the anticyclone. At the top of this layer of warm, humid air (around 2 km), an inversion is usually present, as shown in figure 10.11. (An *inversion* is a layer of the atmosphere in which the temperature rises instead of falls with height, which is an "inversion" of the normal situation.) At higher levels above the inversion, westerly flow brings cool, dry air into the region and over the Gulf air. The temperature in this upper-level flow decreases rapidly with height, which means that the air is unstable for saturated air parcels. However, the thin layer between the base and the top of the inversion (see figure 10.11) is very stable—the temperature increases with height. This stable layer acts as a "lid" to the convective parcels in the lower layer, preventing them from rising into the unstable layer above.

As long as the inversion is maintained, or the convective parcels of air rising from the ground fail to reach saturation, only relatively shallow convection is produced. However, as the ground becomes hotter under the influence of strong solar heating, or as the inversion weakens, some of the warm bubbles of humid air may penetrate the inversion, reach saturation, and extend into the unstable air aloft, thereby triggering explosive thunderstorms.

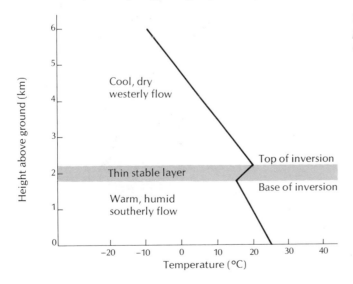

FIGURE 10.11 *Typical vertical temperature structure prior to severe thunderstorms and tornadoes.*

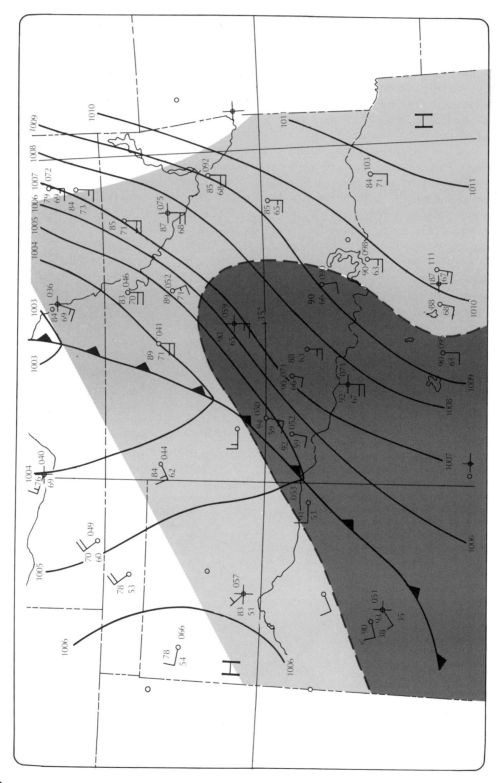

FIGURE 10.12 *Surface weather map (large-scale, pre–squall line) for 12 Noon CST on June 8, 1966.*

In the typical weather pattern, the lifting of the low-level air may be produced by at least two conditions. First, as discussed in section 7.6.5, the dynamical laws of atmospheric motion require a large-scale upward motion ahead of the moving troughs of low pressure. This requirement explains why bad weather generally occurs in the vicinity of fronts, which always lie in troughs of low pressure. Second, radiative heating of the ground by the sun, which is quite strong in spring, heats the air next to the ground and produces rising thermals of hot air, which may eventually become buoyant enough to break through the inversion. These two conditions favor the development of squall lines, which form parallel to and ahead of the advancing cold front. These lines then move eastward at 15–60 km/h, frequently maintaining their identity for several hours. The movement of the cells and the strong vertical wind shear (increase of wind with height) allow the

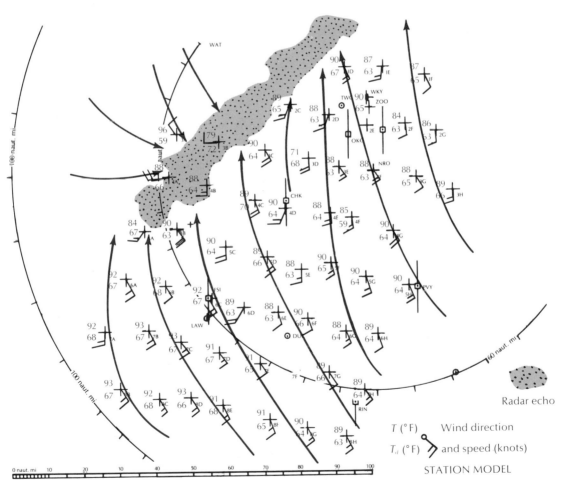

FIGURE 10.13 *Mesoscale surface analysis showing inflow into squall line and radar echoes for 6:30 P.M. CST on June 8, 1966.*

281

raindrops to fall out of one side of the thunderstorm instead of uniformly throughout the entire cell. Thus, the updraft may continue in another part of the cloud, keeping the storm alive.

Within the squall line, individual thunderstorm cells grow, nurtured by the supply of water vapor and warm air accelerating toward the squall line. Under the most extreme cases, a "super" rotating thunderstorm cell may form. These rotating cells are likely to spawn tornadoes, which will be discussed in the next section.

Although idealized histories of squall line development are useful conceptual models, it is perhaps more interesting to consider a real occurrence. For this purpose, we consider the squall line of June 8–9, 1966, which crossed most of Oklahoma. A special observational network, consisting of eleven radiosonde stations in southwestern Oklahoma plus many surface stations staffed by

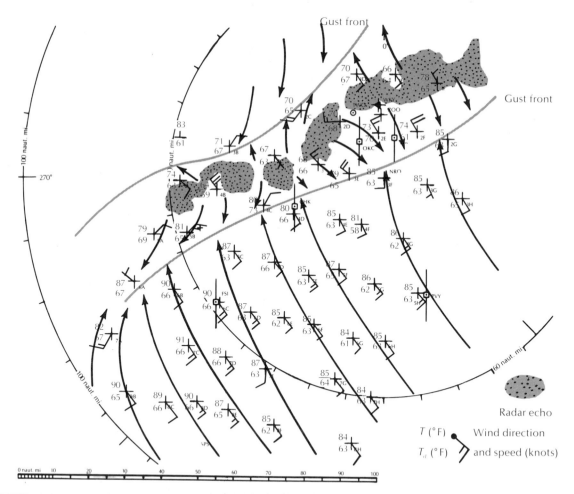

FIGURE 10.14 *Mesoscale surface analysis showing the early stage of a gust front at 8:00 P.M. CST on June 8, 1966.*

volunteers provided a reasonably complete picture of the life cycle of the intense squall line.

Figure 10.12 shows the large-scale weather map at 12 noon CST on June 8, 1966. An advancing cold front separates cooler, drier air to the west and northwest from hot, moist air to the southeast. Notice the shift from southerly winds flowing around the high ahead of the front to westerly winds behind the front. The converging of these surface winds provides a large-scale lifting that was favorable for thunderstorm formation.

Figure 10.13 shows a mesoscale surface weather map $6^{1}/_{2}$ hours later, at 6:30 P.M. CST. This mesoscale map was constructed using the special data network and shows details in the flow that would escape the coarse network of the regularly reporting stations. A line of thunderstorms about 50 km wide and 300 km long had formed ahead of the cold front. The position of the precipitation cells, as revealed by radar, are superimposed on figure 10.13. The smooth, larger-scale wind flow of the earlier map has been perturbed by local circulations associated with the squall line. In general, air is flowing into the squall line at low levels and away from the line at higher levels.

At the ground, the passage of the squall line is marked by an abrupt change in wind direction from south to northwest and a rapid drop in temperature. The surface pressure also varies strongly across the squall line, rising to a maximum directly under the squall line, then falling slightly as the squall passes. Behind the line of thunderstorms, the air becomes drier, the winds become northwesterly, and the pressure begins a slow rise as the next high-pressure system drifts eastward.

As the thunderstorms within the squall line reach maturity, the combined downdrafts reach the ground and cool, moist air spreads out horizontally, forming the gust front. Figure 10.14 shows the beginnings of the gust front associated with the Oklahoma squall line at 8 P.M. CST. Notice that brisk northerly winds at the surface are now outrunning the line of thunderstorm cells, marked by the precipitation echoes. An hour and a half later, at 9:30 P.M. CST, the gust front is well developed (figure 10.15) and is located about 40 kilometers ahead of the line of precipitation echoes. Plate 11 and figure 10.5 show photographs of gust fronts.

TORNADOES 10.5

Without a doubt, the most terrifying weather phenomenon is the tornado, a twisting vortex of winds that can exceed 450 km/h. These winds rotate about a center of pressure so low that buildings explode outward if the tornado passes overhead. This region of low pressure causes the air to expand and cool below its dew point, producing the ominous twisting funnel that identifies the tornado (figure 9.2). No artificial structure is capable of withstanding a direct hit by a strong tornado; railroad cars have been lifted from their tracks and carried hundreds of meters away, and whole brick houses have been demolished, the debris swept cleanly upward into the sky. In rather bizarre instances, small tornadoes cause freakish events such as the stripping of feathers off live chickens without killing them, or the lifting of roofs and walls from houses without disturbing their contents.

Figure 10.16 shows some aerial views of the damage left by tornadoes of varying intensities. As an aid in classifying the intensities of tornadoes for climatological purposes, Professor T. Fujita has developed the F-scale, summarized in figure 10.16. The weakest tornadoes, designated F 0, correspond to wind speeds of from 35 to 63 knots. These do minimum to light damage. At the other end of the F-scale, F 5 tornadoes wreak complete destruction with winds ranging from 227 to 276 knots.

Tornadoes are extremely variable, and because of their small size (usually less than a kilometer in diameter) and short lifetime (a few minutes), are impossible to forecast precisely. However, the large-scale weather conditions that are favorable to severe thunderstorms and tornado formation are well known, as discussed in the preceding section. When these conditions are expected, the Na-

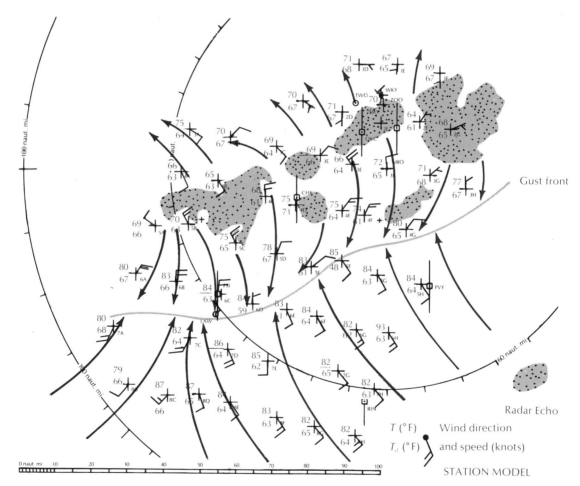

FIGURE 10.15 *Mesoscale surface analysis showing a well-developed gust front at 9:30 P.M. CST on June 8, 1966.*

284

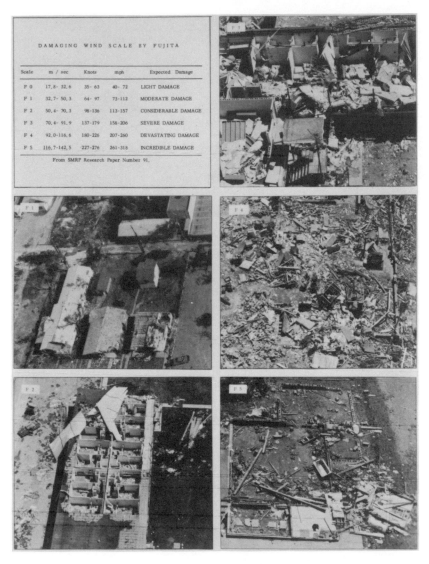

FIGURE 10.16 *Damaging wind scale by Fujita and photographs of damage corresponding to scales F1–F5.*

tional Weather Service issues watches for the areas affected. Fortunately, even under conditions favorable for tornado formation, any one geographical point is unlikely to be hit because of the tornado's small diameter.

The large-scale atmospheric conditions that are favorable for the production of tornadoes are the same as those that produce squall lines and severe thunderstorms (section 10.4). Because of the unique geographical accident of high mountains to the west, the warm, moist Gulf of Mexico to the south, and the cold, dry source region to the north, the midwestern United States has more tornadoes than anywhere else in the world. Plate 12 shows the paths of 19,189 tornadoes that occurred in the United States between 1930 and 1974. The colors correspond

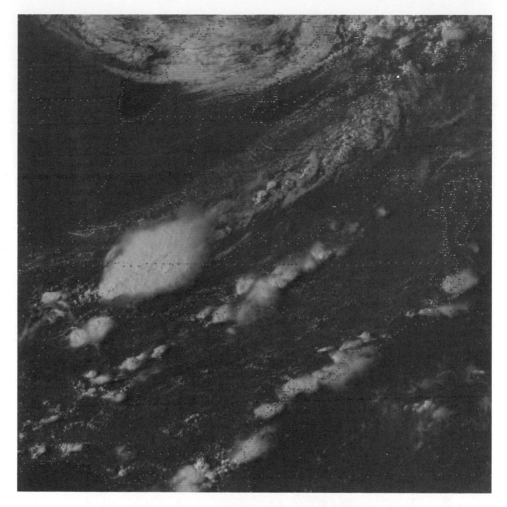

FIGURE 10.17 *Visible satellite photograph of a severe thunderstorm over Tennessee at 4 P.M. on August 16, 1978.*

to different intensities: blue refers to weak tornadoes (F 0 and F 1), green refers to strong tornadoes (F 2 and F 3), and red refers to violent tornadoes (F 4 and F 5). One of the striking features of this figure is the overwhelming predominance of tracks oriented southwest to northeast, at least for the stronger tornadoes. This orientation clearly shows the role of the large-scale circulation in setting up the conditions favorable for tornado formation and then steering the tornado along its subsequent path. Because most of the tornadoes form in the moist, southwesterly flow ahead of an advancing trough, the tornadoes in general move from southwest to northeast.

Although the details of tornado formation are not fully understood, the conservation of angular momentum plays a fundamental role, as it does in hurricane formation. A violent updraft caused by a severe thunderstorm and a concomitant fall of surface pressure would initiate a rapid convergence of low-level air toward the center of the updraft. Because the air in the vicinity of severe thunderstorms already possesses considerable wind speeds (e.g., 60 km/h), it is not hard to see how tornadic velocities could be rapidly generated. For example, a 60-km/h tangential wind at a radius of 5 kilometers from a low-pressure center could achieve 320 km/h if it were accelerated inward to a radius of 1 kilometer while conserving its angular momentum. Indeed, many tornadoes appear to be spawned by a parent cyclone, a rotating severe thunderstorm several kilometers in diameter.

Evidence of the extraordinarily rapid updraft in tornado-producing thunderstorms can be seen in satellite photographs. Figure 10.17 shows a visible photograph of a severe thunderstorm over Tennessee. The strong vertical currents associated with severe thunderstorms carry vast amounts of water vapor into the upper tropopause, where it spreads out horizontally and moves with the winds at this level. From the satellite's elevation, the thunderstorm resembles a smoking chimney with cloud droplets and ice crystals streaking downstream. Such a picture is a good indication of an extremely rapid updraft and suggests the likelihood of violent weather below.

The electrical properties of tornadoes are interesting, even though they are probably results rather than causes of the tornado funnel. In some cases, lightning flickers eerily inside the funnel itself. Tornadoes also emit electromagnetic radiation which may be picked up on ordinary television receivers.

The possibility of tornado modification has attracted considerable interest, but any theories are speculative at this time. Because of the large rate of energy production in a severe thunderstorm (roughly equal to the total power-generating capacity of the United States in 1970), any modification will have to be directed at triggering natural changes in the circulations of these storms by application of relatively small amounts of energy. Perhaps cloud seeding could beneficially alter the tornado-producing clouds, or increase the growth of nearby clouds, thus providing competition for the finite amount of available water vapor. On the other hand, if a possibility for decreasing the frequency or intensity of tornadoes exists, a possibility for an increase must also be admitted. Some of the answers may eventually come from numerical modeling of tornadoes and their parent thunderstorms, but the answers are not likely to come easily or soon.

WATERSPOUTS 10.6

Waterspouts are not produced in cold weather.
[Aristotle]

A sailboat skims along the surface of the emerald waters off the Florida Keys on a bright June morning. The blue skies are interrupted by a few towering cumulus clouds which speckle the water with shadows, but the cloud tops are not very high, and no rain is falling. The winds blow steadily in from the southeast at 10 km/h.

Suddenly, the base of one of the towering cumulus clouds lowers, and a distinct whirling is visible. The water underneath the clouds is disturbed by some unseen force. The whirling cloud material builds downward from the cloud base, forming an unmistakable funnel. Now, the water underneath the lengthening funnel whips around in a frenzy; spray is thrown upward and outward. The funnel finally reaches the water, merging freshwater cloud droplets with saltwater spray. The cloud and its appendage drift toward the sailboat, which frantically jibes to escape the ominous whirlwind. A few minutes later, the funnel passes the boat barely 100 meters to port. At the sailboat's position, the wind and water are scarcely disturbed and the sun is shining; a near miss—as good as a mile.

Although waterspouts are not usually nearly as intense as their big brothers, the tornadoes, they can nevertheless produce wind speeds of over 150 km/h. The average waterspout, however, is much smaller than the tornado, perhaps only 50 meters in diameter, and has maximum wind speeds on the order of 80 km/h. Boats occasionally have passed through the weaker waterspouts with little damage, but the larger spouts can completely destroy small craft.

Waterspouts are of interest because of their dynamic similarity to tornadoes. Both are caused by a rapid updraft and associated inflow of low-level air, which, under the principle of conservation of angular momentum, spins faster and faster until it reaches a point close to the center of low pressure. In moderate to intense waterspouts, the pressure at the vortex center is low enough to cause air to cool by expansion below its dew point, so a visible funnel appears, much like the tornado (plate 13). Thus, the waterspout funnel is not water sucked up from the sea, but merely cloud droplets.

Tornado research in the past has been slowed by the very understandable fact that no one wants to take direct observations of well-developed tornadoes, at least not from very close range. Because the dynamics of tornadoes and waterspouts are similar, however, there is hope of gaining insight into tornadoes through understanding the far less dangerous waterspouts, which are amenable to direct observation.

There are some significant differences between tornadoes and waterspouts, however. One striking difference, obvious to those who have witnessed waterspouts, is that they frequently occur in basically fair weather, at least compared to the severe weather associated with tornadoes. Frequently, the cumulus clouds from which waterspouts descend are not even raining and may extend to only 2.5 kilometers or so, well below the freezing level. In comparison, severe tornado-producing thunderstorms extend to 15 kilometers and higher, well into freezing temperatures aloft. Waterspouts are also much more frequent than tornadoes, especially over the warm, shallow waters surrounding the Florida Keys. For example, Key West alone observes nearly one hundred waterspouts per month during the summer.

Most Florida waterspouts are short-lived, the average duration being about 15 minutes. Their formation is favored by high surface temperatures and humidities (low densities) and, therefore, unstable conditions in the lowest kilometer. In this respect, they are like dust devils, which form over desert regions when the ground is heated to a very high temperature.

Waterspouts are common over coastal waters from Florida to the Gulf Coast and occur most often in the summer, when the water temperatures are high. Waterspouts also occur in other parts of the world, including the Indian Ocean, the equatorial Atlantic, and the Mediterranean Sea.

Occasionally, a tornado which forms in the usual manner over land moves over water and becomes a waterspout. In this case, the waterspout can be very severe, causing the same degree of damage to structures in and on the water as tornadoes do over land.

Questions

1 Describe at least three lifting mechanisms that can trigger thunderstorms.

2 What is the source of energy in a thunderstorm?

3 Discuss the typical structure of a thunderstorm in the early, growing stage, the mature stage, and the decaying stage. Mention the height of the cloud, up- and downdrafts, existence of ice and water, and precipitation.

4 What force initiates the downdrafts in thunderstorms?

5 What is a pileus and how is it formed?

6 How is the gust front formed?

7 What causes some severe thunderstorms to rotate?

8 Why is there a relative maximum of thunderstorms over western North Carolina (see figure 10.8)? Why another over south-central Florida?

9 Why can it be said that a single lightning flash originates from both the cloud and the surface object?

10 Distinguish among the following: stepped leader, dart leader, return stroke.

11 Approximately how fast must an updraft be to suspend a hailstone?

12 If you hear thunder 9 seconds after you see a flash of lightning, how far away was the flash? (Answer: 3 km)

13 In figure 10.12, why are the winds ahead of the cold front from the south when the isobars indicate a geostrophic flow from the southwest?

14 Why do most tornadoes over the U.S. move from southwest to northeast?

15 What produces the visible waterspout funnel?

Problems

16 An inversion is a very stable layer of air, whereas thunderstorms occur in unstable air masses. Why then are inversions often associated with the development of severe thunderstorms?

17 Estimate the kinetic energy $mV^2/2$ in a tornado with a radius 200 m, height 5 km, mean speed 50 m/s, and mean density 0.8 kg/m^3. (Answer: 6.29×10^{11} J) What is the release of latent heat by condensation of water vapor in a typical thunderstorm? Assume an average rainfall of 2 cm over a circle of radius 5 km. (Answer: 3.9×10^{15} J) If all this latent heat were used to produce tornadoes of the intensity and size given above, how many tornadoes could be produced? Assume a latent heat of condensation of 597 cal/g, where 1 cal = 4.187 J. (Answer: 6200)

ELEVEN

Climate

climate • continentality • monsoon • maritime • rain shadow •
valley breeze • chinook • foehn • potential vorticity • general
circulation • Hadley cell • doldrum • horse latitudes • thermals • net
radiation • diurnal • microclimate • urban heat island • urban plume

11.1 CLIMATE IS MORE THAN AVERAGE WEATHER

The climate of a region describes the overall character of the daily weather that prevails from season to season and from year to year. It does not describe the weather on a particular day; rather, it provides a synopsis of the typical weather over a long period of time. Thus, we say that the tropical climate is generally hot and humid, and that the Siberian winter is dry and cold.

But climate not only describes average or typical values of meteorological quantities; it also includes such statistics as the annual range of temperature and the extreme values likely to be reached in any given time period. For example, the largest amount of rainfall in a single 24-hour period in one hundred years, or the strongest wind ever recorded at a station, is part of climate. Climate is the total of all statistical weather information that helps describe the variation of weather at a given place or region.

We need to understand the climate of the various regions of the earth because climate affects all biological processes and human activity, as will be discussed in chapter 16. For example, if we wish to build a house in a certain location, we need to consider several climatic factors. First, the coldest and hottest temperatures likely to be reached during the year should be used to determine the insulation required, the power of the heating system, and the desirability of air conditioning. Wind speed would influence these decisions, too, because wind increases the cooling power of the atmosphere. The likelihood of flooding should be considered in selecting sites near bodies of water or in low-lying areas. The possibility of occasional very strong winds, such as the winds over 160 km/h (100 mi/h) that sometimes blow down the slopes of the Rocky Mountains, has to be considered in determining the strength of the walls. Wind directions also may be important but are often ignored. For example, in much of the northeastern United States, the prevailing wind is from the southwest in summer and from the northwest in winter. From this consideration alone, most windows should face the southwest to take advantage of the summer breezes. There should be few, if any, windows to the northwest to minimize the effect of the chilling winter winds.

Other variables that comprise the climate may also be considered. The amount of snowfall determines the kind of roof that can be used in a certain location without danger of collapse. Thus, the houses in alpine villages have steeply sloped roofs to allow the snow to slide off harmlessly, whereas completely flat roofs are common in Miami. The amount of sunlight, which is strongly influenced by meteorological variables, influences the choice of color for the home. White and other strongly reflecting paints are preferable in climates with strong, hot sunshine, whereas deeper hues may be used effectively for cloudy, cool climates.

Climatological information can be obtained in a number of ways. First, meteorological data are collected at many weather stations, and statistics such as averages, ranges, and extremes that are useful in planning our activities are computed. Such statistics are available for many places in the world. In the United States, the National Ocean and Atmospheric Administration (NOAA) publishes climate summaries by states and by individual cities. Table 11.1 is an example of such a summary for State College, Pennsylvania, for the years 1926–62. This summary gives the monthly average maximum and minimum temperatures, and the extremes. The mean heating *degree-days* are also given. (The number of heating degree-days in a day is obtained by subtracting the average daily temperature from 65°F. Therefore, a day with a high of 60°F and a low of 40°F has 15 heating degree-days. As might be expected, fuel consumption is very closely related to the number of heating degree-days.) The climatological summary in this table also gives the average and extreme amounts of precipitation. In describing precipitation at a point, it is especially important to know the extreme amounts that are possible. Thus, the mean snowfall in March is about 28 centimeters (11 inches), but nearly 127 centimeters (50 inches) fell in March, 1942. Appendix B gives a monthly summary of the climate of 16 cities in the United States. These summaries include values for mean monthly temperatures, percentages of cloudiness, precipitation, and snowfall.

One trouble with climatological summaries is that they apply only to the weather station where the data were observed. However, since we do not build houses or grow gardens right at the weather station, we must understand the factors that cause climate to vary over small distances in order to estimate the climate between. The difference in climate between one location on the southern slope near the foot of a mountain and another a kilometer away near the top and facing north, may be as great as the difference in climate between two weather stations separated by hundreds of kilometers. Therefore, we need to study local as well as global variations in the climate. In this chapter we will first consider the major factors that determine the climate over the Earth. We will then summarize the global climate, the climate of North America, and finally the processes influencing small-scale variations in climate—microclimate.

PHYSICAL FACTORS INFLUENCING CLIMATE 11.2

Climate is determined by many factors—the intensity of sunlight, the proximity of an ocean, the origin of air masses arriving at the locality, and topography. Lakes, mountains of all sizes, and even cities influence the climate. Because there are so many factors influencing climate, we will first separate them into large-scale and local causes. Large-scale controls include the seasonal distribution of sunlight and the global distribution of continents, oceans, and the major mountain chains such

TABLE 11.1 *Climatological summary: Means and Extremes for*

1927–1962
Latitude: 40°48'
Longitude: 77°52'
Elevation (ground): 1175 feet

Station: State College, Pennsylvania

Month	Temperature (°F)*							Mean degree-days †	Precipitation totals (in.)**								Mean number of days				
	Means			Extremes								Snow, sleet					Precip. 0.1 in or more	Temperatures (°F)*			
	Daily maximum	Daily minimum	Monthly	Record highest	Year	Record lowest	Year		Mean	Greatest daily	Year	Mean	Maximum monthly	Year	Greatest daily	Year		Max 90° and above	Max 32° and below	Min 32° and below	Min 0° and below
(a)	37	37	37	37		37		37	37	37		37	37		37		37	37	37	37	37
Jan	35.5	20.9	28.2	71	1950	−14	1936	1141.	2.66	1.60	1952	8.9	28.7	1936	14.0	1936	7	0	12	27	1
Feb	37.5	21.2	29.3	73	1954	−17	1934	1007.	2.39	1.47	1961	9.7	23.6	1961	16.5	1961	6	0	8	25	1
Mar	45.7	27.8	36.7	82	1938	1	1943	876.	3.55	2.10	1936	10.8	47.5	1942	17.5	1942	8	0	3	22	0
Apr	58.6	38.0	48.3	89	1942	17	1950	509.	3.67	2.08	1937	2.3	18.1	1928	17.3	1928	8	0	††	9	0
May	70.9	48.3	59.6	92	1962	28	1931	207.	4.06	2.66	1953	T	0.2	1947	0.2	1947	9	††	0	1	0
June	78.8	56.7	67.8	96	1952	35	1929	49.	3.69	2.28	1957	T	0.3	1954	0.3	1954	8	2	0	0	0
July	83.1	60.7	71.9	102	1936	40	1929	8.	3.71	2.11	1933	0.0	0.0	—	0.0	—	7	4	0	0	0
Aug	80.9	59.2	70.0	101	1930	39	1934	20.	3.40	3.95	1955	0.0	0.0	—	0.0	—	6	3	0	0	0
Sept	73.9	52.3	63.1	98	1953	28	1947	132.	2.69	2.15	1939	0.0	0.0	—	0.0	—	6	1	0	0	0
Oct	62.8	42.2	52.5	90	1941	18	1936	396.	3.03	3.42	1954	0.1	1.0	1962	1.0	1962	6	††	††	4	0
Nov	48.8	33.0	40.9	81	1950	1	1929	723.	3.04	3.21	1950	3.2	14.0	1953	10.0	1953	6	0	2	15	0
Dec	37.3	23.2	30.2	65	1933	−6	1942	1077.	2.60	1.52	1941	8.0	24.4	1950	10.0	1960	6	0	10	26	1
Year	59.5	40.3	49.9	102	July 1936	−17	Feb 1934	6145.	38.49	3.95	Aug 1955	43.0	47.5	Mar 1942	17.5	Mar 1942	83	10	35	129	3

(a) Average length of record., years.
* Number of degrees Celsius = (5/9) (number of degrees Fahrenheit − 32).
** One inch = 2.54 centimeters.
† Base 65°F
†† Less than one-half.
T = trace, an amount too small to measure.

294

as the Rockies or Alps. These global features produce climatic characteristics that dominate large regions of the earth; therefore, their effects can be interpolated easily between weather stations.

11.2.1   LATITUDE

The single most important variable affecting climate is latitude. In fact, the word climate itself comes from the Greek word *klimas,* meaning angle of inclination. The relevant angle of inclination is, of course, the angle of the sun above the horizon; on a yearly average, this depends almost entirely on latitude. In chapter 4 we saw that the solar radiation received per unit area was proportional to the cosine of the angle that the sun makes with the zenith. Thus when the sun is 60° above the horizon (30° from the zenith), a given area receives 1.73 times as much radiation as when the sun is 30° above the horizon (cos 30 ÷ cos 60 = 1.73). Over the year, the sun is much higher near the equator than the polar regions. Since most of the solar energy is absorbed and converted to heat at the surface of the earth, tropospheric temperatures show a general maximum at the equator and a decrease toward the poles (figure 11.1). The difference in temperature between the poles and the equator is greatest in winter, because the variation of solar radiation with latitude is greatest then. In summer, the long days at high latitudes compensate for the sun's low angle so that the differential heating between the polar and equatorial regions is much less (figure 4.20).

11.2.2   CONTINENTS AND OCEANS

Although latitude is the dominant control on climate, even a cursory glance at global temperature averages shows that other factors are also quite important (figure 11.1). For example, the average January temperature over central Siberia at a latitude of 60°N is about −40°F (−40°C), whereas at the same latitude northwest of Ireland, the temperature is higher than +40°F (4°C). This extreme temperature difference at the same latitude is caused by the second most important geographic factor affecting climate—the distribution of continents and oceans.

As shown by the global temperature patterns in January and July (figure 11.1), there are great differences in temperature between maritime (oceanic) and continental climates. In winter the continents are much colder than surrounding oceans; in summer they are considerably warmer.

The seasonal contrast in temperatures between continents and oceans is caused by the great difference in what happens to solar radiation when it strikes water and land. For several reasons, a given amount of solar heat input will warm the ocean surface much less than a land surface. First, the water in the ocean can overturn vertically, so warm surface waters mix with colder water below. Second, the ocean is partly transparent, so a thick layer of water has to be heated. Third, some of the incoming heat is used to evaporate rather than warm the water. Finally, water has a greater heat capacity than land, which means that it takes

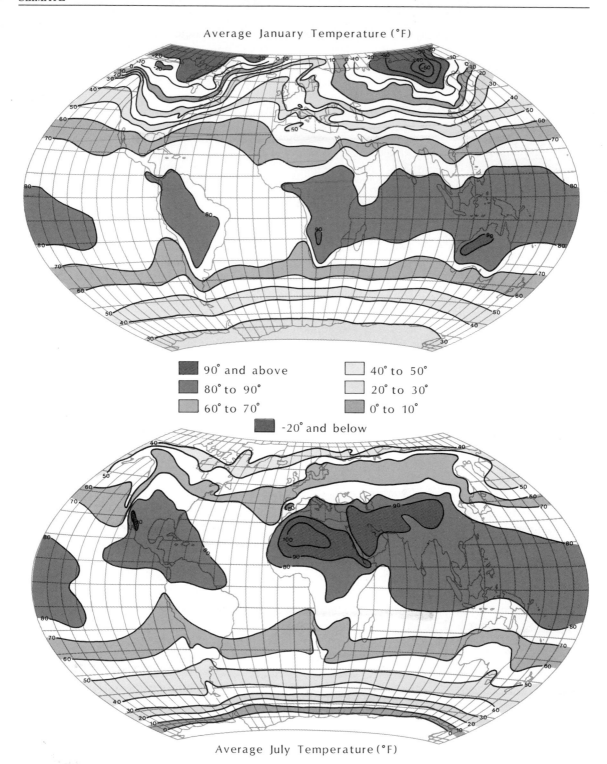

FIGURE 11.1 *Average January and July temperatures (°F).*

more heat energy to warm a mass of water by one degree than an equal mass of land.

Just as the surface temperature of water increases during the day and spring season more slowly than the temperature of nearby land surfaces, water cools more slowly at night and during the autumn. The higher heat capacity means less of a temperature fall for a given amount of heat loss. Furthermore, as a thin layer of water at the surface begins to cool, it becomes denser than the underlying layers and convection (vertical overturning) begins, with cool water mixing downward and warmer water rising to the surface. The maximum density of fresh water occurs at 4°C. Only when the entire depth of water is cooled to this temperature does the vertical exchange diminish. Thus the cooling of water occurs over a much deeper layer than cooling over land.

For these reasons, diurnal and annual variations of temperature over and near water are small compared to the variations over land. The diurnal and annual ranges of temperature are thus useful measures of the continentality of the climate.

Seasonal differences in temperature between continents and oceans produce important variations in pressure, winds, and precipitation between summer and winter. These seasonally varying circulations are called monsoons. To see how a typical monsoon works, consider figure 11.2 showing a monsoon circulation associated with an idealized square continent in the Northern Hemisphere in summer and in winter.

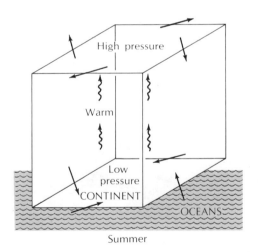
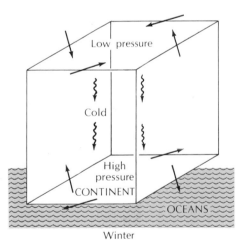

FIGURE 11.2 *Monsoon circulations around continents in the Northern Hemisphere. Warm air rises over the continent in the summer, and air from the ocean moves in to take its place. In winter, cold air sinks over the continent and spreads outward over the ocean.*

In summer, the continent is hot and the continental air rises. To replace this rising air, other air rushes in from the oceans toward the land. However, the Coriolis force deflects the air to the right, so a counterclockwise (cyclonic) circulation results. In winter, the circulation reverses, as the land is then cold.

297

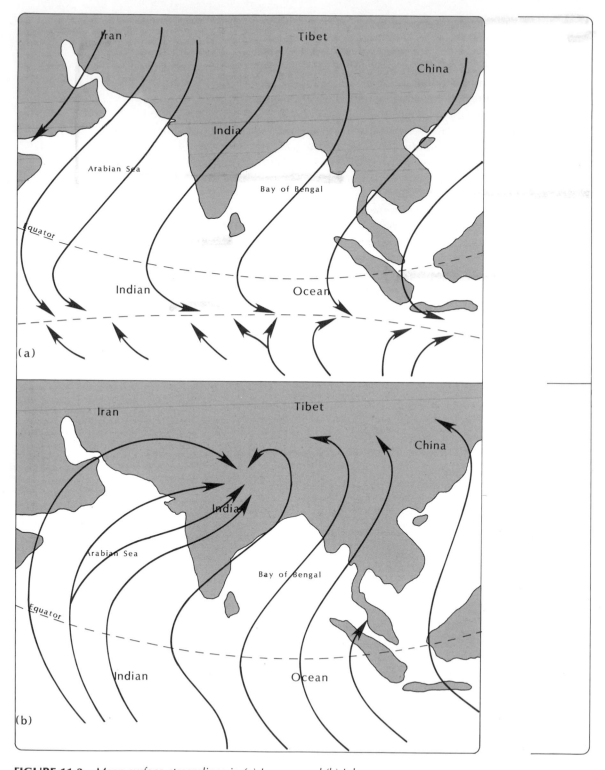

FIGURE 11.3 *Mean surface streamlines in (a) January and (b) July.*

Cold, heavy surface air flows off the continent over the seas, acquiring a clockwise (anticyclonic) circulation under the influence of the earth's rotation.

Real monsoons are not quite so simple, because continents are not squares, and mountains interfere. The bigger the continent, the better the idealized patterns work. Monsoons are therefore especially prominent in Asia. The surface flow over India and China in winter is generally from the north (figure 11.3), bringing cold air out of Siberia, which makes winter much colder than if the monsoon were not present. In summer, however, the mean surface flow reverses dramatically (figure 11.3b). The humid southwesterly flow from the Indian Ocean flows up over the Himalaya Mountains and releases huge amounts of moisture in the form of rain and snow. Some of the largest rainfalls in the world have been measured in summer at Cherrapunji in northeast India, which holds the world's record for rainfall in one year (2647 centimeters, or 1042 inches) and in one month (930 centimeters, or 366 inches). No wonder that the term *monsoon* has come to mean heavy rainfall; but it is, strictly, a seasonal wind.

It is apparent that maritime climates are quite different: Because water loses and gains heat slowly, maritime climates are mild and have their extreme temperatures late in the seasons. We have both maritime and continental climates in the United States. Because prevailing winds blow from west to east, weather in California, western Oregon and western Washington is brewed over the Pacific and the climate is maritime, with late seasons and little temperature contrast from winter to summer. On the other hand, the central and eastern United States have continental climates, with early seasons and large seasonal contrasts. Thus, the annual range in temperature in San Francisco is only 6C° (11F°), compared with a range of 29C° (53F°) in Chicago (see appendix B). On a day-to-day basis, it is the great polar and tropical air masses that produce the range in Chicago.

Another difference between continental and maritime climates is the difference in annual distribution of precipitation. In continental climates, most precipitation occurs in summer. At this time, the ground gets so hot compared with the air aloft that strong convection currents are produced in the afternoon, leading to thunderstorms and large amounts of precipitation.

11.2.3. MOUNTAINS

Mountains play a major role in modifying the climate of continents.

First, air ascending mountains cools and deposits precipitation on the upwind side. Thus, where the winds are westerly, most rain falls on the western slopes. A conspicuous example is the west side of the Olympic Mountains in Washington state, where the annual precipitation is over 305 centimeters (120 inches), allowing rain forests to grow. On the east slopes, some towns get only about 38 centimeters (15 inches) a year. Figure 11.4 shows the strong control of orography on precipitation across the state of Washington. The dry zone east of the Cascades is an example of what is called the *rain shadow effect.*

In most mountainous regions, there is a very strong correlation of precipitation with elevation. In addition to forced lifting of air when horizontal winds encounter the sloping mountain surface, heating of the air over the moun-

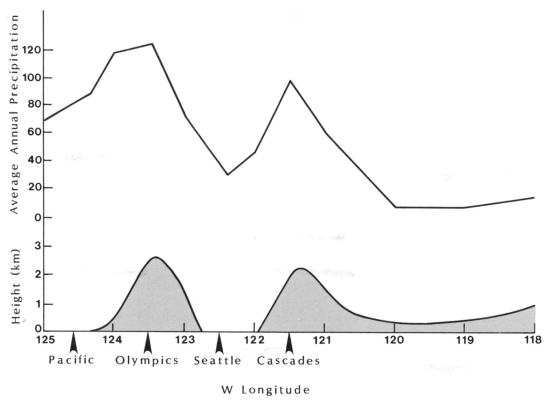

FIGURE 11.4 *West-to-east cross section of terrain at 47.5°N and average annual precipitation (in inches).*

tains relative to air at the same level over lower terrain results in upward motion and often convection. Figure 11.5 shows the strong relationship between terrain elevation and rainfall over the island of Oahu, Hawaii. The average annual rainfall of 1234 centimeters (486 inches) on Mt. Waialeale, which is on the Hawaiian island of Kauai, is the highest in the world. Along the slopes of Mt. Waialeale, annual rainfall varies by as much as 101 centimeters (40 inches) per kilometer.

Mountains affect the winds in a variety of ways depending on the height and shape of the mountains, the direction and speed of the prevailing wind at various levels, and the vertical temperature structure of the atmosphere. Under light wind conditions, heating the air over mountains results in air rising over the mountain peaks, with surface air blowing up the slopes of the mountains (a *valley breeze*). Sinking motion occurs over lower terrain. If the air is sufficiently moist, clouds will form where the air is rising and dissipate where the air is sinking. Figure 11.6 shows a satellite photograph of the eastern United States when a high pressure system was centered over Tennessee. After a clear sunrise over the entire Southeast, cumulus clouds have formed over the high terrain of West Virginia, western North Carolina, South Carolina, and northern Georgia. A conspicuous lack of clouds is evident over the lower terrain of the Tennessee valley.

When the air is blowing perpendicular to a long ridge, a variety of mountain wave effects are possible. Under extremely stable conditions (figure

300

11.7a) the low-level flow may be completely blocked. For faster winds and less stable conditions, strong waves may be formed with variations in wind speed by 100 percent or more across the mountain (figure 11.7b). These waves tilt upstream and extend to great heights into the stratosphere. On the lee side of the mountains the sinking air warms as it compresses. The dry, warm wind that descends on the lee side of the Rockies in the winter is called the *chinook*. In other parts of the world, similar winds are called *foehns*.

Occasionally the mean wind and stability structure will allow the mountain waves to develop extremely high amplitudes, so that surface winds in the lee of the mountains exceed hurricane force (73 mi/h or 117 km/h). Boulder and other Colorado cities along the eastern foothills of the Rockies experience these destructive storms several times each winter.

A close relative of the chinook is the Santa Ana of southern California, which occurs most often in autumn. When a strong anticyclone is centered over the central Rockies, east winds on the south side of the high descend over southern California to the Pacific Coast. Already warm after their passage over the southwest deserts of Arizona and Nevada, air is warmed still further by compression, and temperatures over 38°C (100°F) are quite common—even on the Pacific coast.

An important property of large mountain ranges which modifies the climate is the blocking effect on large-scale air masses. Thus the Himalayas often confine polar air masses to their north over Tibet and Mongolia. In North America, the westward extent of cold air masses in winter is often limited by the Rocky Mountains.

Because of their effect on perturbing the large-scale currents of air, mountains also play a major role in the cyclogenesis process. As air ascends larger

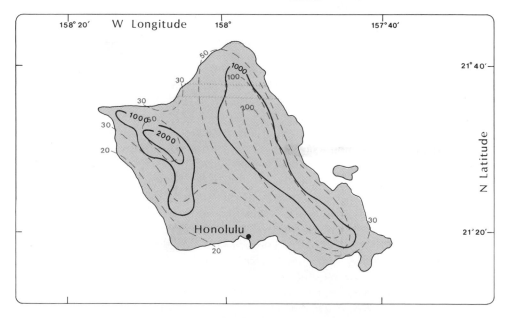

FIGURE 11.5   *Terrain elevation (in feet) and annual rainfall (in inches) on Oahu, Hawaii.*

FIGURE 11.6 *Satellite photograph for 1:30 P.M. EST on April 20, 1976, showing cumulus clouds over high terrain and clear skies, indicating sinking motion over Tennessee Valley.*

mountains like the Rockies, it is compressed vertically (figure 11.8). According to the conservation of potential vorticity (see section 5.3.4), air undergoing horizontal divergence spins more anticyclonically (clockwise). Thus straight westerly flow crossing the Rockies turns in a clockwise direction (toward the south). As the air descends the mountains on the east side, it is stretched vertically and its spin becomes more cyclonic (clockwise). Because it is now at a lower latitude (figure 11.8), the earth's vorticity represented by f in equation (5.6) is less than the original value. In order to keep the potential vorticity constant, the relative vorticity must be greater than its initial value (more cyclonic). The air turns counterclockwise back toward the north and a trough is formed.

In summary, when westerly wind encounters a major north-south range of mountains such as the Rockies, changes of the depth of air columns and changes in latitude produce anticyclonic flow (a ridge) over the mountains and

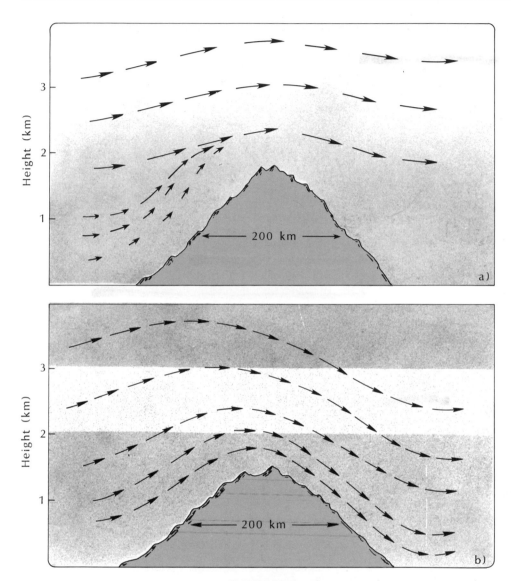

FIGURE 11.7 *Flow over a large mountain with two different vertical temperature structures. (a) Stable at all levels with very cold air near surface. (b) Stable in low levels, less stable in middle levels. Length of arrows represents wind speed; continuous lines denote paths of air parcels.*

cyclonic flow (a trough) to the lee of the mountains. Because upper-level troughs are favorable for development of surface lows, cyclogenesis is favored along the eastern slopes of the Rockies and other high mountains.

High terrain, whether mountains or plateaus, can be heated strongly during daytime in the warm seasons. Because the surface absorbing the solar radiation is elevated, the air temperature near the surface at places like Denver or

Salt Lake City can be much higher in the afternoon than the temperature in the free atmosphere at the same elevation (1 to 2 kilometers). We think of such ground as *high-level heat sources*. The heating of mountain peaks often generates afternoon thunderstorms, which can lead to dramatic and potentially dangerous changes in the weather. Following a calm, sunny morning with temperatures in the 60s (° F), a stunning chill associated with evaporational cooling of rain in thunderstorms and gusty winds can attack the hiker in the Rockies. Temperature decreases of 30 F° are not uncommon in summer thunderstorms over high terrain, even without a change in air mass. With a change to a colder airmass, the temperature decrease can be even greater. Caution is the word in anticipating mountain weather!

11.2.4 LAKES

Continental regions in the vicinity of lakes have their climates modified by the tempering effects of water. Over the United States and Canada, the Great Lakes significantly affect the climate throughout the year (unless the lakes are frozen). In the summer, the water is cooler than the land. During hot days the air over the warm land adjacent to the lakes rises and cool air from the water flows onshore, producing a welcome cooling *lake breeze*. Often Chicago's Midway Airport, only 15 kilometers from the Lake Michigan shore, will have temperatures in the 60s (°F), whereas O'Hare Airport, at a distance of 25 kilometers from the lake, will have temperatures in the 90s (°F).

FIGURE 11.8 *Change of vorticity (spin) as a column of air moves over mountain. Through conservation of potential vorticity, air spins more anticyclonically as it crosses ridge and depth (D) decreases. As the air column stretches on its lee side, cyclonic spin increases again. These changes of vorticity result in streamlines across long ranges of mountains as shown in (b).*

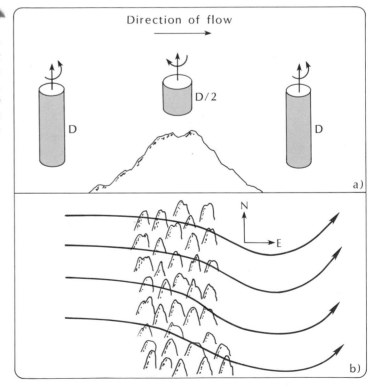

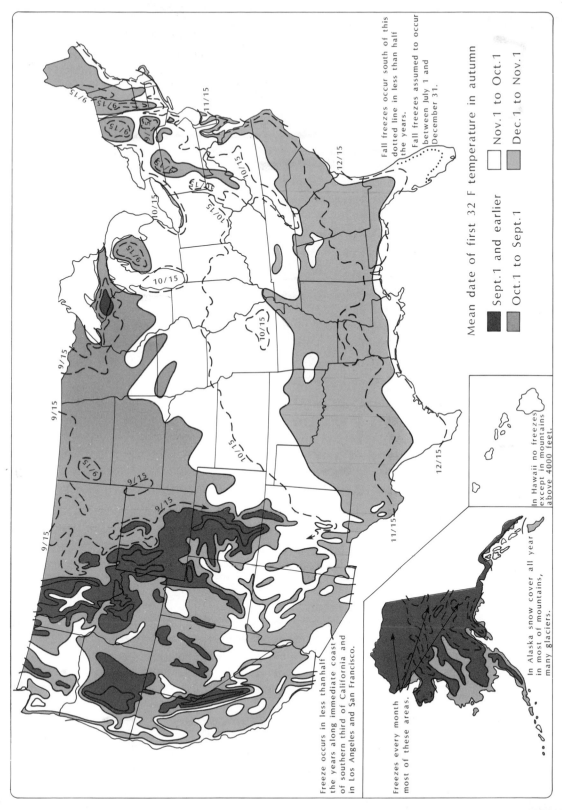

Mean date of first 32 F temperature in autumn

- Sept.1 and earlier
- Oct.1 to Sept.1
- Nov.1 to Oct.1
- Dec.1 to Nov.1

Fall freezes occur south of this dotted line in less than half the years.

Fall freezes assumed to occur between July 1 and December 31.

Freeze occurs in less than half the years along immediate coast of southern third of California and in Los Angeles and San Francisco.

Freezes every month most of these areas.

In Alaska snow cover all year in most of mountains, many glaciers.

In Hawaii no freezes except in mountains above 4000 feet.

FIGURE 11.9 *Mean date of first 32°F temperature in autumn.*

The relatively cool water of the Great Lakes during the summer is responsible for less cumulus cloudiness over the water. Furthermore, thunderstorms which form over the warm land often weaken or dissipate as they move over the Lakes.

During autumn the role of the Great Lakes in modifying the climate begins to change. As the land cools more rapidly than the water, the Lakes become relatively warm. Thus the average date of the first freeze is delayed by as much as a month compared to areas well inland (figure 11.9). The mild autumns near the shores of Lake Erie allow for wine grapes to be grown along a narrow zone near the shore when the average climate of the region is far too harsh to permit the ripening of the grapes. Because of the milder, more humid winters on the east shore of Lake Michigan, peaches can be grown in western Michigan but not in eastern Wisconsin.

As the air masses which follow the passage of cyclones become colder in late autumn and winter, their movement across the relatively warm Great Lakes results in intense vertical mixing and convective clouds. These clouds, though only a kilometer or two thick, often produce significant rain and snow showers over the Lakes and on adjacent shores. The mean annual snowfall map for the northeastern United States and southeastern Canada (figure 17.1) attests to the efficiency of the Great Lakes in enhancing the seasonal snowfall. Although the significant increases in precipitation are usually confined to fifty kilometers or so from the coast, the clouds often persist for several hundred kilometers downwind, and significantly affect the wintertime climatology of low stratocumulus clouds in Pennsylvania and New York. Similar snowbursts can occur downwind of Hudson Bay and on the shores of the North Sea and the Baltic.

Finally, in spring and early summer, the land again becomes warmer than the Great Lakes. Spring is thus delayed by several weeks along the shores. Trees along the shore of Lake Michigan, for example, do not produce leaves until early June, several weeks later than their inland counterparts.

11.2.5 CHARACTERISTICS OF THE GROUND

Even after the effect of latitude, oceans, continents, mountains, and lakes have been separated to explain the variability of climate, there are additional factors associated with the properties of the ground surface which shape a region's climate. These properties include albedo, availability of water at the surface, heat capacity of the soil, and roughness of the terrain. Soils with high heat capacities, such as clay, are associated with somewhat less diurnal temperatures variation than soils with lower heat capacities, such as sandy loam. Low-level winds over a rough surface (such as forests) are slower than over smooth surfaces (such as prairie grass or sand). The properties of the surface that are most important in regional climatology, however, are albedo and water availability, because these parameters affect the amount of energy absorbed by the region and thereby the horizontal and vertical motion and precipitation. Variations in vegetation are extremely important in modifying the albedo and water availability. Over a sandy desert the surface albedo may be as high as 40 percent, which means that 40 percent of the incoming solar radiation is reflected from the surface. The water availability of such a surface may be nearly zero, so that evaporation is negligible. Thus the Sahara desert, in spite of its low latitude, can actually be a net heat sink for the planet because of its high albedo. On the other hand, a forest may have an albedo of less than 10 percent, and hence absorb 1.5 times as much solar energy as an unvegetated desert surface. Part of the absorbed energy is used to evaporate water from the vegetation, so the actual temperature may not be any higher over a vegetated area than a nearby unvegetated region that is brighter. However, the moisture content and the rainfall is likely to be substantially greater.

Unlike many of the factors that control climate, vegetation is relatively easy for humans to modify. It has been so easy, in fact, that humans have destroyed an estimated 15×10^6 km^2 of forests, which corresponds to 3 percent of the earth's area. Because of the feedback associated with replacing vegetated with unvegetated surfaces (less energy absorbed, less rainfall, and less growth of vegetation), there is strong evidence that humans have changed the climate significantly on a regional basis (as in the African Sahel Desert and the Rajasthan Desert of India). There is less strong, but still significant evidence that these anthropogenic changes in vegetation have affected global climate over the past 10,000 years to a similar degree as other natural causes (see chapter 12). The cumulative effect of human activities on changing the surface of the earth has been an increase in mean global albedo of about 0.6 percent. Although this change does not appear large at first, theoretical calculations indicate a decrease of mean global temperature by about 2°C for a 1 percent increase in mean global albedo. As we shall see in chapter 12, this change is quite significant.

GLOBAL CLIMATE AND THE GENERAL CIRCULATION 11.3

If we average the wind and temperature perturbations associated with passing cyclones and anticyclones on a daily basis over many years, a mean temperature and wind structure of the atmosphere would emerge. This highly smoothed atmospheric state, showing gradual variations from day to day as the seasons changed, is known as the *general circulation*. It includes persistent features, such as the trade winds, middle-latitude westerlies, and the polar jet streams. The properties of the general circulation, especially the regions of mean upward and downward motion, reveal much about the global variations of temperature and precipitation.

11.3.1 THE GENERAL CIRCULATION

General-circulation theories start with air at a uniform temperature. We imagine that the sun is suddenly turned on and then use the basic laws of physics to describe the development of the wind and temperature distributions. These laws may be expressed mathematically, and the resulting equations solved on fast electronic computers. Such computer models of the atmosphere (see section 9.3 for a further discussion of numerical models) correctly describe many

FIGURE 11.10 *Mean vertical circula-tion cells in the general circulation.*

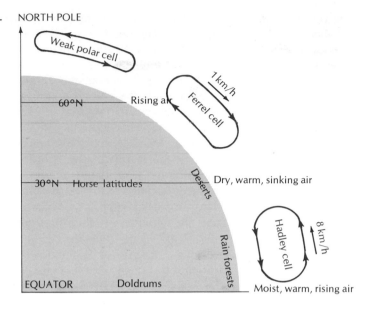

NORTH POLE

of the observed features of the real atmosphere. In particular, mathematical models clearly show the tendency for the atmosphere to concentrate winds into narrow jets and produce large temperature changes over short distances. They not only explain the development of fronts and jet streams, but also correctly describe the tendency for air near the ground to move from the tropics toward the equator, rise at the equator, move poleward above the ground, and sink at about latitude 30 degrees north or south.* This mean circulation—averaging about 8 kilometers per hour and, therefore, being much slower than the west-east motions described above—has been known for a long time (since 1735); it is called the *Hadley cell,* after its discoverer, George Hadley (see figure 11.10). This circulation cell, with its rising motion in equatorial regions, explains why there is so much rain in this area. (Rising motion leads to condensation and precipitation). Most of the rainforests of the world are located in the rising branch of the Hadley cell, where surface winds are light and variable. This region is called the *doldrums,* meaning "much rain and light winds." Figure 11.11 shows the global distribution of annual precipitation.

At latitude 30 degrees north or south, the air sinks and warms, and skies are clear most of the time. Again, there is little organized motion near the ground; sailors have often been becalmed in this area. Because of the mean down-ward motion, there is also almost no rain, and most of the world's deserts are found here. We call the latitude band near 30 degrees north or south the *horse latitudes,* because sailors sometimes threw overboard (or perhaps ate) horses they could not feed. Between the horse latitudes and the doldrums, the air near the

---

*This simplified circulation applies only over the oceans. Over the continents, mon-soon circulations dominate (see section 11.2.2)

ground flows toward the equator where the surface pressure is low. The Coriolis force deflects the wind to the right in the Northern Hemisphere and to the left in the Southern Hemisphere (figure 11.12). In both hemispheres, therefore, the wind blows toward the west, with a strong component toward the equator. These winds (northeast in the Northern Hemisphere and southeast in the Southern Hemisphere) are very steady and reliable and are called *trade winds*. Sailors made good use of the trades to reach America from the Old World.

The Hadley cell emerges clearly from wind observations. But observations and general circulation models also suggest two other, much weaker cells (figure 11.10). These are difficult to detect from observations because the small average north-south motions (less than 1 km/h) are masked by much larger motions associated with circulations around traveling high- and low-pressure centers. The more important of these two weak cells is the *Ferrel cell* (named after William Ferrel), where winds near the ground tend to blow from west to east. Thus, in

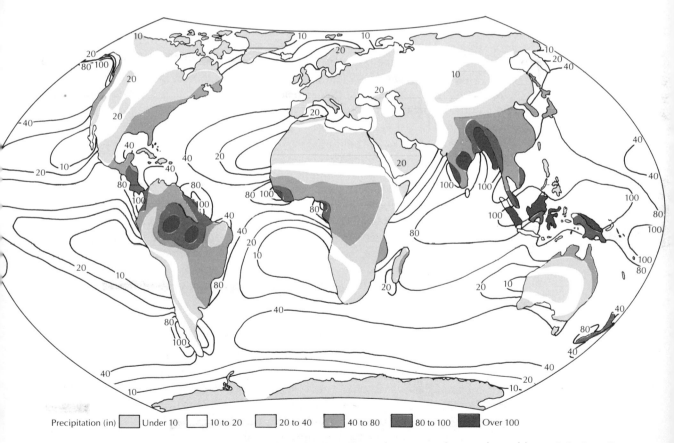

FIGURE 11.11 *General pattern of annual world precipitation (in inches).*

FIGURE 11.12 *Mean surface winds in the general circulation.*

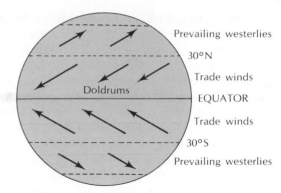

Prevailing westerlies
30°N
Trade winds
EQUATOR
Trade winds
30°S
Prevailing westerlies
Doldrums

regions between roughly 30°N and 60°N (latitudes of Jacksonville, Florida, and Anchorage, Alaska, respectively), most of the surface winds have westerly components; winds from the southwest, west, and northwest are much more common than easterly winds. However, these westerly winds are not as reliable as the trade winds in the tropics and are frequently disrupted by moving storm systems.

The general storminess between latitudes 30°N and 60°N is a consequence of the strong north-south temperature difference in these latitudes. This temperature difference is usually concentrated along a relatively narrow zone which marks the surface boundary between warm air flowing northward and cold air flowing southward. This zone of converging air currents and strong horizontal temperature contrasts is the *polar front.* As we have seen, the fastest large-scale wind currents on earth occur aloft over the polar front in the polar jet stream. These fast winds and the horizontal temperature differences that accompany them are sources of energy for the cyclones and anticyclones that travel eastward through the middle-latitude belt.

Sharp fronts that separate cold air and warm air and fast and narrow jet streams seem to be fundamental properties of the atmosphere and can be duplicated well by computer models. Satellite pictures often show such fronts because they are associated with narrow regions of clouds and precipitation (figure 3.6). These fronts and jet streams meander north and south and, in an average picture of the atmosphere, are quite indistinct, since averages eliminate the extremes. Because averages smooth out characteristic features of great practical importance, it is inaccurate to consider climate as simply average weather.

11.3.2 WHY THE WEST WINDS?

To understand the reason for the westerly winds that prevail throughout the middle and upper troposphere in middle latitudes, we need to understand the relationship between horizontal pressure differences and horizontal temperature differences. The key to this relationship is that the pressure decreases with height more rapidly in cold air than warm air, as illustrated in figure 11.13. Cold air is denser than warm air because it is more concentrated in the vertical. Thus, if the surface pressures of a warm column and a cold column of air are equal, at any level above the surface the pressure will be higher in the warm column.

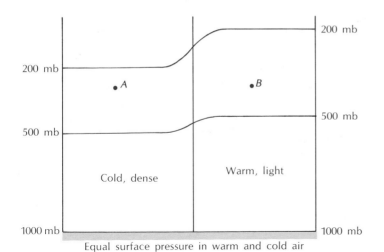

Equal surface pressure in warm and cold air

FIGURE 11.13 *Schematic diagram showing that pressure decreases with height more rapidly in cold air than in warm air. Note that the pressure is higher at point* B *in the warm air than it is at the same level, point* A, *in the cold air.*

In the troposphere, the warmest air is located near the equator; the pressure several kilometers up is high over the equator and decreases toward both poles. It is important to note that this horizontal pressure variation aloft results from the horizontal temperature variation at lower levels. In general, the larger the horizontal temperature difference below, the larger the pressure difference above. Again, this horizontal pressure variation is not large, but is sufficient to produce strong winds.

To see how the large-scale winds develop, let us consider what might have happened after the creation of the atmosphere. The air near the equator would have first become warmer than the air near the poles, and horizontal pressure differences would have developed in response to this temperature difference. If the atmosphere was initially at rest, the pressure gradient aloft would have caused air in the northern tropics to accelerate toward the low pressure at the North Pole, and the air in the southern tropics toward the South Pole (as shown in figure 11.14). However, the rotation of the earth produces a deflection of the air to the *right* in the Northern Hemisphere and to the *left* in the Southern Hemisphere. The force which causes this deflection is called the *Coriolis force* (see section 5.3.3). Therefore, in both hemispheres, the air flowing toward the poles would have been deflected toward the east, and west winds would have prevailed all over the earth, except right around the equator. The strength of these winds is proportional to the magnitude of the pressure gradient, which, as we have seen, depends on the temperature gradient below. Therefore, the larger the temperature gradient, the stronger the west winds aloft. Also, the more the temperature varies horizontally from one place to another, the more the pressure gradient and the wind change with height. This is a general meteorological principle with many applications.

There are many consequences of the general westerly flow aloft in the middle latitudes. Clouds generally move from west to east, as do cyclones and anticyclones. Airplane trips are shorter from west to east than from east to west. With a good tail wind, a flight from Los Angeles to New York takes four hours, but the return trip takes over five hours.

311

FIGURE 11.14 *Schematic diagram showing differential heating, expansion of atmosphere, horizontal accelerations, and geostrophic balance.*

(a) Solar radiation reaching the spherical earth produces differential heating, warming the equatorial regions more than the poles.

(b) Warm air over the equator expands upward, creating higher pressure aloft and lower pressure at the surface.

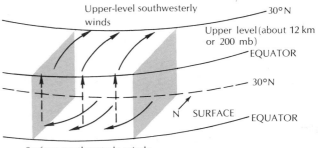

(c) Air rises near the equator and returns poleward aloft. Rotation of the earth deflects poleward-moving air to the right in the Northern Hemisphere (Coriolis effect), producing southwest winds aloft. Air flows toward the equator in low levels. The Coriolis effect deflects air toward the right of motion, producing northeast winds near the surface.

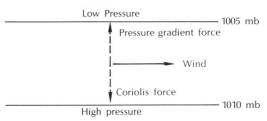

(d) A balanced state of motion is achieved when the Coriolis force exerts a force equal in magnitude but opposite in direction to the pressure gradient force. This balance is called geostrophic balance.

We have seen the fundamental result of west winds existing aloft because the temperature decreases from the tropics toward the poles. This increase in westerly wind speed continues up to the tropopause. However, the horizontal temperature gradient generally reverses in the stratosphere, where cold air is located over the equator (consider figure 3.3 at 150 millibars). Thus, west winds in the stratosphere generally decrease upwards, which means that somewhere near the tropopause the west winds reach a maximum, averaging about 130 km/h (80 mi/h) in winter and 65 km/h (40 mi/h) in summer. These strong west winds are by no means uniformly distributed from tropics to poles. They occur in relatively narrow ribbons of especially strong winds, called *jet streams,* which may be only 60 kilometers wide. In such bands, the west winds may blow over 320 km/h (200 mi/h). These jet streams are often blamed by the media for unusual or bad weather. This is somewhat like blaming the blizzard on the snow and high winds. The jet stream, the general circulation, and the daily weather are so entwined that it is impossible to separate them into cause and effect.

11.3.3 SEASONAL VARIATION OF CLIMATE PATTERNS

The warmest and coldest times of the year lag behind the times of the maximum and minimum radiation, for it takes time to warm and cool the surface of the earth and the air. How much time depends on the properties of the surface. The oceans take more time than land, for reasons given in the discussion of the diurnal variation of temperatures over water (section 11.2.2). Land heats more rapidly in spring and cools more rapidly in autumn than the sea. Thus, continents in the Northern Hemisphere in middle latitudes are warmest in July and coldest in January; the oceans, on the other hand, are coldest in March and warmest in September.

The intensity of the seasons also changes with latitude from the arctic to the tropics. In the tropics, the sun is always high at noon, and the days and nights always last about twelve hours. In high latitudes, the sun is low in the sky in winter or does not shine at all. Therefore, Minnesota has much more severe seasonal contrasts than Texas, even though both have continental climates.

There is little annual variation in temperature at the equator, but much in the arctic, and so we find another important difference between summer and winter. In summer, the north-south temperature difference is small, but in winter it is much larger; compare figures 11.1a and 11.1b. For example, in winter there is a large temperature difference between New York and Miami, which is why so many New Yorkers go to Miami in winter. In summer, however, the temperatures at New York and Miami are about the same, so relatively few Miamians go to New York.

We noted earlier that temperature differences in the horizontal are related to vertical differences in wind speed, known as *vertical wind shear.* Therefore, vertical wind shears are much greater in winter than summer; west winds aloft are about twice as fast in winter. Weather patterns in middle latitudes also move faster from west to east in winter than summer, and airplane schedules have to be adjusted for this seasonal wind change. Strong wind shear also tends to stir up the air and make it turbulent. This turbulence, which is most common and severe in the winter months, shakes planes and passengers and is quite common at the flight level of conventional jets.

A less pronounced, but still important, seasonal change is that most weather patterns, including cyclones and anticyclones, shift southward in winter and northward in summer (as do the sun, birds, and some people). These weather-pattern migrations alter the position of the doldrums and other basic features of the general circulation. For example, the doldrums are about 8 degrees of latitude north of the equator in summer, but are 4 degrees south of the equator in winter. Similarly, the dry belt at about 30 degrees latitude drifts about 5 degrees north of its winter position to its summer position. In California, a belt of relatively high precipitation moves southward in fall and is responsible for the heaviest precipitation of the year in winter.

Up to now, we have been concerned mainly with annual changes near the ground, but it turns out that some huge variations occur in the high atmosphere. For example, at the North pole, 30 kilometers up, the temperature decreases from a "warm" $-35°C$ ($-31°F$) in summer to about $-80°C$ ($-112°F$) in winter. Also, at high levels, most of the world's wind circulation outside the tropics completely reverses in direction from summer to winter; 55 kilometers up, 300-km/h west winds in winter are replaced by 150-km/h east winds in summer.

A most peculiar thing happens in the stratosphere above the equator. Here, the wind direction does not change from summer to winter, but reverses, on the average, every year. Such a year-to-year change is called a quasi-biennial oscillation, that is, an oscillation with a period of about two years. A very complicated theory of this reversal has been given, having to do with atmospheric waves from the ground traveling upward and interfering with the stratospheric flow. Since the discovery of this phenomenon, meteorologists have tried to find such oscillations in weather near the ground. One two-year cycle is a tendency for warm, pleasant summers in northern Europe to occur every other year (good wine years), with cold, rainy summers (and poor grapes) in between.

We saw earlier that seasonal variations are smaller over sea than land. This difference produces great differences in climate between the two hemispheres; the Southern Hemisphere, consisting of 80 percent water, is more maritime than the Northern Hemisphere, which is only 60 percent water. Thus, seasonal variations are generally smaller in the Southern than in the Northern Hemisphere, in spite of the slightly greater contrasts between the amount of radiation received by the Southern Hemisphere in its summer versus its winter season. Only the polar regions are exceptions. The North Pole is covered by water and ice (the Arctic Ocean). In contrast, the South Pole is surrounded by the antarctic continent, an ice-covered landmass of considerable height. Therefore, the coldest ground temperatures observed anywhere in the world (around $-87°C$, or $-125°F$) occur near the South Pole in winter.

314

The climate of North America is primarily continental, owing to the nearly unbroken chains of mountain which are oriented in a general north-south direction near the western boundary. These mountains confine maritime air masses from the Pacific to a narrow strip along the west coast and remove much of the moisture from the Pacific air as the westerly winds lift air over the mountains. Thus the Great Plains of Canada and the United States are colder in winter, warmer in summer, and drier all year than would be the case if the mountains were not present.

The Appalachians in the eastern United States affect the climate of that region in more subtle ways. They are not high enough to block the deep, cold northwest flow in winter, so the climate of the East Coast is more continental than maritime.

Figures 11.15 and 11.16 show the mean daily temperatures for January and July. The tremendous influence of the western mountains in both seasons is evident. In January, temperatures along the West Coast range from the 50s (°F) in southern California to the 40s in British Columbia. Only 50 kilometers inland, mean temperatures decrease by 20 to 40F°. Because of the prevailing westerly flow in the East, the temperatures along the East Coast are not very much higher than inland temperatures.

In July, the temperature gradient along the West Coast reverses, with cooler air on the coast and warmer air inland. Again, there is only a slight coastal effect in the East.

With the rapid increase in the price of energy, home heating costs have become a larger fraction of most people's income. The cost to heat a house for a year is closely related to the number of heating degree days. One heating degree day is given for each degree F that the daily mean temperature departs below the base of 65°F. Thus if the mean temperature on a given day is 45°F, that day contributes 20 heating degree days to the yearly total. Figure 11.17 shows the mean annual number of heating degree days for North America. In general, the isopleths of heating degree days follow the pattern of the January mean isotherms (figure 11.15); the major difference is a somewhat greater north-south variation in heating degree days than January temperatures.

A comparison of the north-south temperature difference from 60°N to 25°N in January and July indicates a much smaller difference in July (about 30F°) than January (90F°). We have seen that this seasonal variation in north-south temperature difference is caused by variations in solar radiation. Because horizontal temperature differences in the lower atmosphere are associated with horizontal pressure differences aloft, which in turn are directly related to the strength of the winds, the westerlies are much faster in winter than in summer over central North America.

The average surface winds over North America show evidence of a monsoon circulation, though not as pronounced as the Asian monsoon. In January (figure 11.18) the mean surface winds show an anticyclonic spiral out of the center

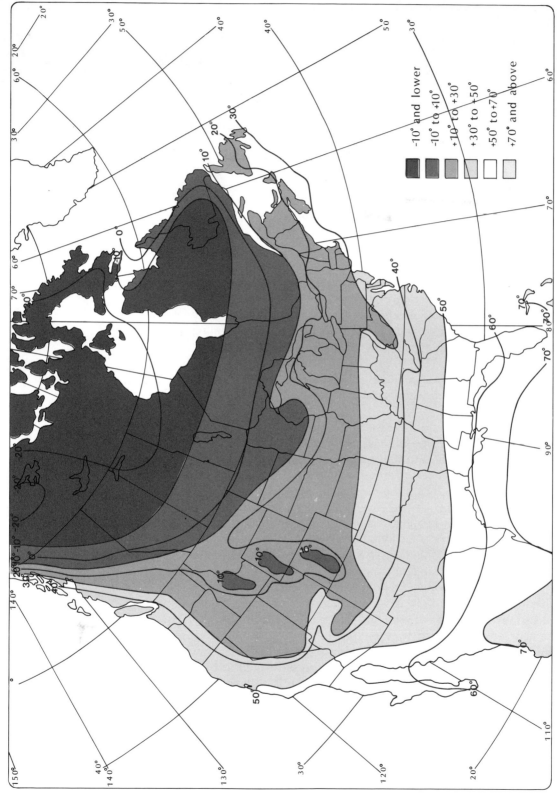

FIGURE 11.15 *Mean daily temperature (°F) in January.*

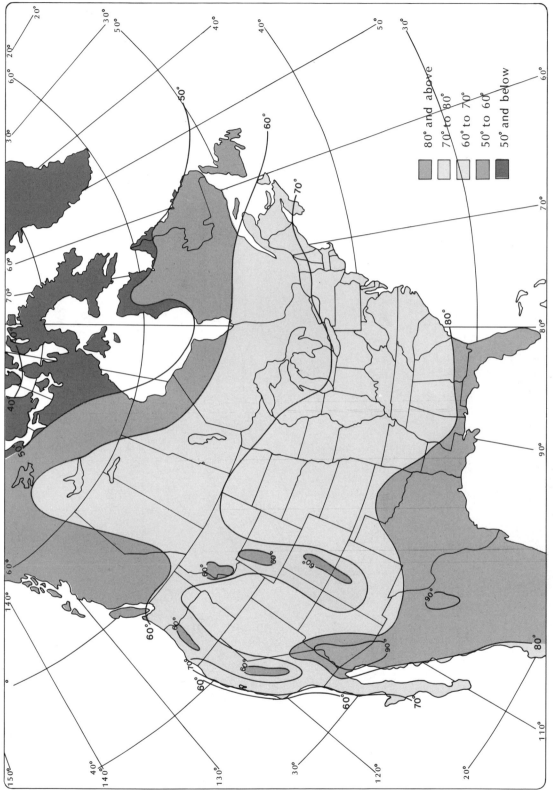

FIGURE 11.16 *Mean daily temperature (°F) in July.*

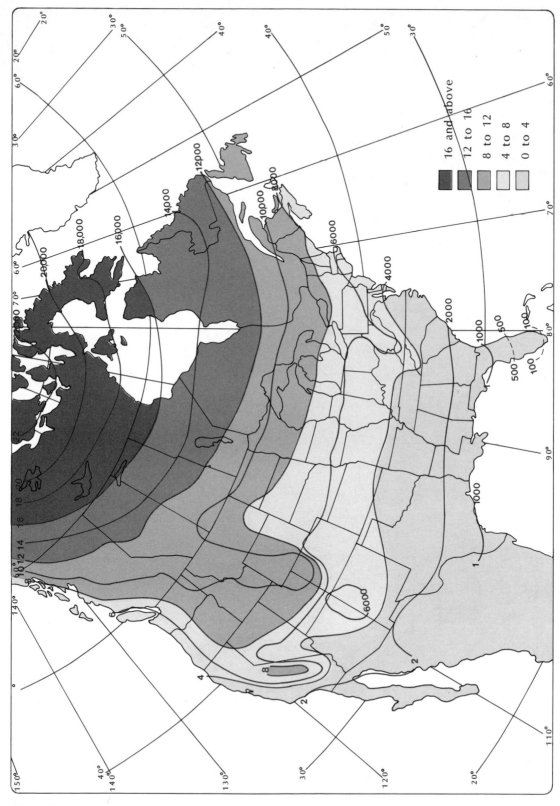

FIGURE 11.17 *Mean annual heating degree days (base, 65°F) in thousands of F°.*

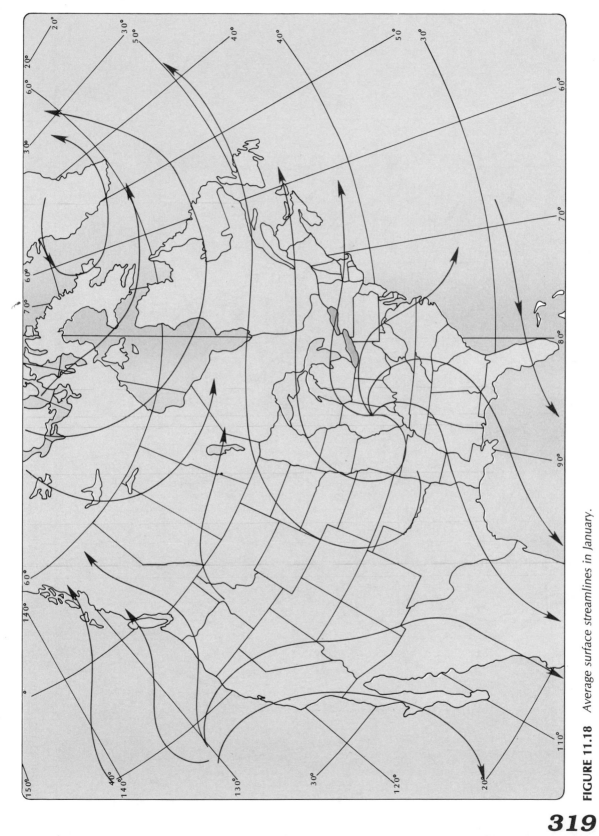

FIGURE 11.18 *Average surface streamlines in January.*

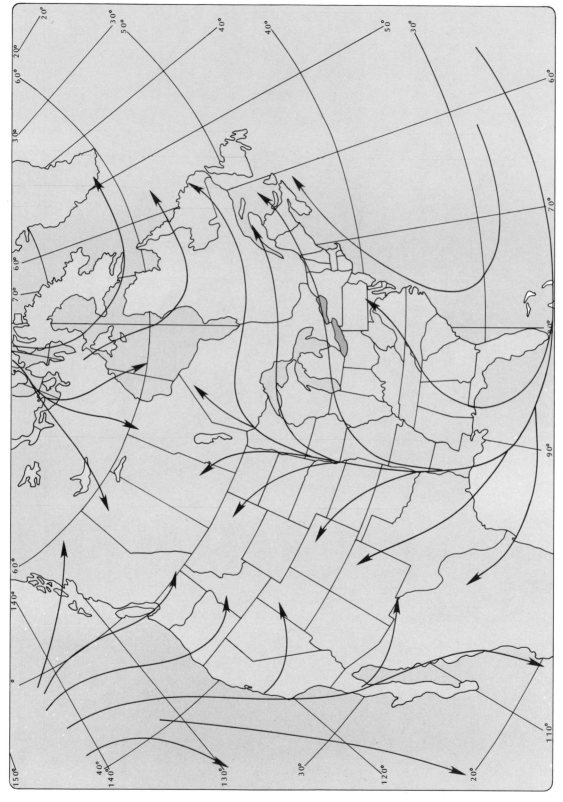

FIGURE 11.19 *Average surface streamlines in July.*

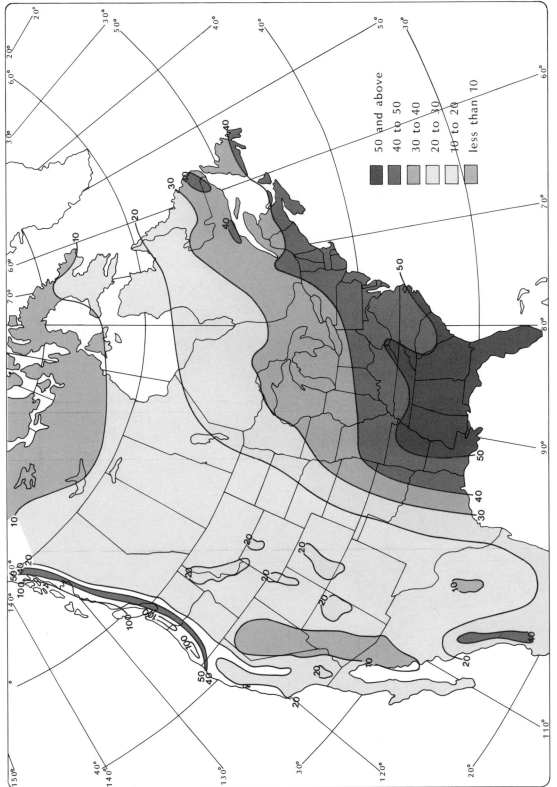

FIGURE 11.20 *Mean annual precipitation (in inches).*

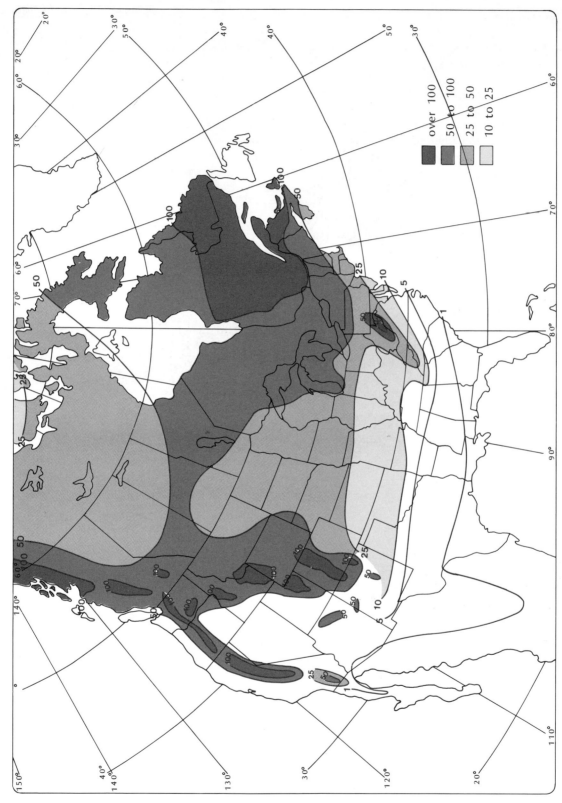

FIGURE 11.21 *Mean annual snowfall (in inches).*

of the United States. Strong northerly flow occurs across the Gulf of Mexico and Mexico. In July, however, this flow reverses with southerly winds flowing into the United States across the Gulf Coast (figure 11.19). Thus, a general convergence of low-level air occurs over the warm continent in July, in contrast to a general divergence away from the cold continent in January.

The mean annual precipitation and snowfall patterns (figures 11.20 and 11.21) show the influence of mountains and the direction of the prevailing flow. The heaviest precipitation occurs over the high terrain along the West Coast where moist Pacific air is lifted over the Cascades and the Sierra Nevada range. East of the Rockies, the large-scale precipitation increases toward the Gulf of Mexico and the Atlantic Ocean. In winter, most of this precipitation comes from large-scale cyclone systems which obtain their moisture from the Gulf and Atlantic. During the summer, cyclones are considerably weaker, and much of the precipitation is associated with thunderstorms triggered by weak disturbances aloft and surface heating over the mountains.

The average snowfall chart (figure 11.21) shows that both the western mountains and the Appalachians produce local maxima which stand out against the general increase of snow toward the north. Mesoscale maxima are located to lee of the Great Lakes (see also figure 17.1). It is noteworthy that, after the effects of the mountains and Great Lakes are removed, a large-scale maximum in snowfall occurs between latitudes 40°N and 60°N, with the greatest amount (over 200 inches) occurring over Newfoundland. This latitude is far enough south to receive precipitation associated with migratory cyclones, and yet far enough north for that

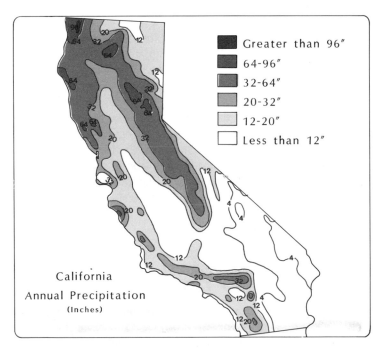

FIGURE 11.22   *Annual precipitation in California (in inches).*

precipitation to be snow. Northward of 60°N the precipitation (and snowfall) decreases, and south of 40°N much of the wintertime precipitation is rain. The maximum snowfall over Newfoundland is a result of the large number of strong cyclones which pass just south of this region.

The temperature and precipitation patterns shown in figures 11.15, 11.16, 11.20, and 11.21 are highly smoothed over mountainous terrain. A more accurate picture of the true variability in precipitation is shown by the average precipitation in California (figure 11.22). Precipitation maxima of over 162 centimeters (64 inches) occur over high terrain of the coastal range and the Sierra Nevadas. In the low interior basins precipitation is much less, in some locations less than 10 centimeters (4 inches) per year.

11.5 CLIMATE CLOSE TO THE GROUND (MICROCLIMATE)

As was already pointed out, local geographic features—lakes, cities, and hills—vary greatly over small distances and cause considerable differences in the local climate from one place to another. For example, on a cold night, there may be a difference in temperature of 10C° between a valley and the adjacent ridge only a few miles away. On a cold night in State College, Pennsylvania, the authors measured −28°C at the bottom of a hill and −18°C at the top, merely two blocks away. Tabulated data at weather stations are not much help in estimating local variations in climate if such differences occur frequently. To interpret properly these local variations in climate, we must understand the small-scale meteorological processes that produce them.

Because most of our activities are carried out in the lowest three meters of the atmosphere, there is special interest in the climate next to the ground. This lowest layer is a rather complicated region which shows tremendous variation from one height to another and from one horizontal location to another. For example, we have all experienced surfaces so hot that we cannot walk on them barefoot—a macadam road on a sunny July day or the sand on a beach. In these cases, the surface temperature may be 60°C (140°F) or more. Yet, merely 2 meters above the surface, where our noses are, the temperature may be only 30°C (86°F). Along with such large temperature contrasts immediately above the ground, we find correspondingly vast variations of wind, moisture, and other variables. We might wonder how such great changes are possible over these short distances. For an explanation, we will start with the temperature.

11.5.1 VERTICAL TEMPERATURE DISTRIBUTION AT LOW LEVELS

The atmosphere is largely transparent to incoming solar radiation. Hence, most of the sunlight heats the ground, not the air. How hot the ground gets depends on how much sunlight is reflected and on how much heat is conducted into the ground. If the ground is dark (a poor reflector) and conducts heat poorly, the temperature can be extremely high. Sand is a poor heat conductor.

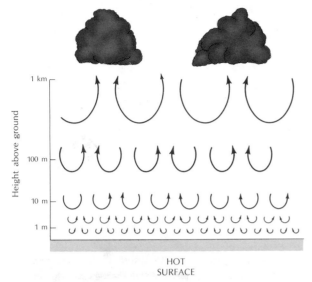

Cumulus clouds form where air is lifted high enough to cool below its dew point.

Large eddies are able to mix heat upward more efficiently.

Eddies increase in size away from ground.

Size of eddies limited by proximity to ground; vertical mixing is inefficient here.

FIGURE 11.23 *Vertical eddies and convection currents.*

mainly because air is trapped among the grains. Air is an especially poor conductor, which is why we use double-paned windows with an air space between to keep heat in buildings.

After the ground gets hot, the air in contact with the ground begins to warm by conduction. However, since air is such a poor conductor, it would take a very long time to heat the air a few meters above the ground if conduction were the only mechanism. Conduction does heat the air very close to the ground, perhaps 0.04 centimeter above it. This heated air expands and becomes so light that the air above the next layer is actually denser. This heavy air begins to sink in certain places, while the warm, light air rises in others. These upward- and downward-moving bubbles of warm and cool air, called *convection currents* (figure 11.23), heat the air higher up by carrying hot air upward in thermals with the air aloft. The convection currents can actually be seen because the associated variations in density cause shimmering of light.

As the ground continues to warm, convection currents rise higher and higher, eventually reaching about one kilometer on a typical day. If there is enough moisture in the air, the rising air becomes visible in the form of cumulus clouds. The updrafts and downdrafts of the thermals are also felt by airplane passengers. Birds and sailplanes use the updrafts to stay aloft without effort.

From our point of view, the function of convection or thermals is to mix the hot air in contact with the ground with the cooler air higher up. The result is that the air at 5, 10, and 100 meters, and eventually at thousands of meters, is being heated. At the same time, if the sun continues to shine, the ground is heated more and more. The question is, What eventually will be the temperature distribution? To answer this, consider again figure 11.23. Thermals close to the ground

325

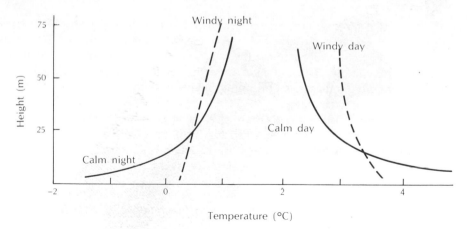

FIGURE 11.24 *Vertical temperature distributions near the ground on calm and windy days and nights.*

are small because they have little vertical space in which to grow. Higher up, away from the constraining ground, the eddies are larger; the larger the eddies, the more efficiently they mix air of different properties. It is because the thermals close to the surface do not mix air efficiently that large vertical temperature gradients can exist. Thus, we get temperature distributions such as those shown in figure 11.24, with huge temperature differences near the ground and smaller differences higher up. A 10C°-temperature difference in the lowest meter is quite a common occurrence on a hot day.

Thermals produced by heating are not the only eddies able to mix air with different properties. Eddies can also be produced mechanically, by rapid changes in wind speed or direction with height (wind shear). We already have met this type of eddy in connection with jet streams at high levels. Such mechanical turbulence also occurs near the ground when the wind blows. At the ground itself, the speed is zero, but just a few meters above the ground, the wind speed can be considerable, which means that the wind shear near the ground is usually strong and will produce eddies. Again, because of space limitations, these eddies are small close to the ground and larger and more efficient higher up. Such mechanical eddies can help the thermals carry the heat upward. Therefore, on windy, sunny days, the air above the ground is heated more rapidly than on calm days. At the same time, the surface will not get quite as hot, since some of the heat is carried aloft. Figure 11.24 compares temperature distribution on calm and windy days.

So far, we have considered air motions near the ground during the daytime. At sunset, the ground cools rapidly, and eventually gets quite cold, as anyone who has slept outside on the ground knows. This is because the ground is an excellent radiator (it radiates heat away easily). It gets especially cold if the air is dry and cloudless. How does this cold ground cool the air at higher levels? No thermals are generated, because the air just in contact with the surface is colder and denser than the air above. If there is enough wind, though, mechanical mixing can cool the air aloft and simultaneously prevent the surface from cooling as strongly, producing the "windy-night" distribution shown in 11.24.

326

If there is not enough wind to cause turbulence, the air above the ground slowly cools by radiation. Some of the radiation from the air is returned to the ground. On clear, dry nights, this returned radiation is relatively small, so the ground cools rapidly. The radiation from the air to the ground increases as the relative humidity increases, so on humid nights the ground cools less than on dry nights. The maximum return of radiation from the nocturnal sky occurs when low clouds are present. These clouds may return nearly as much radiation to the ground as the ground is losing to the air and so prevent any significant cooling. Thus, the coldest mornings (which hold the greatest possibility of frost near the beginning and end of the growing season) are calm, dry, and clear.

11.5.2 VERTICAL DISTRIBUTION OF WIND AT LOW LEVELS

In some respects, the vertical distribution of wind near the ground is similar to that of temperature. There is much more wind shear from the ground to a height of one meter than from one meter to two meters. You can observe this characteristic on a windy day at the beach; by lying down, you will feel much less wind than when standing. A typical wind distribution is shown in figure 11.24. The theory of such a distribution is well understood, and a logarithmic equation fits it well. The reason for this behavior is, again, that eddies near the surface are much smaller than they are higher up; hence, large differences of wind can occur only close to the ground, where the eddies are small and mixing is inefficient. If the surface is rough—if it is the top of a forest or a city—bigger eddies are possible right at the surface, so large gradients in wind speed cannot develop without being destroyed by mixing. Typical wind distributions over rough and smooth terrain are contrasted in figure 11.25.

11.5.3 DIURNAL VARIATIONS IN TEMPERATURE, CLOUDS, AND WIND

Many meteorological variables undergo daily, or *diurnal*, cycles which are caused by the earth's rotation. The earth rotates on an imaginary axis that connects the North and South Poles, and the time from local

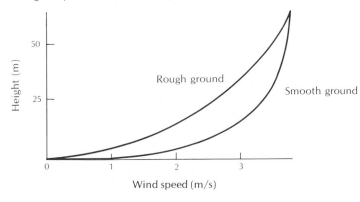

FIGURE 11.25 *Vertical wind distributions near the ground over rough and smooth surfaces.*

327

noon to the next local noon is twenty-four hours on the average. The rotation is counterclockwise, as seen from above the North Pole (figure 11.26.) The result of this rotation is obvious. As point A rotates counterclockwise after noon, the angle between the surface of the earth and the rays of the sun becomes smaller, and the incoming radiation decreases. As everybody knows, sunlight increases in intensity during the morning, reaches its peak at noon, and decreases in the afternoon; however, the temperature reaches its peak after the radiation does. The ground and the air in contact with it warm as long as the incoming solar radiation exceeds the loss from longwave radiation. The difference between the incoming and outgoing radiation at the surface is the net radiation, which remains positive for several hours after the maximum solar radiation.

Thus, the temperature is usually highest two to three hours after noon; it begins falling in the late afternoon, continues through the night, and reaches its lowest point about sunrise. This daily cycle is called the diurnal variation of temperature.

The diurnal variation of temperature is influenced by many factors: latitude, season, cloudiness, wind speed, and proximity of water bodies. Cloudiness, of course, reduces the temperature variation, since clouds block incoming radiation during the day, making the daytime temperatures cool. In contrast, at night, the clouds stop the radiation loss by the ground and reradiate heat earthward, helping to keep nighttime temperatures mild. We might consider clouds as insulators which shelter us from extreme temperatures.

Cloudiness can also change the time of the temperature maximum. Normally, as we have seen, it is warmest between about 2 P.M. and 3 P.M. In Denver, Colorado, however, on typical summer days it is clear in the morning only. By 11 A.M., it becomes quite hot, perhaps 32°C (90°F). The hot ground starts convection currents over the mountains to the west, which form cumulus clouds. After noon, these clouds drift eastward and cover most of the sky, preventing the sunshine from reaching the ground. The air begins to cool, so the temperature reaches its peak about noon.

The variation of cloudiness just described is quite typical; clouds have their characteristic diurnal variations, too. For instance, cumulus clouds are always an indication of strong convection currents, both updrafts and downdrafts, which are usually strongest about midday. Flying through a region of such clouds is likely to be quite bumpy. These strong vertical motions (typically 10 km/h) have

FIGURE 11.26 *Rotation of the earth as seen from above the North Pole.*

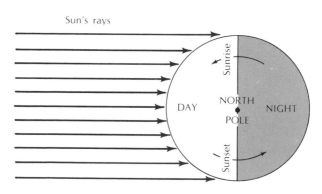

another important effect in that they mix air near the ground with air aloft. Under this vertical mixing, polluted air near the ground is exchanged with the cleaner air aloft, so the pollution near the ground is diluted. Generally, air quality is best in the afternoon at the time of maximum convection.

The vertical convection currents also produce a diurnal variation of wind speed. As we shall see, the horizontal air motions (winds) are normally fast aloft and slow on the ground. If there is strong vertical mixing in the middle of the day, the fast air is carried to the ground, and the wind speed increases. Therefore, we observe fast winds in the daytime and slow winds at night.

All of these diurnal variations are basically caused by the diurnal variation of temperature. Because the magnitude of the diurnal temperature change varies over the earth, some places have stronger diurnal cycles than others. For example, as seen on figure 4.14, points near the poles keep the same low angle of the sun all day and so have no noticeable diurnal variation. In general, the diurnal variation decreases from the equator toward both poles.

Since the diurnal variation of temperature is caused by absorption of radiation near the ground, it decreases with increasing height. Above about 1 kilometer, there is almost no diurnal cycle. Paradoxically, in the stratosphere, the daily temperature variation becomes large again, with perhaps a 10C° difference between maximum and minimum temperatures. The variation at this level is caused by ozone, which exists above an altitude of 25 kilometers and absorbs ultraviolet radiation from the sun during the daytime.

11.5.4 HORIZONTAL VARIATIONS OF TEMPERATURE NEAR THE GROUND

So far, we have emphasized the enormous vertical variations of meteorological variables close to the ground; huge horizontal variations are also possible. For example, we have seen that the surface temperature depends on the characteristics of the surface. A dark, poorly conducting surface becomes very hot, but lighter surfaces remain cool. Different types of surfaces are frequently located very close to each other, with the result that large changes in temperature can occur over very short distances. Large differences can also be produced by even small changes in topography. Cold air at night collects in low spots, which, on calm nights, may be 10°C or more colder than higher elevations. Sometimes, the air cools below the dew point, so that such low spots often fill with fog, which is absent at higher elevations.

Proximity to bodies of water can also produce large horizontal temperature variations near the ground. On cold nights, oceans, lakes, and even ponds retain enough heat to keep the nearby shore areas warmer than inland areas. Conversely, on hot days, the water heats slowly, and the temperatures are frequently 10°C cooler on the beach than inland. This latter effect is more pronounced near large bodies of water in the spring, when water temperatures may be close to 5°C (41°F), while air temperatures away from the water may stand at 20°C (68°F) or higher.

There are many aspects of microclimate which are important to specialists, for example, the special character of the climate of cities. Cities are warm

329

compared to the country, and this temperature difference (recall the importance of differential heating) produces peculiar wind-flow patterns around cities. Because these small-scale circulations play an important role in understanding phenomena such as the dispersion and chemical reactions of air pollution, we will describe them here.

11.5.5 URBAN CLIMATE

Human activities, especially when concentrated in large cities, affect the local and even regional climate in numerous ways. Some of these changes, expressed as percent variations from rural conditions, are summarized in table 11.2. The greatest changes are in air quality, with atmospheric contaminants showing an increase of five to ten times over rural levels. The temperature of the cities is also higher than that of rural areas, so cities are referred to as urban heat islands. The combination of increased contaminants (which provide abundant cloud nuclei) and warmer temperatures tends to increase the cloudiness and precipitation while reducing the incoming solar radiation and the visibility.

The warmth of the city has been recognized for over one hundred years. Figure 11.27 shows the average minimum temperatures in the Washington D.C., metropolitan area for the winter season (December-February) of 1946–1950. The warmest minimum temperatures (31°F) occurred near the Potomac River in the heart of the city. In the suburbs and surrounding countryside of Maryland and Virginia, the *average* minimum temperature was 6F° lower than that in the city. A difference of 6F° in average minimum temperature is very significant, because on many nights (cloudy, windy nights), the differences are much less. In fact, on nights which are clear, dry, and calm, the temperature difference between city and country may be 20F° or more.

Because of the warmer temperatures (heat-island effect), the growing season in cities is three to eight weeks longer than in the country. For example, Figure 11.28 shows the average date of the last freezing temperature for different

TABLE 11.2 *Percentage differences in meteorological parameters in urban area compared with rural area*

	Annual difference (percent	Cold-season difference (percent)	Warm-season difference (percent)
Contaminants	+1000	+2000	+500
Solar radiation	−22	−34	−20
Temperature	+3	+10	+2
Relative humidity	−6	−2	−8
Visibility	−26	−34	−17
Fog	+60	+100	+30
Wind speed	−25	−20	−30
Cloudiness	+8	+5	+10
Rainfall	+11	+13	+11
Snowfall	±10	±10	—
Thunderstorms	+8	+5	+17

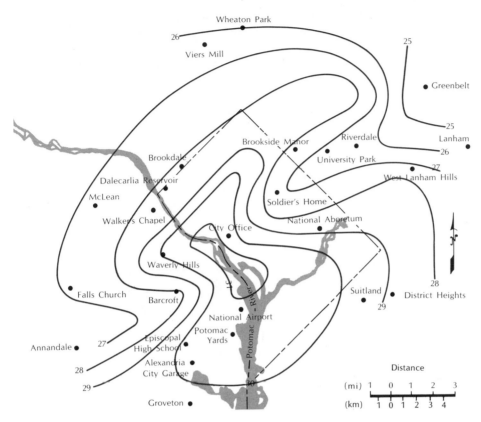

FIGURE 11.27 *Average minimum temperatures (in °F) for the Washington, D.C., area during the winter season.*

sections of the Washington, D.C., area. The last freeze of the winter occurs nearly a month earlier in the center part of the city than in the surrounding countryside.

The differences between urban and rural climates shown in table 11.2 are caused by a variety of factors, including modifications of surface characteristics and emission of waste heat and pollution into the atmosphere. Cities consist of tall buildings made of materials such as brick or stone which generally have higher heat capacities than natural materials found in rural areas. The buildings and roads are able to absorb and store vast amounts of heat, both from natural and human sources. Thus while rural areas cool rapidly after sunset, these materials release heat slowly all night and keep the urban temperature several degrees Celsius higher than in urban areas.

The replacement of soil and vegetation by buildings and roads in cities contributes to more rapid runoff of precipitation, and therefore the air tends to be drier over cities than over the countryside. The reduced evaporation (and associated cooling) is another reason for the urban heat island. Warmer air and surface temperatures in cities also lead to less snowfall and less accumulation of snow.

331

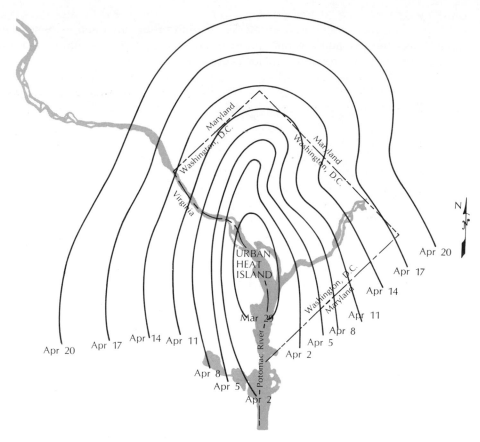

FIGURE 11.28 *Average dates of the last freezing temperatures during spring for the Washington, D.C., area.*

Pollutants emitted by automobiles, heating systems and industry in the cities contribute to much lower air quality, including visibility, in urban areas. Some of the particulates may serve as cloud nuclei and increase the amount of fog or even clouds and precipitation in and downwind of cities. Modification to clouds and precipitation have been observed far downwind of major sources of pollution such as fires and paper mills.

One of the more famous examples of apparent modification of precipitation downwind of an urban/industrial area is the increase in precipitation by 30 to 40 percent since 1925 in La Porte, Indiana. La Porte is located downwind (east) of major mills and industries of South Chicago. Precipitation increases in La Porte have closely followed the upward trend in iron and steel production in Chicago and Gary, Indiana. There is strong circumstantial evidence that these increases were caused by excess particles as well as waste heat and moisture emitted by industries upstream.

A major study of the climate of St. Louis, *METROMEX*, indicated urban effects produced significant increases in rain, thunderstorms and hail in and downwind of this city. Warm polluted air over the city apparently aids the development of convective storms. The relative warmth of the city often leads to rising motion over the city and a gentle inflow of air into the city. This lifting and low-level convergence is probably a major cause of enhanced precipitation over cities.

An advantage of the urban heat island is the decreased frequency of surfaced-based inversions. Thus vertical mixing and dispersion of pollutants is enchanced compared to surrounding rural areas.

Questions

1   List at least five factors that affect the climate.

2   Why does the Northern Hemisphere show a greater annual range of temperature than the Southern Hemisphere?

3   Give four reasons why lakes are slower to warm than land surfaces.

4   Why are Santa Ana winds so warm?

5   Explain why middle-level flow often shows a ridge over mountains and a trough to the lee of the mountains.

6   Why would it be dangerous to estimate the date of the first freeze in the mountainous states from figure 11.9?

7   Which two of the following surface parameters are most important in affecting regional climatology: albedo, water availability, heat capacity of soil, roughness of terrain.

8   Why are cold temperatures in the lower atmosphere associated with low heights of constant pressure surfaces aloft?

9   Why is the temperature near the ground higher on a windy night than on a calm night, other factors being the same?

10   Describe ways in which human activities in urban areas affect relative humidity, temperature, precipitation, and visibility.

Problems

11   Suppose the obliquity of the ecliptic decreased from $23\frac{1}{2}°$ to $10°$. How would the seasons be affected? When would the winter solstice occur? How would the length of a June day in Chicago be changed? A December day? A March day?

12   Sketch a typical graph of net radiation over a 24-hour period. Assume a maximum of 300 W/m^2 and a minimum of -50 W/m^2.

TWELVE

Climate Changes and the Changing Climate

climate change • volcanoes • year without a summer • glaciers • $\delta 0^{18}$ ratio • obliquity of the ecliptic • precession of the equinox • eccentricity of orbit • climatic optimum • tree rings • Sahel desert • Little Ice Age • sunspots • autovariation • carbon dioxide • particles • albedo • deforestation • climatic transition

12.1 *INTRODUCTION*

In beginning a discussion of climate change, a topic top-heavy with theory and speculation, it is useful to note two uncontested facts: (1) The climate has always been changing over periods ranging from tens of millions of years; (2) For most of the food-producing regions of the world, the climate at present is better than it has been over 92 percent of the last million years. These two facts say nothing about what has caused the climate to fluctuate, nor do they predict what will happen in the future. They are sobering, however, when we reflect upon the unprecedented expansion of the human population over the earth (figure 12.1) and upon the implications of a possible return to "normal" climatic patterns of the harsh past.

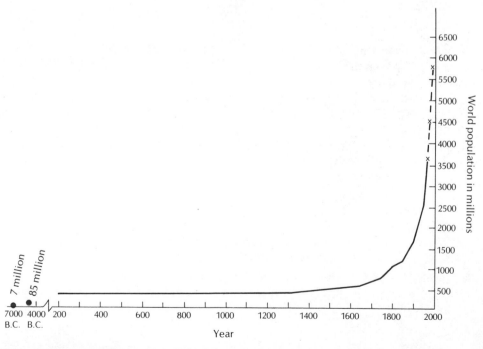

FIGURE 12.1 *World population from 7000 B.C. to the present.*

Actual time scale (time before present)	Compressed time scale
4500 million years (age of earth)	86 years
1 million years (Pleitocene epoch)	1 week
100,000 years	17 hours
10,000 years (civilization began)	1.7 hours
1000 years	10 minutes
100 years	1 minute
10 years	6 seconds
1 year	0.6 second
1 season	0.15 second

TABLE 12.1 *Scaled geologic times*

In discussing climate change over the age of the earth (approximately 4500 million years), most minds boggle at the tremendous time scales involved. Because the climatic record of more than a million years ago is very fuzzy, most of our attention will be focussed on the past million years. Let us compress this time scale into a week to express the relative magnitude of the various cycles in familiar terms. Accordingly, we let one million years equal one week and obtain the compressed time scale shown in table 12.1.

From the data given in table 12.1, we see why we must guard against interpreting a bitterly cold winter or a stifling hot summer as an indication of climate change. One anomalous season is only a flicker against the time scale of a million years. If we were interested in measuring the variation of weather over a million years, we would need extremely fast-response intruments to record the "turbulence" of one season.

In answer to the question "Are we well-off?" one may be tempted to say, "Compared to what?" Such relativity applies well when we evaluate our present climate against the climates of the past 1000 million years (approximately one-fourth of the earth's age). Compared to the average climate during this 1000 million years, we are presently not well off. In fact, we are suffering through the fourth ice age of this period. During the long warm periods that separated the four glacial epochs, global temperatures averaged about 8C° warmer than at present. Dinosaurs ruled the land during the previous warm period (about 70–200 million years ago), the time in which our present coal and oil supplies were produced.

The warm spell did not last forever, and about 2 million years ago, the earth entered its present ice age, which is known as the Pleistocene epoch. Temperatures during the last million years have fluctuated over a range of about 8C°. At present, we are in one of the warmest periods of the Pleistocene. Therefore, considering the climate of the past million years, we are doing very well indeed.

The changes in climate over the last million years are often summarized by changes in the mean surface air temperature over the earth, and, indeed, this measure of the global climate amply illustrates our favored position in time. Figure 12.2 shows the estimated variation of mean global temperature over several time scales.*

---

*We will discuss some of the methods by which these estimates were made in section 12.2.

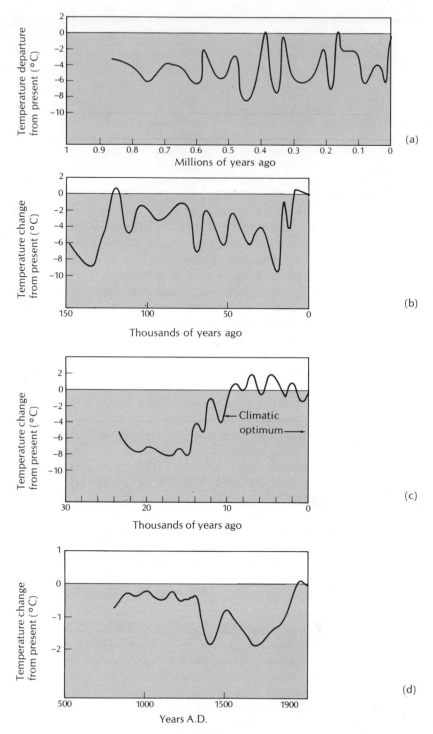

FIGURE 12.2 *Variation in temperature (in C°) over the past (a) 1 million years, (b) 150,000 years, (c) 25,000 years, (d) 1200 years.*

338

If we accept the temperature records shown in figure 12.2, it is evident that only a small fraction of the past million years has had temperatures as warm as those we are experiencing now. However, noting the small magnitude of the changes (approximately 2C° over the past 10,000 years), a skeptic might ask, "So what?" After all, the variation in daily temperatures from the "average" are often 10–15C° in the middle latitudes, and monthly departures from normal are frequently 5C°. Is a 2C° change in mean temperature really important?

The answer depends in part on the location and the climate of that location. On a tropical island, a drop of the mean annual temperature by 2C° might not be damaging at all. In fact, it may even represent an improvement. A decrease of this magnitude would lessen the need for air conditioners and perhaps permit the raising of crops that cannot stand extreme heat.

However, consider the effect of a mean temperature drop of 2C° on a continental location in middle latitudes. Suppose that the "normal" monthly minimum temperature behaved as shown in figure 12.3. The average minimum is given by the solid curve. The dotted curve shows the minimum temperatures that likely would be experienced 20 percent of the time, or once every fifth year. For example, on June 15 and September 15, the minimum temperature would likely fall below 0°C once every five years. Now, if a typical crop requires 120 frost-free days from sprouting to maturity, the "normal" climate of this location would be marginal for the growth of this crop. Farmers would have to accept the fact that they would harvest a crop only four out of five years.

In contrast to the tropical example, a drop of 2C° would alter the agriculture possibilities of this continental station drastically if the temperature decrease were spread uniformly over the year and affected the maximum and minimum temperatures equally. As shown in figure 12.3, the period of frost-free days on four out of five years has decreased from 120 to 90 days—a season too short for many crops.

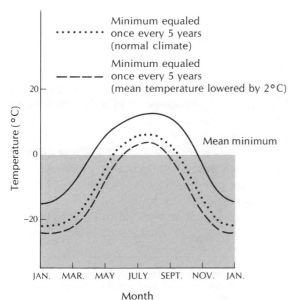

FIGURE 12.3 *Mean monthly minimum temperatures (in °C) at a station in a marginal agricultural area.*

The above example illustrates the importance of small temperature changes to the agriculture of marginal areas. Unfortunately, much of the world's grain is grown in such regions—in the United States, Canada, and Russia. Furthermore, the hypothesized change in the probability of a killing freeze was relatively minor compared to other reasonable hypotheses. For example, months in which the average temperature is 2C° below normal often are composed of many days of near-normal, or even above-normal, temperatures and a few days of much-below-normal temperatures.

The infamous year without a summer, 1816 (also called "eighteen hundred and froze to death"), is a classic example of how simple departures from normal fail to tell the importance of the change. In the spring of 1815, the volcano Tambora, located in the East Indies, exploded in a fury of fire, smoke, and dust. Thrown high into the stable stratosphere where the fallout rate is slow, the enormous amounts of dust spread horizontally and colored skies around the Northern Hemisphere during the fall and winter of 1815. This dust layer reflected enough solar radiation to cause a temperature decrease of several degrees Celsius around the Northern Hemisphere.

The summer of 1816 was remarkable for its frequent incursions of arctic air into the northeastern United States and adjacent Canadian provinces.* Diaries of the period noted June snows and July and August frosts throughout the Northeast. During the four-day period from June 6 through 9, killing frosts occurred as far south as Virginia; ice 2 $^1/_2$ centimeters thick covered water in Vermont, and the green leaves of summer blackened and withered in the cold. Snow flurries fell in Massachusetts and in the Catskills. In July and August, additional outbreaks of cold air killed vegetables in northern New England. On September 27, a widespread killing frost ended what was left of the growing season in the Northeast.

Frost in every month, yes—but a year without a summer, hardly. Temperatures often reached 80°F, and, in fact, "normal" summer weather occurred during the month between the June and July cold waves and between the July and August freezes. During these short interludes, farmers replanted and crops grew normally, only to be struck down again with the next cold wave.

In fact, the temperatures in New England averaged only 2–4C° below normal in June and July and 1–2C° below normal in August. The significant fact is that these departures were caused by a few days with temperatures extremely far below normal in a region where minimum temperatures often approach freezing even in normal years. The effects were a reduction of about 80 percent in the harvest and widespread famine during the next winter.

The preceding examples illustrate the importance of the variability as well as the mean of temperature in marginal areas. The importance of the variation of precipitation is even more significant because much of the agriculture in North America occurs in areas of marginal "normal" rainfall.

Figure 12.4 shows the total precipitation that fell in Columbia, Missouri, during July for the years 1900–1974, together with the three-year running

---

*Although the occurrence of the cold summer the year after the Tambora volcanic eruption is suggestive of a cause-and-effect relationship, none can be proved. Other volcanic eruptions have been followed by normal summers.

Plate 13   Waterspout in the Florida Keys (courtesy of Joseph H. Golden).

Plate 14   Cumulus clouds over a power
station (courtesy of Frank Schiermeier).

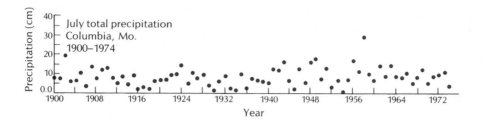

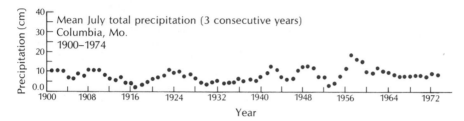

FIGURE 12.4 *Mean July precipitation at Columbia, Missouri.*

mean. In the Midwest, July is a month when adequate rainfall is critical to the growth of corn and other crops. As shown in figure 12.4, the average, or "normal," precipitation is about 10 centimeters. However, many years have had less than half this amount, and even when the amounts are smoothed to give the three-year means, significant variability remains. Other places show much greater year-to-year variability. In general, drier areas have the greatest variability and therefore the greatest likelihood of drought years.

The year without a summer illustrates that climatic changes of enormous importance may occur without large changes in mean temperatures. This is because there is a variety of fairly persistent weather patterns that may occur with the same mean global temperature. The main determinants of whether the weather for a given period at a given location is above or below normal in temperature and precipitation are the position and amplitude of the long-wave troughs and ridges. As we saw in earlier chapters, troughs aloft coincide with cold air, whereas ridges are warm. Furthermore, air to the east of the trough axis is rising on the average while air to the west of the trough axis is sinking (figure 12.5). Therefore, when the long-wave-pattern of the trough and ridge is stationary, regions under the trough axis experience below-normal temperatures, and areas that are under the ridges experience above-normal temperatures. Areas under the southwesterly flow aloft have above-normal precipitation, and areas under the northwesterly flow will have below-normal precipitation.

The magnitudes of the above anomalies depend on the amplitude of the long-wave troughs and ridges. If the amplitude is slight, as illustrated in figure 12.5a, the temperatures will average only a few degrees from normal and the precipitation will depart only slightly from normal. Intense cold and heat will be confined to the far north and tropics, respectively, while middle latitudes will have moderate weather. However, when the amplitude is large, as illustrated in figure 12.5b, temperature departures as large as 10C° may occur at the surface, and there may be flooding ahead of the trough while droughts occur just east of the ridge

FIGURE 12.5 *Typical long-wave positions of troughs and ridges in (a) moderate weather patterns and (b) high-amplitude weather patterns. Solid lines indicate direction of flow; dashed lines indicate temperature departures in C° from normal.*

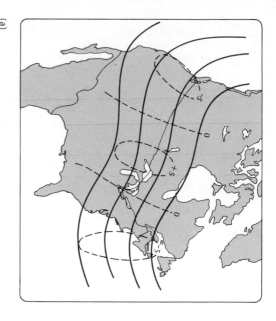

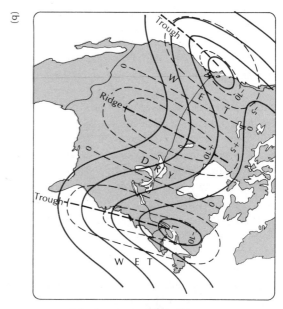

axis. These high-amplitude patterns are responsible for record-breaking extremes in temperature and precipitation in the middle latitudes.

The importance of the position and amplitude of the long-wave troughs and ridges can be seen by comparing the mean 700-millibar-height contour maps for January 1974 and 1977, which are shown in figures 12.6a and 12.6b, respectively. The corresponding departures from normal in the temperatures are shown in figures 12.7 and 12.8.

In January, 1974, the mean trough position was over western North America while a weak ridge persisted off the west coast. The air over the East

TABLE 12.2   *Record and near-record monthly mean temperatures observed during January, 1974*

| Station | Temperature | | | | Anomaly |
	°C	°F	C°	F°	Remarks
Key West, Fla.	25.1	76.8	+3.4	+ 6.1	Warmest Jan. on record
Miami Beach, Fla.	23.6	74.2	+3.2	+ 5.7	Warmest Jan. on record
West Palm Beach, Fla.	23.1	73.2	+4.3	+ 7.7	Warmest Jan. on record
Lakeland, Fla.	21.8	71.0	+5.7	+10.2	2nd warmest Jan. since 1915, 1st Jan. with no heating degree-days
Fort Myers, Fla.	23.0	73.0	+5.3	+ 9.5	2nd warmest Jan. on record
Tampa, Fla.	21.9	71.1	+5.9	+10.7	2nd warmest Jan. on record
Pensacola, Fla.	24.5	65.7	+7.6	+13.6	2nd warmest Jan. on record
Jacksonville, Fla.	19.4	66.7	+6.7	+12.1	3rd warmest Jan. on record
Tallahassee, Fla.	19.4	66.6	+7.8	+14.0	Warmest Jan. on record
Mobile, Ala.	18.0	64.0	+7.1	+12.8	2nd warmest Jan. on record
Columbus Ga.	15.5	59.6	+7.1	+12.7	2nd warmest Jan. on record
Augusta, Ga.	13.9	56.8	+6.1	+11.0	Warmest Jan. on record
Charleston, S.C.	16.7	61.8	+7.3	+13.2	Warmest Jan. on record
Columbia, S.C.	15.2	59.2	+7.7	+13.8	Warmest Jan. on record
Wilmington, S.C.	15.0	58.8	+6.9	+12.4	Equaled 2nd warmest Jan.
Greensboro, N.C.	7.7	45.8	+3.9	+ 7.1	Warmest Jan. since 1950
Charlotte, N.C.	10.0	49.8	+4.3	+ 7.7	Warmest Jan. since 1950
Bristol, Tenn.	8.4	47.0	+5.9	+10.6	Warmest Jan. since 1950
Norfolk, Va.	9.3	48.6	+4.5	+ 8.1	Warmest Jan. since 1950
Richmond, Va.	7.7	45.8	+4.6	+ 8.3	Warmest Jan. since 1950, also 4th warmest Jan. on record
Beckley, W. Va.	5.3	41.4	+5.6	+10.0	Warmest Jan. since 1950
Cleveland, Ohio	0.0	32.0	+2.8	+ 5.1	Warmest Jan. since 1953
Cairo, Ill.	3.4	38.0	+0.9	+ 1.7	Warmest Jan. since 1937

generally came from the west rather than from the north. The East therefore enjoyed unusually mild weather.*. Table 12.2 lists some of the records that were set. Precipitation was heavy over the Ohio Valley which was under the main storm track ahead of the trough.

If the East got a break in 1974, they paid back more than enough in 1977. People in the East will be able to tell their children and grandchildren for years to come about the severity of the winter of 1976–1977, and without the need to exaggerate. Table 12.3, which is practically a mirror of table 12.2, lists a few of the records that were set. Probably the most remarkable event during this month was snow in Miami, Florida, and over the Bahamas—an unprecedented occurrence.

The mean 700-millibar chart reflects the extremely cold month in the East. In January, 1977, a very high-amplitude, trough-ridge system persisted with the trough in the East and the ridge in the West. Extreme patterns such as this satisfy almost no one—certainly not those in the Northeast, where over half a mil-

---

*And saved money and energy: a typical home in the Northeast saved $30 during January, 1974 because of reduced heating bills.

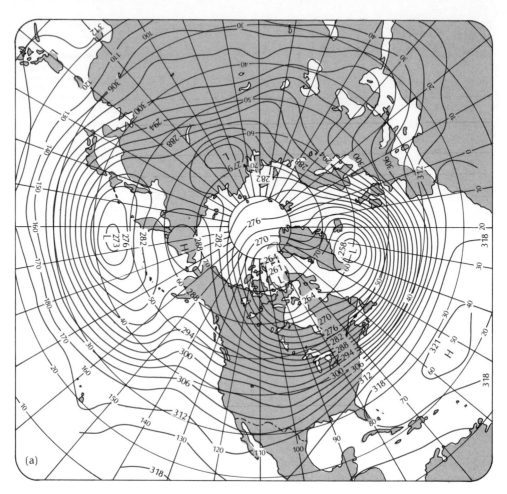

FIGURE 12.6 *Mean 700-millibar height contours (in tens of meters) for (a) January, 1974 and (b) January, 1977.*

lion jobs were lost, some schools were closed for the entire month, and heating bills averaged $50 above normal.* The vacationers who went to Miami and found snow were not overjoyed. In addition, ski operators and skiers in Colorado were frustrated by the cloudless skies day after day. Perhaps most important of all, the West Coast received virtually no rain or snow in a month that normally produces 25 percent of its total annual precipitation.

It is important to emphasize in the above examples that major departures from "normal" climate can occur with little or no change in the mean temperatures or amounts of precipitation over the globe. In figures 12.7 and 12.8, colder-than-normal areas are largely offset by warmer-than-normal conditions in other areas. The average over the United States is quite close to "normal," but the climate during January was drastically abnormal almost everywhere.

---

*With approximately 41 million households east of the Mississippi, the total extra cost for fuel was about $2000 million for this one month.

344

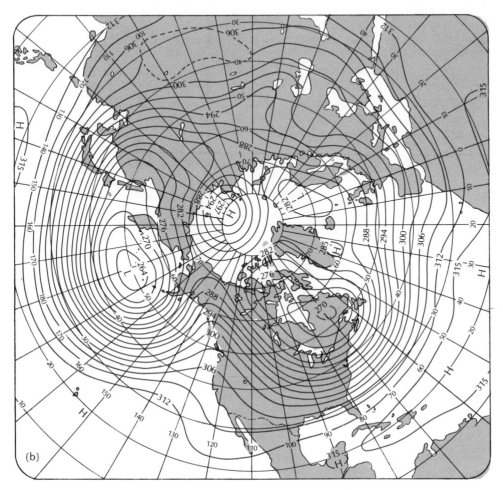

(b)

Another important thing to note from the anomaly patterns associated with the long-wave trough positions is the difficulty in getting representative mean global data from a limited number of observations. Weather stations, for example, are naturally concentrated over continents. If the long-wave, trough-ridge position is such that the troughs lie mainly over the continents and the ridges lie over the water, most stations over the globe will report below-normal temperatures. Unless care is taken, therefore, mean global temperatures may simply reflect the position of the troughs and ridges.

The possibility of incorrectly inferring global temperature trends from local data is not confined to weather-station data, which, after all, are comparatively recent (last 300 years). As we shall see, past climate on longer time scales is inferred from biological and geological evidence. For example, the size of tree rings gives an estimate of the yearly temperature and precipitation. By this method, we know from tree growth records of the bristlecone pines that California experienced a very cold period around 1440–1450. Without additional evidence, this may simply represent a persistent pattern in the long-wave troughs and ridges during this period rather than a global trend.

345

TABLE 12.3 *Record and near-record low monthly mean temperatures observed during January, 1977*

Station	Temperature °C	Temperature °F	Anomaly C°	Anomaly F°	Remarks*
Topeka, Kans.	− 9.4	15.2	− 7.1	−12.8	2nd coldest Jan. (1940)
Shreveport, La	3.0	37.3	− 5.5	− 9.9	2nd coldest Jan. (1940)
Columbia, Mo.	−10.3	13.6	− 8.7	−15.7	Coldest month
Dubuque, Iowa	−16.0	3.4	− 7.9	−14.3	Coldest month
Waterloo, Iowa	−18.0	− 0.1	− 9.1	−16.4	Coldest month
Rochester, Minn.	−19.0	− 1.8	− 8.2	−14.7	2nd coldest Jan. (1912)
Minneapolis, Minn.	−18.1	0.3	− 6.3	−11.9	2nd coldest Jan. (1912)
Madison, Wisc.	−15.9	3.7	− 7.3	−13.1	Tied for 2nd coldest (1912)
Green Bay, Wisc.	−16.2	3.1	− 6.8	−12.3	2nd coldest Jan. (1912)
Milwaukee, Wisc.	−13.3	8.3	− 6.2	−11.1	2nd coldest Jan. (1912)
Rockford, Ill.	−14.6	5.9	− 7.9	−14.3	2nd coldest Jan. (1912)
Moline, Ill.	−13.3	8.2	− 7.4	−13.3	2nd coldest Jan. (1912)
Peoria, Ill.	−13.2	8.5	− 8.5	−15.3	Coldest month
Springfield, Ill.	−12.2	10.3	− 9.1	−16.4	Coldest month
Chicago, Ill. (Midway Airport)	−12.3	10.1	− 7.9	−14.2	
Ft. Wayne, Ind.	−12.8	9.2	− 8.9	−16.1	Coldest month
Evansville, Ind.	−9.6	14.8	−10.0	−17.8	Coldest month
Cincinnati, Ohio	−11.2	12.0	−10.6	−19.1	Coldest month
Columbus, Ohio	−11.5	11.4	− 9.4	−17.0	Coldest month
Akron, Ohio	−11.5	11.4	− 8.3	−14.9	Coldest month
Youngstown, Ohio	−12.2	10.3	− 8.6	−15.4	Coldest month
Cleveland, Ohio	−11.8	11.0	− 8.8	−15.9	Coldest month
Erie, Pa.	−10.9	12.5	− 7.0	−12.6	Coldest month
Pittsburgh, Pa.	−11.5	11.4	− 9.3	−16.7	Coldest month
Avoca, Pa.	− 9.5	15.0	− 6.1	−11.0	Coldest month
Philadelphia, Pa.	− 6.7	20.0	− 6.8	−12.3	Coldest month
New York, N.Y. (Central Park Observatory)	− 5.5	22.1	− 5.6	−10.1	2nd coldest Jan. (1918)
Concord, N.H.	−12.0	10.6	− 5.6	−10.0	Coldest month
Worcester, Mass.	− 8.8	16.2	− 4.1	− 7.4	2nd coldest Jan. (1970)
Trenton, N.J.	− 5.8	21.7	− 5.8	−10.4	2nd coldest Jan. (1918)
Richmond, Va.	− 3.8	25.3	− 6.8	−12.2	
Greensboro, N.C.	− 3.0	26.7	− 6.7	−12.0	
Asheville, N.C.	− 4.0	24.8	− 7.2	−13.1	
Lexington, Ky.	− 8.0	17.8	− 8.4	−15.1	Coldest month
Louisville, Ky.	− 7.5	18.6	− 8.2	−14.7	Coldest month
Knoxville, Tenn.	− 2.7	27.2	− 7.4	−13.4	2nd coldest Jan. (1940)
Nashville, Tenn	− 4.2	24.5	− 7.7	−13.8	Coldest month
Jackson, Miss.	1.8	35.3	− 6.6	−11.8	2nd coldest Jan. (1940)
Columbia, S.C.	2.1	35.8	− 5.3	− 9.6	2nd coldest Jan. (1940)
Athens, Ga.	− 0.4	31.2	− 6.8	−12.2	Coldest month
Atlanta, Ga.	− 1.5	29.3	− 7.3	−13.1	
Rome, Ga.	− 1.7	28.9	− 6.8	−12.2	Coldest month

346

TABLE 12.3 *(continued)*

Station	Temperature		Anomaly		Remarks*
	°C	°F	C°	F°	
Tallahassee, Fla.	6.7	43.9	− 4.8	− 8.7	2nd coldest Jan. (1940)
West Palm Beach, Fla.	14.8	58.5	− 3.9	− 7.0	Coldest month
Lakeland, Fla	11.2	52.0	− 4.9	− 8.8	Coldest month

*All are coldest January on record except as otherwise indicated. When shown, a date indicates the year of the coldest January on record other than the current one.

So, we have seen that persistent anomalies in the westerly winds aloft may cause changes in the temperature and precipitation patterns of enormous magnitudes. Therefore, the question of climate change is more than simply determining the average temperature and precipitation over the earth; it must also consider the physical factors that cause the waves in the westerlies to assume the position and amplitude that they do. It may come as a surprise that only a little is known about these factors, although sea-surface temperature anomalies appear to play a role.

In the next sections, we will examine a number of hypotheses governing climate changes of the past. These may be classified into two basic types: those that explain changes in climate by changes in the external physical processes affecting the earth (such as variation in solar radiation), and those that attribute climate change to internal changes and feedbacks within the earth-atmosphere sys-

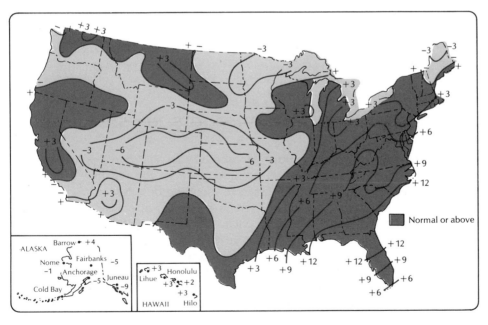

FIGURE 12.7 *Temperature departure (in F°) from normal in January, 1974.*

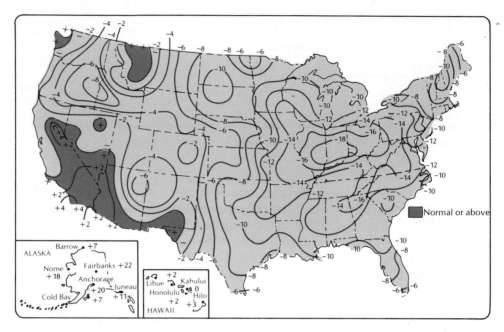

FIGURE 12.8 *Temperature departure (F°) from normal in January, 1977.*

tem. We stress that no one theory can explain all of the changes. One fascinating thing about the study of climate change is that all of the theories may contribute to the observed changes. Another is that there may be theories not yet considered that explain part of the changes. The enormous complexity also leads to a certain degree of frustration because it is unlikely that any one theory can ever be "proven" correct.

12.2 CLIMATIC CLUES IN THE GEOLOGIC RECORD

Although we will not attempt to prove why the climate has changed, it is beyond dispute that it has. It is interesting to consider some of the possible reasons for change and see which among them begin to meet the only test that science demands—that of supporting evidence. There is no dearth of plausible (and sometimes implausible) ideas. For instance, a potato bug in a big potato wrapped in foil and plunged into a hot oven might come to believe in Bug Hades, or he might rightly speculate that his climate was being altered by surface effects, by heat transported along a shiny metal grid, by changing radiation from a distant heat source, and by a progressively warmer environment. He might even note reversals in the heat flow as the oven door is opened before throwing it all in and joining an end-of-the-world vigil.

Like the potato bug, we too do not understand all of the reasons for the changes in our climate. We do, however, have some useful tools for testing the hypotheses that have been advanced to explain previous climatic variations. Geological dating techniques applied to cores drawn from the sea beds, lake bottoms, glaciers, and soils have revealed much about what has happened. They have also revealed a good bit of the sequence of the happenings, and even a modest notion of when some things happened. However, we must always remember one rule: The longer ago an event is believed to have occurred, the less reliable our information about it.

Accordingly, though the earth is well over 4500 million years old, we will not even discuss the climate of most of its life, and even what we say about the last 1000 million years is very sketchy. Only for changes during the most recent million years is it possible to apply any scientific tests, and even then, only in a very tentative way. But whatever the time frame, and whatever the degree of uncertainty, it is stimulating to think about climate change and its potential causes. At the least, we can test our understanding of the physical phenomena of our planet.

THE LAST MILLION YEARS (ONE WEEK)—IN AND OUT OF (MOSTLY IN) THE DEEPFREEZE 12.3

During the most recent 1000 million years, the earth's climate has been substantially warmer than it is now. However, on at least three relatively brief occasions, separated by 300 million years or so, cooling led to substantial glaciation. Evidence of one such occasion, about 600 million years ago, can be found in and on pre-Cambrian rocks in northern Europe, Africa, Australia, and Asia. But inasmuch as the continents move about on the face of the earth, and have done so very substantially since then, we can only say that the glaciation was apparently widespread.

Three hundred million years later, approximately 300 million years ago, glaciation struck again, during the Carboniferous period. This period of glaciation lasted about 50 million years and was followed by a very warm interglacial time which lasted from approximately 250 million years ago to 50 million years ago. After this last interglacial, a cooling trend set in and has culminated in the extensive glaciation of current times, meaning the last few million years.

Why have these dramatic climatic events occurred? One hypothesis holds that a necessary condition for extensive glaciation is a suitable arrangement of the continents as they shift their positions in response to massive convective movements in the earth's interior. The arrangement that appears to be most favorable has substantial amounts of land in middle and high latitudes. We had that arrangement 300 million years ago and we have it now, and cold periods have occurred at these times. From the point of view of what we have learned, the idea makes some sense. Because continental air masses contain less water vapor than air masses exposed to the oceans, almost as a matter of definition, they permit greater losses of surface heat to space during the night, which means nights are relatively cool. When the continents are at middle or high latitudes, the nights are long and daytimes short during the winter half-year, allowing very cold air masses to form, especially near the ground.

However, cold air is not enough; the critical factor, at least in the initial stage of glaciation, is the excess of snowfall over losses to evaporation and melting. Very cold air masses cannot produce copious snowfalls. Remember the old saying, "It's too cold to snow." Bitterly cold air, though easily saturated, simply cannot deliver large volumes of precipitation. As a result, most climatologists look for relatively warm winters and cool summers as the most favorable conditions for glaciation. Heavy snows delivered in any season have their best chance to survive in relatively cool summers.

We should not dismiss the importance of cold, dry air in the maintenance of glaciers, however. Greenland is now practically a desert in terms of new snowfall, but its snowpack survives easily in the cold, dry air which covers it most of the time. The heat balance on a glacier at a moderately high latitude is influenced by two factors that prevent the glacier from vanishing without the necessity of large amounts of new snow. First, in the summer half-year, much of the heat from the sun is reflected away by the white surface. Second, if the air over the glacier is both cold and dry, that heat which is not reflected tends to evaporate the snow and ice rather than melt it. Inasmuch as evaporation uses about 675 calories per gram of ice lost, where melting uses only 80, the same amount of heat will consume about eight times as much ice if it melts the ice rather than evaporates it. This is one reason the dry, warm days are far less damaging to ski resorts than moist, warm days.

An existing glacier may thrive for a long time on cold, dry air, which probably means that a certain minimum amount of continent be at middle or high latitudes. There are reasons to believe that additional factors are needed for glacial periods. Other important influences on glaciation are orogeny (mountain building) and volcanic activity, which could dramatically alter the atmosphere's radiation exchange with space. Both processes have occurred on a grand scale at the surface of this active, evolving planet.

Mountains help ice survive the summer. The temperature decrease that we find in the atmosphere away from the ground is ordained by the fact that the ground is the heat source for the troposphere: The nearer the stove, the warmer the hands. But over mountainous terrain, the effective level of this radiation surface is raised, on the average, by the mountains themselves and ends up somewhere between the deepest valleys and highest peaks. The upshot is that the peaks are above that level and are thus cooler, by an amount that is determined by the ventilation effects of the wind. Therefore, glaciers have an excellent chance to survive on high mountains. In fact, the great glaciations have been associated with extensive mountain glaciers. Whether the mountains played any causative role is unclear. Extensive mountain building has occurred without concurrent glaciation, but when other factors are right, it ought to be useful for a glacier builder to have some high mountains around.

There is rather suggestive recent evidence that volcanoes cause surface cooling by absorbing sunlight in the stratosphere, as though there is a tenuous ground there. Brief periods of volcano-related cooling may have occurred around 1816, 1883, and in 1912 after sensational eruptions. Furthermore, the geologic record amply documents extensive volcanic ash deposits, but the timing is not always right for glaciation. Perhaps vulcanism is another of those factors that can exacerbate a cooling trend over a short period. If the earth is in a period of glaciation, as it is at the present, volcanic eruptions may be associated with glacial advance.

Still another factor that may encourage glaciation is the manner in which the arrangement of continents controls the path of the ocean currents. Ocean currents transport much heat poleward, and impediments to that poleward transport may contribute to drastic changes to climate in high latitudes. It has been argued that relatively warm-water transport into a partially open Arctic Ocean could increase glaciation by increasing snowfall around the arctic rim. It appears that this effect would have to be associated with very long-term climatic changes inasmuch as the arctic has been at least as ice-covered as it is now for the last million years or so. A related suspected long-term cause of climate change was the breakdown of the land bridge from Australia to Antarctica and the subsequent establishment of a current around Antarctica about 30 million years ago. This is thought to have contributed to the development of the antarctic glacier about 10 million years ago by cutting off a heat supply to that high-latitude continent.

In the preceding paragraphs, we have emphasized the major glaciations of 300 million and 600 million years ago as evidence of major changes in the earth's climate, but the potential causes we have discussed, as well as others that are mentioned below, operated at times during the long-ice-free periods as well. They may well have produced extensive climatic changes of much shorter periods, but the geologic record is obscure on the matter of such short-term fluctuations so long ago. For that reason, we concentrate next on the more specific clues in the record of the great period of glaciation that set in "yesterday" during the last few million years, and especially on the last few hundred thousand years of the Pleistocene glaciation which is still with us. However, to understand the sleuthing that rivals any by Inspector Poirot or Columbo, we have to think first about what happens to the waters of the earth when glaciation increases.

At the present time, about 97 percent of the earth's water is stored in the oceans and about 2 percent is interred in the glaciers and icepacks. The amount that is stored as water vapor is trivial. If all of it were condensed out, it would raise the sea level about 3 centimeters. During most of the last 1000 million years, all of the water was stored in oceans, because there were no glaciers. The atmospheric storage was also trivial during these times. We can be sure of this last fact because we know that the atmosphere's storage is limited by its temperature. While temperatures were somewhat higher then than at present, even if we imagine the impossible situation of a totally cloudy, warm, saturated atmosphere, there still would have been only 10 centimeters or so of water stored. By contrast, putting several percent of the earth's water "on ice" would lower sea level well over 100 meters. And putting it back would just as surely raise sea level. The present glaciers have about 70 meters of water locked up.

The formation or melting of extensive continental glaciers lowers or raises sea level dramatically. Material along the old shorelines can be dated by radioactive-decay methods, yielding the dates of the changes in sea level.

In addition, the glaciers themselves have an interesting property, namely that they contain less of the heavy isotope of oxygen, O^{18}, in their meltwater than water from the warmer oceans. Thus, periods of large stands of ice on the

351

continents were marked by relatively large ratios of O^{18} to "regular" oxygen, O^{16}, in the open sea. This ratio decreased during periods of glacial melting as the lighter glacial water mixed with the surface water of the rest of the earth.

The ratio of the two isotopes of oxygen is measured as the departure, called δO^{18}, from a standard, with positive values denoting an excess of the heavy isotope. This value may be calculated from cores taken from sea beds, which contained calcium carbonate ($CaCO_3$) in skeletons of fossil plankton which lived near the sea surface at the time. The $CaCO_3$ in these fossils was formed from the oxygen in the water. Accordingly, the cores present a chronological geologic record of the melting and forming of the glaciers, with the oldest cores buried the deepest. A layer with a large value of δO^{18} would come from a glacial period. There is very encouraging agreement in such core samples, although the time sequencing depends on unsatisfying assumptions about the sedimentation rates that produced the cores.

Independent estimates of sea-surface temperatures have been obtained from sea-bed cores by comparing the types of fossils in these cores with the fauna that live in the sea today. Other estimates of past temperatures can be made from δO^{18} ratios in existing ice cores (which reveal the temperature at which the snow fell), and in coral reefs.

While all of these techniques require assumptions, and even shrewd guesses, most provide independent estimates of the timing and magnitude of the same events. Agreement is surprisingly good among them, especially over the last 150,000 years or so. We turn now to see what they tell us and to speculate why the climate has been changing.

12.4 THOUSANDS OF YEARS AGO—THE LAST FEW HOURS

Having discussed the changes in climate over the past million years (one week on our modified time scale), we now come to the changes during the last 18,000 years. This period corresponds to only the last three hours of our week of weather. In spite of this relatively short time, the climate has not been at all constant. Figure 12.2c shows the variation in middle-latitude temperatures over the last 25,000 years. The temperatures have varied by about 10C° (18F°) over this period. Only 18,000 years ago, extensive regions of the Northern Hemisphere were covered with glaciers and the mean sea level was 140 meters lower than it is at present. The lower sea level made it possible to cross the Bering Strait by foot, and it was around this time (25,000 years ago) that the first people migrated from Asia to the Americas.

Figure 12.9 shows the climate during the Northern Hemisphere summer 18,000 years ago, near the height of the most recent glaciation as deduced by the CLIMAP* project members. The sea-surface temperatures were constructed

---

*CLIMAP means Climate: Long-range Investigation Mapping and Prediction. The CLIMAP project is part of the National Science Foundation's International Decade of Ocean Exploration program.

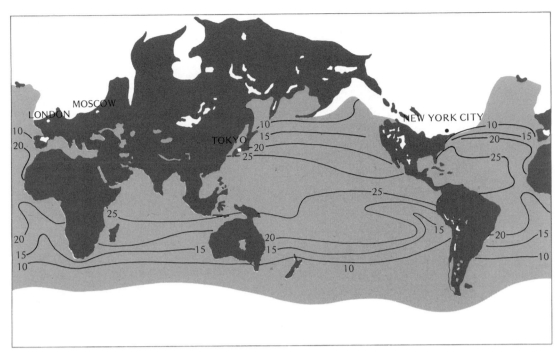

FIGURE 12.9 *Climate during August in the Northern Hemisphere 18,000 years ago. Isotherms are given in °C.*

from geologic data including ocean and lake sediments, ancient soil types, and ice cores. For comparison, the current conditions are shown in figure 12.10.

A comparison of figures 12.9 and 12.10 indicates relatively small differences in oceanic temperatures in the tropics. Both the equatorial Atlantic and the Pacific show temperatures slightly above 25°C. The differences in middle latitudes, however, are dramatic. Ice covers almost all of Canada and most of the Northeastern United States. In Europe, much of the British Isles, Scandanavia, and Northern Russia are glaciated.

Figure 12.11 shows the map of North America for August 18,000 years ago depicting the coverage of ice together with the sea-surface temperature pattern. At that time, what is now Canada was not habitable; the site of the present-day city of Toronto, which now enjoys delightful summers and points the tallest artificial structure in the world some 600 meters skyward, lay under more than 1500 meters of ice. Sea level was about 85 meters lower than at present (note the increased amount of "Florida"), and the weather over the United States must have been spectacularly different only 18,000 years ago, a veritable blink in the lifetime of the earth.

Let us consider what the weather maps might have looked like on a typical August day. To do this we draw on the results of a computer simulation of the ice-age atomsphere.* This simulation used the lower boundary conditions

---

*Simulation from W. L. Gates, "The Numerical Simulation of Ice-Age Climate with a Global General Circulation Model," *Journal of the Atmospheric Sciences,* 33:10 (October, 1976), pp. 1844–73.

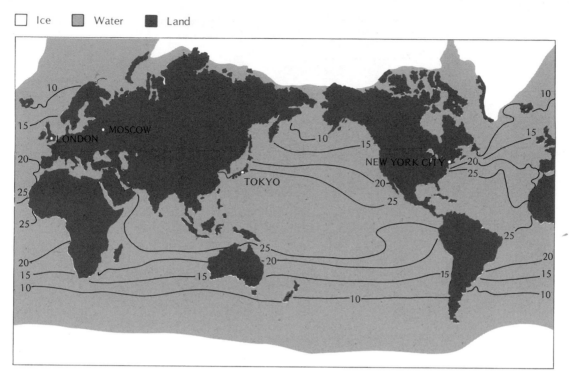

Ice   Water   Land

FIGURE 12.10   *Climate during August in the Northern Hemisphere today. Isotherms are given in °C.*

imposed by the ice coverage and sea temperatures depicted in figure 12.11. (It is worth noting that computer simulations by numerical models are powerful tools for diagnosing climatic behavior. The ultimate goal of the current and very active research effort using computer models is to study the stability and trends of the present climate as well as human impact on it. Important work of this type is being performed at several U.S. government laboratories and elsewhere.) We will add our own meteorological reasoning to check both our own understanding and the model results.

The pattern of packed isotherms in the Pacific in figure 12.11 (see also figure 12.9) trend mainly east-west, suggesting that atmospheric isotherms must have been similar, inasmuch as the ocean temperature exerts a strong influence on the air temperature. Thus, an atmospheric frontal zone would trend across the Pacific near the warm boundary of these isotherms, with a jet stream just on the cold side of the front, blowing more or less parallel to the isotherms. We recall that cyclones form and move along the surface fronts, under the jet stream. Several cyclones of varying intensity are depicted in figure 12.12. The simulation indicates that the atmospheric temperature gradient was not, on the average, a great deal stronger 18,000 years ago but that the temperatures were rather uniformly cooler, and the strong temperature gradient was further south over the Northern Hemisphere. Thus, we would expect the Pacific cyclones to have been about as strong as at present and to have had a favored track in the

354

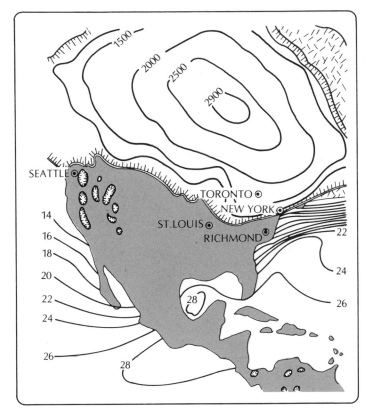

FIGURE 12.11 *Sea-surface temperatures (in °C) and depth of ice (in meters) for August 18,000 years ago.*

monthly mean low-pressure trough of figure 12.13a*. These would have weakened as they encountered the high ground and high sea-level pressures over Canada and the eastern Pacific, but not before delivering rains to the western and southern ends of the ice mass over Canada and Alaska, and snows where the ice was deeper. Some thick fogs must have occurred there as well.

The monthly sea-level pressure chart (figure 12.13a) also shows a pronounced trough of surface low pressure extending from what is now the southern plains of the United States on southeastward to about Georgia; the trough then extends northeastward just off the East Coast to a point near where the Icelandic low-pressure system now exists. This trough must have been the frontal boundary between the cold arctic masses pulsing out of the 1040-millibar highs over central Canada and Greenland and the warm maritime tropical summer air masses over the Gulf of Mexico, Florida, and the southern United States.

The cold ice and its associated cold air dipping southward into Pennsylvania must have produced a vigorous mean upper-level trough near the U.S. East Coast. As a result, the weak cyclones moving southeastward across the United States must have moved into a much more favorable pattern for intensification

*Figure 12.13b shows the mean sea-level pressure map for July under the current climatic conditions for comparison with figure 12.13a. Note the absense of the 1040-millibar high over Canada and the northern United States under the current conditions.

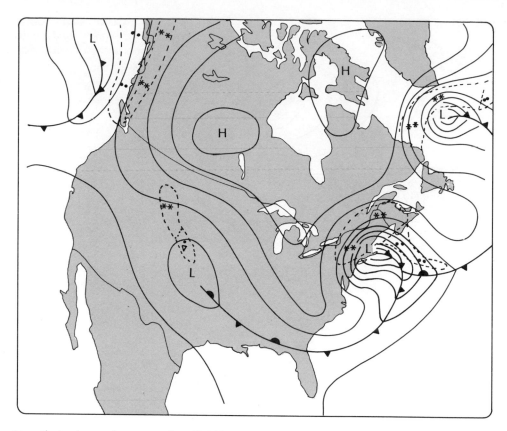

FIGURE 12.12 *Hypothetical weather map for 18,000 years ago showing two summer snowstorms.*

when they passed off the east coast and turned northeastward under the vigorous southwesterly jet stream. This jet stream would have been supported by the strong horizontal temperature gradient between the warm Atlantic air and the cold air over the ice fields. Every cyclone would not have intensified, especially in summer, as the air must have been very stable (in the hydrostatic sense) much of the time with so much cool air near the ground. But in figure 12.12, we depict two in the mean trough position that might have been more vigorous than the summer storms of today. On this day, a terrific summer snowstorm would have added to the ice over New England and the Atlantic provinces of Canada.

How do we know that all this is true? The answer is that we don't know any of the details, and there are some uncertainties in the broad picture. Our vigorous cyclones along the east coast of North America may never have happened; there is just good reason to believe they could have. But the core samples in the ice and the ocean, combined with carbon-14 radioactive dating techniques and the study of fossil pollen in lake sediments, tend to confirm the basic picture

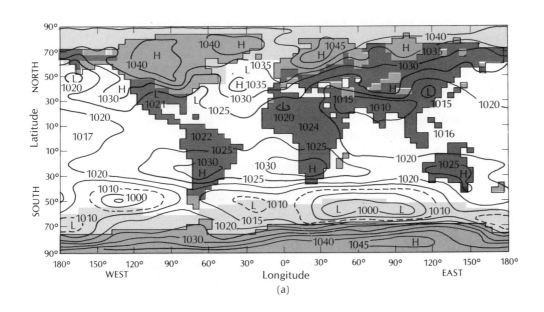

(a)

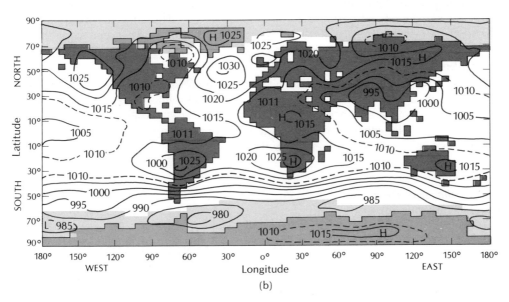

(b)

FIGURE 12.13 *Sea-level pressure (in millibars) for (a) ice-age July (simulated) and (b) July today. The continental outlines are those resolved by the model's grid. The locations of ice sheets and sea ice are indicated by the shading of small and large dots, respectively.*

that we see in figure 12.11. Then the physical laws which govern atmospheric behavior give us valuable help in guessing at some of the details.

Tree ring data, including fossil trees, changes in tree lines, and layering in lake sediments give remarkable details of the warm period that began in earnest 14,000 years before present (B.P.) and peaked at about 6000 years before present. However, this warmup was not steady over all of the earth. One rather spectacular readvance of the glaciers in northern Europe about 11,000 B.P. was very sudden and quite brief as these things go and shows dramatically that the shift from warm to cold or cold to warm need not be steady, nor necessarily very slow. It leads us to ask, Is there any regularity in the glacial advances and recessions at all?

There is evidence in the geologic record of cyclic behavior in the climate. The dominant cycle, or return period, for the last million years or so is about 100,000 years. Other geologic evidence from ancient soils supports that periodicity. Cores from the ice in Greenland and warm-water shoreline features confirm that the last previous severe glaciation before the one that peaked about 18,000 B.P. was centered on 135,000 B.P., a little over 100,000 years earlier. Figure 12.14 shows the estimate of global ice volume taken from a sea-bottom core in the equatorial Pacific.

Both of these recent global ice maxima were followed by dramatic collapses of the ice, as can be seen in figure 12.14. In both of these cases, it is well documented that within a space of 10,000 years or so, the ice went from a maximum to a minimum. Similar events apparently occurred at about 690,000 B.P., 590,000 B.P., 420,000 B.P., and 330,000 B.P. Thus, the ice maxima have tended to occur every 100,000 years, and the minima have come very shortly later, something like 10,000 years. What is more, the warmest periods themselves have been rather short, perhaps 10,000 years long, after which the ice started an irregular buildup that took most of the next 100,000 years to reach the next icy peak. There are those who use this record, and other data, to suggest that the earth faces a new glacial advance within the next few thousand years. *

Notwithstanding our caution about making forecasts on the basis of previous cycles, we are impressed by the evidence that cycles do exist in the sea-

FIGURE 12.14 *Estimate of the global ice volume over the last million years.*

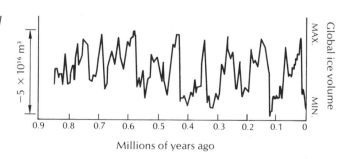

*In these pages, we have tried to introduce the reader to some of the complexities of forecasting tomorrow's weather, so we will refrain from endorsing such long-range forecasts.

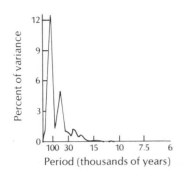

FIGURE 12.15 *Spectrum analysis of deep-sea cores taken from Southern Hemisphere oceans.*

bed fossils, because the length of time between the peaks of the cycles match very well the natural periods of some of the astronomical changes that alter the amount of sunlight reaching the top of the atmosphere, especially at high latitudes and in summer. It is possible to extract, by mathematical methods, the periods of the dominant cycles in the ocean-core variables such as δO^{18}. This technique is called *spectrum analysis*. Figure 12.15 shows a spectrum of δO^{18} for a Southern Hemisphere core. The peaks in the spectra show the dominant cycles in the data; here it can be seen that there is a very strong peak (periodicity) at about 100,000 years, a less strong one at 41,000 years, and still another at 23,000 years.

The 41,000-year cycle in the observed data happens to be very, very close to the frequency of changes in the angle between the earth's axis of rotation and its orbital plane. This angle is called the *obliquity of the ecliptic* and is responsible for the seasons. With all else equal, when the obliquity increases, the seasons are more severe, and when it decreases, the winters are warm and the summers are cool relative to the average. The latter is a condition we are looking for to produce glaciers. The 23,000-year peak is very close to the period of the cycle when the earth makes its closest approach to the sun, which is currently about January 5. For cool summers, we would wish this to occur on the winter solstice for the hemisphere of interest. This variation is called the *precession of the equinox*.

Finally, the 100,000-year peak is close to the period of the cycle when the earth has moved from a more elliptic orbit to a more circular one, that is, to changes in the *eccentricity*.

All of these orbital variations interact to change the sunlight at the top of the atmosphere in very complex ways, and the spectral data suggest that the effects have been felt, to some degree, over the past few hundred thousand years. What is more, while spectral data do not tell anything about when events occur relative to the purported causes, only that they may have similar repetition intervals, other analyses suggest that heavy surface waters, cool sea temperatures, and number of cool-water organisms tend to have maxima at regular intervals after the astronomical controls dictate that the Northern Hemisphere summers should be cool.

Since the above cycles have sound physical bases, it is possible to make tentative predictions from them. Projections of the precession-dominated cycle indicates cooler weather and more ice volume ahead, but not on a time scale that any of us will live to see.

We conclude this section by noting that the astronomically controlled changes in the insolation at the top of the atmosphere may be important in modifying the glaciation, and of these, changes in orbital shape, probably through interactions with the changing date of nearest approach and degree of seasonality, may be the most important. But these variations have been repeating long before the present glaciation. Clearly, other factors, and possibly all factors, play some role in climatic change.

12.5 RESPONSE OF SOCIETIES TO CHANGES IN CLIMATE

After the peak of the glaciation around 16,000 B.C., the glaciers began to retreat. One of their legacies was the Great Lakes, which were scoured out by the shifting ice. During the period between 12,500 B.C. and 4500 B.C., the glaciers continued to melt and the sea level rose to nearly its present value. The warmest period during the past 20,000 years began at the conclusion of this withdrawal, about 8000 B.C., and continued until around 4000 B.C. (figure 12.2c). This period in time has been called the climatic optimum. Since then, the global climate has been gradually cooling off, although irregularly.

One of the most fascinating aspects of studying the climate of the past 10,000 years as compared to the climate before that time is that we can see the role of climate in shaping the development of human civilization. Figure 12.2c shows that there have been short-period oscillations in the mean middle-latitude temperatures for the past 10,000 years, even though the period as a whole has been warm. These relatively minor dips and rises have apparently played a major, if not decisive, role in the history of civilizations. It is probably no coincidence that the development of civilization, which is so dependent upon agriculture for stability, began around 10,000 years ago in the Middle and Near East. The climate during this period became optimal for the growth of wild grains, particularly wheat.

A more detailed picture of the mean temperature variation over the past 5000 years is shown in figure 12.16. Here, the 100-year mean ring widths of

FIGURE 12.16 *Average ring widths (100-year means) of California bristlecone pines from 3500 B.C. to A.D. 1950. Positive departures indicate above-normal growing season temperatures.*

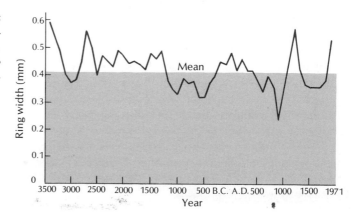

the bristlecone pine tree from the White Mountains of California are plotted. The growth of these trees increases during a warmer-than-normal growing season. The total range in figure 12.16 corresponds to about 2C° (3.6F°). Although we may wonder about the applicability of California data to other parts of the world, there is evidence of a strong correlation between long-term mean temperatures there and other places.

We gave examples earlier in this chapter of the important effect one or two degrees of mean temperature change can have on the agriculture of marginal regions. Furthermore, we saw that the changes in the amplitudes and positions of the long-wave troughs and ridges that cause temperature fluctuations also cause important variations in precipitation.

The regions of civilization during the 3000 years before Christ were generally marginal regions from a climatic point of view. In northern Africa, the climate has always shown varying degrees of aridity. During the wet periods, enough water was available to support a fairly large nomadic population. Caravans were common across the Arabian Desert. For example, around 550 B.C., food supplies were sent by camel from Babylonia to Teima in northwestern Arabia. Today, this route is across a desert that would be impossible to traverse by primitive caravan. And the large lake which bordered Teima then is now gone; the ruins of the formerly prosperous city stand starkly above the ancient depression that once held water.

The above example is not to say that the Arabian Desert was not a desert then, simply that there are varying degrees of desert. Relatively minor changes can cause a great imbalance in the delicate equilibrium that exists between people (or for that matter, animals) and their environment in marginal zones. This equilibrium is worth a brief exploration. Over a long period of time when climate varies significantly, the animal and human populations expand during the "good" times when food is plentiful. In animal or primitive human societies which are not highly developed technologically, an equilibrium is reached relatively soon after an improvement in the climate appears. In this equilibrium state, the further growth of the population is checked by the limitations of the food supply.

When the climate fluctuates toward a less favorable state, as has always happened in the past, the population is suddenly too large for what the land can produce. Animals are faced with two choices—accept a decrease in population through natural processes, such as starvation, or migrate to better climates. When there are no satisfactory places to which to migrate, starvation is the only answer, as was the case with the dinosaurs. Human societies have these same two choices, with the additional ability of reducing their population by birth control or war. Humans also have a third choice—at least sometimes—and that is to develop technology that can increase the capability of the land to support a higher population.

An example of the use of technology to increase the capability of the land is the so-called Green Revolution, which is the increased productivity of agriculture through the use of fertilizers, irrigation, pesticides, etc. Figure 12.17 shows the increase in the corn production in the United States since 1945. Much of this increased output has come at the expense of an increased input. The figure shows that the ratio of energy output to input has actually been decreasing since

FIGURE 12.17 *Energy invested in U.S. corn production* versus *energy returned.*

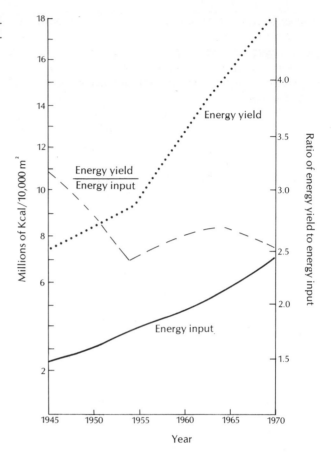

1945; in 1970, it was about 2.6, which means that 1 kilocalorie of energy was required to produce 2.6 kilocalories of corn. Clearly, we are approaching a point of diminishing return. Therefore, the third approach to coping with overpopulation is expensive and limited.

In the hundreds of years before Christ, technology could not keep pace with the climate changes, and the availabile methods of population reduction were undoubtedly quite unpopular. Therefore, the common response to adverse changes in climate was the second choice—migration. Unfortunately, large migrations generally encountered other civilizations, which, naturally enough, resisted the influx of the strangers. There are those who argue that the innocuous bumps and wiggles in figure 12.16 represent climatic changes that were primarily responsible for some of the great migrations of the past (table 12.4). Although a precise timing of the migrations listed in table 12.4 and a correlation with the climate of that time are impossible, there is evidence that the peaks in migrations occurred during periods of climatic stress. We might reconstruct a typical scenario as follows.

After a relatively long (100–200-year) period of more favorable than average climatic conditions, a thriving nomadic civilization exists on the semiarid steppes of central Asia. Although there is not enough rainfall for extensive agricul-

3000–2000 B.C.	The Hellenes migrated from the north into Greece	
2000–1000 B.C.	Aryans entered India from the northwest	
ca. 1195 B.C.	Exodus of Jews from Egypt under Moses; Israelites invaded Palestine	
480 B.C.	Xerxes invaded Greece with 180,000 men from Persia	
330 B.C.	Alexander the Great conquered Persia	
A.D. 630–730	Islam expanded from Arabia into Iraq, Babylonia, Syria, Egypt, Mesopotamia, North Africa, France, Spain	

TABLE 12.4 *Partial list of major migrations of the past*

ture, oases are plentiful, and natural vegetation serves as an abundant food source for animals. Then, the general circulation patterns change, gradually at first. The principal track of rain-producing cyclones shifts northward, and the amount of annual rainfall diminishes slightly. At first, the change is mainly an inconvenience, causing the population to congregate closer to the larger, more stable sources of water.

Because the change in climate occurs gradually, some years have almost the "normal" amount of rain. However, the proportion of "dry" years increases over several decades to the point where the smaller lakes and rivers cannot be replenished by the little rain that falls. Evaporation exceeds rainfall on the average. The large population begins to feel the pressures more and more as they crowd around the remaining water supplies. Overgrazing occurs, and the earth is unable to replace the grass in the overgrazed areas. The political and religious systems become unstable as people seek scapegoats for their harsher lives.

Finally, a series of unusually dry years causes starvation on a large scale. People give up on the land and follow their leaders elsewhere. Families, animals, and personal belongings are gathered together as the entire civilization moves out.

The above scenario was apparently followed by the Arabs around 600 A.D., shortly before the time of Mohammed. Historical records indicate a severe drought in Palestine at this time, which severely disrupted the civilization in this region. The response was typical; the Arabians migrated to the African coast near Tunis. And, with continued unfavorable climatic conditions, the scene was ripe for Mohammed to lead the Islam expansion during the next one hundred years.

Lest we think that the migratory response to adverse conditions is a solution only of antiquity, we need only recall the migrations of people out of the American Dust Bowl to California in the 1930s. Even as late as 1957, when another drought hit Oklahoma, farm families were deserting the land at a rate of 4000 per day. There is little reason to doubt that similar migrations will occur in the future as climates continue to change.

For a recent example of the human response to worsening climatic conditions, we consider the region of the Saharan desert known as the Sahel. The African Sahel stretches from the Sudan to the Atlantic between about 14 and 18

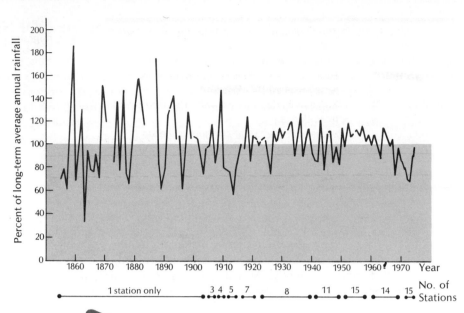

FIGURE 12.18 *Annual rainfall in the Sahel area expressed as a percentage of normal.*

degrees north latitude. The Sahel is a marginal area in terms of rainfall, as indicated by figure 12.18. The variation from year to year is often 50 percent higher or lower than the average. Droughts of two to five years are followed by periods of relatively plentiful rainfall. The five-year period from 1969 through 1973 marked a severe drought, during which the nomadic people lost between 20 and 50 percent of their flocks. Only massive help from the outside limited the deaths to around 100,000 (out of 2 million).

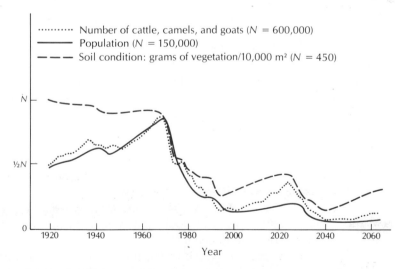

FIGURE 12.19 *Population, number of animals (cattle, camels, and goats), and extent of vegetation in the Sahel area as predicted by a model.*

364

The Sahel experience is summarized in figure 12.19, which shows the soil conditions in terms of amount of vegetation; the number of cattle, camels and goats; and the number of people for a portion of Niger in the Sahel. From 1920 to 1965, conditions were stable, as indicated by the relatively constant amount of vegetation. The rainfall variation during the period was exceptionally small, with the average being above normal (figure 12.18). During this favorable time, the number of animals and people increased rapidly, nearly doubling in forty-five years. Then came the drought in the late 1960s. The combination of a decrease in rainfall and an increase in grazing caused the amount of vegetation to plunge, and with it, the basis for the large population expired. The animal and human populations began to decrease. The continuation of the curves beyond 1975 is based on the assumption that humans will not intervene with improved veterinary services, breeding programs, restocking, and well drilling. The effect of these programs would be to so overtax the region that a total desertification (conversion to desert) would result.

THE LITTLE CLIMATIC OPTIMUM AND THE LITTLE ICE AGE 12.6

In the last 2000 years, two major variations in the climate of the Northern Hemisphere stand out. The first occurred roughly from A.D. 800 to A.D. 1200, which was a relatively warm, tranquil period. It is called the Little Climatic Optimum. During this time, the Vikings settled Iceland and Greenland. There was virtually no ice near Iceland in the summer, and Greenland was really green rather than the present day frozen white. Leif Ericson, in the year 1000, landed somewhere on the North American coast. Grapes were introduced, and wine was produced in England and even in Scotland. Today, grape vines still survive in England,* but the summers are not warm enough to ripen the grapes.

A period of increased storminess began around A.D. 1250 and continued until approximately A.D. 1350. Great storms pounded Europe and the North Sea, flooding large areas in England, Belgium, and Holland. Famines occurred periodically in the 1300s, weakening the population and probably setting the stage for the Black Death which ravaged Europe from 1348 to 1350, killing half the population.

The stormy period signaled the end of the relatively benign climate of northern Europe and wreaked great havoc with cultures that had prospered during the preceding several hundred years. Agriculture in Ireland was particularly hard hit, and the high cultural aspects of the Irish civilization which had flourished in the preceding 500 years (the Golden Age of Ireland) waned dramatically. After the Black Death, the winters became so cold that most of the animals on Iceland were killed. Eventually, the Viking settlements on Greenland and Iceland were abandoned completely (during the fifteenth century) as the climate turned too cold for wheat and ice choked the sailing routes. Greenland itself became covered with

*A famous vine, planted in 1769, still exists in Hampton Court just outside London.

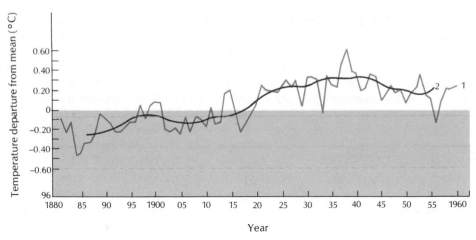

FIGURE 12.20 *Variation of annual temperature (in °C) from 1880 to 1960.*

ice, as it is today. In 1323, sleighs crossed the Baltic Sea between Sweden and Germany.

The period beginning in 1550 and continuing to 1850 is known as the Little Ice Age. Glaciers began to increase again, while famines and sickness plagued Europe. Cyclones stayed far to the south of their present tracks, producing snowy winters and cool, damp summers. Then, for reasons still unknown, the climate began to improve again, and the mean temperature in the middle latitudes rose 1–2 C° toward their highest levels since the climatic optimum around 4000 B.C. The advance of the glaciers was suddenly halted and then reversed. Agriculture prospered in the middle latitudes, and the earth's population grew rapidly. The warming apparently reached its peak in 1940 (figure 12.20). Since then, the mean Northern Hemispheric temperature has been decreasing. However, there is evidence that part of this decrease has been offset by warming in the Southern Hemisphere.

12.7 THEORIES OF CLIMATIC CHANGE ON TIME SCALES OF HUNDREDS OF YEARS

In discussing theories on the changes of climate that occur over periods of hundreds of years, we will concentrate on the

1 natural autovariation of the oceans and the atmosphere;

2 effect of changes in CO_2;

3 effect of particles in the atmosphere;

4 variation in solar radiation (sunspots).

The explanation of climate change by the autovariation of the oceans and the atmosphere does not necessitate any change in the solar output or in the

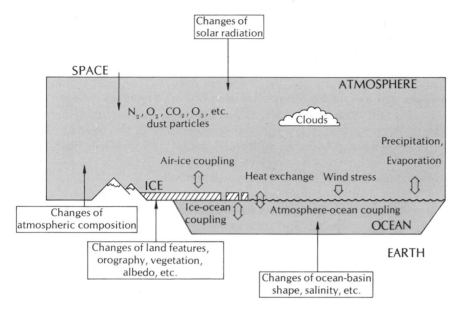

FIGURE 12.21 *Schematic illustration of the components of the coupled earth-ocean–ice-atmosphere systems. The solid arrows are external processes; the open arrows denote internal feedbacks that can cause climate change.*

orbital characteristics of the earth. Crucial to the autovariation theory are the positive feedbacks that can occur with an interacting oceanic and atmospheric system (figure 12.21). Let us explain this theory by giving an example. Suppose that because of several intense volcanic eruptions, the upper atmosphere is filled with dust. The dust blocks some of the solar radiation from reaching the ground, and the surface temperature drops slightly. Several winters in a row are colder than normal, so the polar ice cap begins to grow.

The replacement of a water surface with an ice surface has several effects. First, the solid surface allows snow to accumulate rather than to melt in the open water, thus adding to the accumulation of ice. Second, the albedo of snow-covered ice is much higher than that of water; thus, more solar radiation is reflected back into space and lost to the earth forever. These factors cause additional cooling. The following winter starts early in the arctic because there is less open water to warm the air and offset the loss of radiation to space. The ice pack continues to grow.

The effect of the larger ice mass early in the season is to increase the north-south temperature gradient and enhance the jet stream. Cyclone activity begins earlier and farther to the south than normal. By Thanksgiving, all of Canada and the northern United States lies under a deep snow cover. Air flowing over this snow is modified only slightly in its passage from the arctic to the middle latitudes. The increased cloud cover associated with the frequent cyclones reflects more radiation back to space. With each succeeding cold wave, the ocean temperatures along the coast are lowered. By late January, all of the Great Lakes are frozen, and by Febuary, the Baltic freezes over for the first time in 300 years.

367

Because of the tremendous amount of ice and snow, spring comes late to the Northern Hemisphere. Summer is cool and cloudy because of the frequent cyclones that still exist. The cyclones are more vigorous than during previous summers because of the increased north-south temperature gradient associated with the snow-covered tundra and the still-warm tropics. Thus, less melting than normal occurs, and ice persists all summer in some high mountain valleys.

As this sequence is repeated year after year, the accumulation of ice allows the sea level to drop, increasing the continentality of the Northern Hemisphere. The polar ice cap becomes wider and thicker and thus less likely to be broken up by the action of the waves at its boundary. Cyclones follow a track farther to the south, bringing rainfall to semiarid zones.

After several hundred years of the mini ice age, a number of checks and balances begin to slow down the runaway growth of the ice. First, the cooler temperatures mean that less infrared radiation is lost to space, and the balance of radiation is more easily triggered in favor of an excess of short-wave incoming radiation over outgoing long-wave radiation. Second, the cold interior of the continents relative to the oceans modifies the general circulation so that the troughs and ridges are more highly amplified, with strong northerly winds on the west side of continents and strong southerly winds on the east sides. This stronger north-south flow exchanges heat between the tropics and poles more efficiently.

The modified atmospheric circulation also causes a slow change in the ocean currents. These currents also act to transport more heat poleward. Finally, the cooler temperatures and the increase in the cloud cover in middle and high latitudes reduce the evaporation, and snowfall decreases, especially at high latitudes. Eventually, the ice pack stops growing. In a particularly favorable year, melting exceeds snowfall. The decreased albedo trips the radiation balance toward the positive side, and the higher latitudes begin a gradual warmup, forced by the radiation surplus and the increased northward transports by the atmosphere and ocean. Eventually, the climate returns to its state of 400 years ago, and the cycle is complete.

The above hypothetical example is just one of many reasonable scenarios that could be developed without involving external changes. The difficulty is that the interacting system is very complicated; changing one variable immediately changes other variables, even if only slightly. The change quickly spreads through the system, eventually touching all the variables. If the situation is unstable and positive feedbacks exist, the perturbation may grow and a climate change occur. In a stable situation, the perturbation introduces negative feedbacks and the perturbation dies out.

12.7.1 EFFECT OF ICE ON THE CLIMATE

The growth of the polar ice caps played an important role in the preceding hypothetical sequence of events. Indeed, the presence or absence of ice plays a major role in determining global climate. The high albedo of snow-covered ice (around 0.9) compared to that for water (about 0.03) means that once ice forms, most of the incoming radiation is reflected back to space rather than being absorbed by the water. So powerful is the reflecting ability of ice that

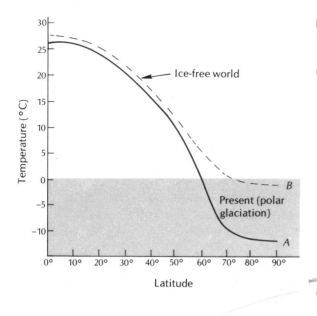

FIGURE 12.22 *Two possible mean temperature distributions (in °C) for an ice-free world and the present state of polar glaciation.*

completely ice-free poles (as has occurred in the past) with today's orbital and solar characteristics are possible. If the ice could be melted at the poles, increased solar absorption could keep them melted. This line of reasoning is the basis for climate modification proposals to close the Bering Strait and pump water out of the Arctic Ocean into the Pacific to permit more warm water to flow in from the Atlantic and melt the ice. Another proposal is to spread soot over the Arctic Ocean ice to decrease the albedo. Figure 12.22 shows the possible variation of temperature with latitude that could occur under conditions of polar glaciation and no glaciation. Both states are theoretically possible with energy balance between the incoming short-wave radiation and the loss of long-wave radiation.

12.7.2 EFFECT OF CARBON DIOXIDE AND PARTICLES IN THE ATMOSPHERE

We will see in chapter 13 that carbon dioxide (CO_2) pollution, caused by the burning of fossil fuels, may cause the earth to warm up by the atmospheric effect. The CO_2 content is approximately 13 percent higher now than in 1850 and is projected to increase to 32 percent by the year 2000. Calculations have estimated a global warming of about 0.3°C per 10 percent change of CO_2, assuming that other factors such as humidity and cloud cover remain constant and that there is no ocean feedback.

Acting to offset the effects of increasing CO_2 is the increase of particulate matter in the atmosphere. Particles less than 5 micrometers in diameter remain suspended in the atmosphere for a long time. Certain particles absorb and reflect solar radiation, thereby reducing the amount reaching the ground. Cooling may result at the surface as a result of these particles. At times, the cooling has been dramatic following volcano eruptions, as in the year without a summer following the eruption of Tambora.

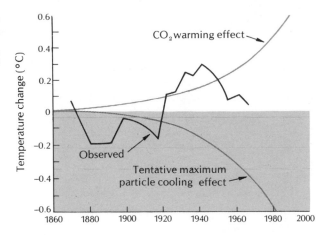

FIGURE 12.23 *Trends of mean temperature (in °C) from 1860 to 1960, with estimated warming effect due to CO_2 and cooling effect due to particles in the atmosphere.*

At present, about 4×10^7 tons of particles exist in the atmosphere; about a quarter of this due to human activities. The present-day amount in the stratosphere exceeds the average amount over the past one hundred years due to volcanoes. However, it is only about 20 percent of the total that existed following the eruption of Krakatoa in 1883.

Figure 12.23 shows the calculated warming effect due to CO_2 and the maximum cooling effect due to particles. The observed temperature changes are given for comparison. Figure 12.23 shows that the calculated CO_2 and particle effects have nearly canceled each other in the past. The observed global temperature trend fluctuates, first decreasing, and then increasing.

Climatologists disagree strongly on the interpretation of figure 12.23 and its implication for the future. Some believe that the observed cooling since 1940 is not caused by particulates but is part of a global cycle that is temporarily offsetting the CO_2 effect. They believe that when the warm portion of the cycle resumes, the rise in global temperature due to the CO_2 increase will continue, warming the earth to the highest temperature in 1000 years. Other scientists say that the particulate effect began to override the CO_2 effect around 1940 and will continue to be dominant, causing another period of extensive glaciation. Finally, other scientists question whether the global mean temperature has fallen at all since 1940. They cite recent evidence from the Southern Hemisphere that suggests that temperature rises there may be offsetting temperature decreases in the Northern Hemisphere. It is possible that particulates, which are greatest in the Northern Hemisphere where industrial activity is concentrated, may be causing a cooling there while the CO_2 effect is causing a warming in the Southern Hemisphere. And so the controversy goes.

12.7.3 CYCLES IN SUNSPOTS

Perhaps no single theory of climate change has been so widely studied as the solar variability theory. The most commonly mentioned solar cycle is the sunspot cycle. Every eleven years or so, there is an increase in the

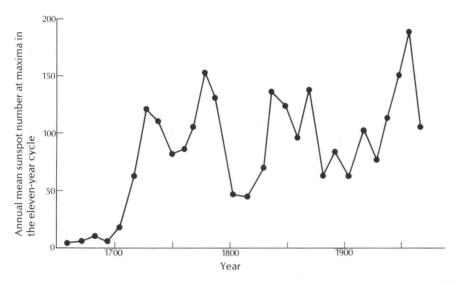

FIGURE 12.24 *Annual mean sunspot numbers at maxima in the 11-year cycle from 1645 to the present.*

storm activity on the surface of the sun. The storms are relatively cool and appear as dark spots on the sun. During the periods of sunspot maxima, radiation in the wavelengths corresponding to X rays increases significantly. However, the percent of total radiation .in these wavelengths is very small, and so the total amount of radiation reaching the earth is not significantly changed during the sunspot cycles.

There is evidence that significant solar variability may occur on time scales of hundreds of years. Between about 1645 and 1700, for example, there was a pronounced minimum of sunspot activity. This minimum, called the Maunder minimum after its discoverer in the 1890s, apparently coincided with a reduction in the amount of radiation reaching the earth. As we have seen, the Little Ice Age occurred at this time. On the other hand, a maximum in sunspot activity occurred between A.D. 1100 and 1250—a time of relatively warm conditions over the Northern Hemisphere. If these two points represent extremes in a cycle, the next warm period will occur around 2700. And, indeed, sunspot activity has been increasing over the past 300 years (figure 12.24).

"Also"

CLIMATE MODIFICATION BY HUMAN ACTIVITIES 12.8

The saying "Everybody talks about it but nobody does anything about it" is no longer a trite truism. It may be as trite as ever, but it is no longer true. By their very numbers, people are significantly changing the weather and the climate, at least on a local scale. The burning of fossil fuels and the resulting pollution by carbon dioxide is very likely contributing to a global warming, and particles are probably causing the opposite effect. The emission of sulphates has lowered the pH of precipitation to as low as 3 in some locations, enough to dissolve marble statues and buildings. And pollutants emitted by supersonic transports and ground sources may threaten the ozone layer. This section considers some additional effects of human activity on the climate.

In general, people affect the weather and climate in three ways:

1. By operating machines that emit gases or particles into the atmosphere, thereby changing the composition and radiation characteristics of the atmosphere;

2. By altering the surface of the earth by deforestation, farming, or by some other activity so that the energy budget at the earth's surface is changed;

3. By simply adding heat to the atmosphere by burning fossil fuels.

TABLE 12.5 *Summary of inadvertent climate changes due to human activity*

Changes	Probable effect(s) (first order)
Changes in composition of atmosphere	
Increased amount of CO_2 by burning fossil fuels	Increased average temperature of 0.3°C for 10 percent increase in CO_2
Increased number of particulates	Decreased average surface temperature; increased local precipitation
Addition of sulphur	Increased acidity of precipitation
Addition of fluorocarbons	Decreased ozone; warmer earth's surface
Addition of H_2O in contrails	Increased number of cirrus clouds; cooler earth's surface
Oxides of nitrogen (NO_x)	Decreased ozone; warmer earth's surface
Modification of surface and subsequent changes in surface energy budget	
Urbanization	Urban heat island; reduction of wind in cities
Deforestation	Increased albedo; decreased evaporation; increased winds
Overgrazing in semiarid regions	Desertification
Thermal pollution	
Urban waste heat	Urban heat island; increased convective precipitation; longer growing season
Cooling tower heat disposal	Increased convective precipitation; increased fog

Plate 15a (above)   Cumulus clouds over an Everglades grass fire (courtesy of Ronald Holle).

Plate 15b (below)   Circumhorizontal arc over Mt. Olympus, Washington (photo by Steven Hodge; © Alistair B. Fraser).

Plate 16   Aurora borealis over Alaska (courtesy of Gustav Lamprecht).

Some of our activities that belong in these categories, along with their probable effects on climate, are listed in table 12.5. The changes due to altering the composition of the atmosphere through various pollutants are discussed in Chapter 13. Therefore, we will discuss here only those climatic changes due to modification of the surface and to the addition of heat.

12.8.1 MODIFICATION OF THE SURFACE ENERGY BUDGET

The short- and long-wave energy balances presented in chapter 4 were global averages. Local balances can be quite different, depending on the characteristics of the surface and the time of year. The important surface parameters are albedo (reflectivity), heat capacity (ability to store heat), heat conductivity, and availability of water. For a given amount of solar radiation, the surface temperatures decrease with increasing albedo. High values of heat capacity and conductivity are associated with lower daytime and higher nighttime temperatures. Greater water availability lowers the temperature by day and raises it at night. Thus, we see why diurnal (and annual) temperature variations are so much less over water than over land—water has a low albedo, high heat capacity and conductivity, and, of course, maximum water availability. Very little of the incoming solar radiation warms the air above water directly; most of it is used to heat and evaporate the water. Therefore, the greatest contrast in the surface energy budget occurs between land and water.

Important contrasts do exist over land, however, Anyone who drives across the semideserts of New Mexico in the summer to visit the White Sands National Monument will notice a cooling as the bright white sand replaces the much darker ground cover surrounding the monument. The high albedo of the white sand reflects much of the light back upward, allowing local cooling.

Important changes in the surface characteristics occur when we try to "remake" natural areas into something else. An example is the sad story of the American expansion into the Great Plains which began in the 1870s. As news spread back East about the unlimited "free" grass for grazing and buffalo for hunting, settlers swarmed into west Texas. In 25 years, the "unlimited" grass was gone, and the cattle were dying by the thousands on an artificial desert. The fragile ecology of the grassland region had been destroyed by overgrazing. A similar tragedy occurred in the western Great Plains where rainfall averages 25–50 centimeters (10–20 inches) a year—not enough to support the animals and agriculture that were thrust upon the region around 1900. When the farmers dug up the prairie grass that had held the soil together for centuries, the dry winds ravaged the exposed soil, setting the stage for the infamous Dust Bowl of the 1930s.

Abuse of semiarid areas through overpopulation and overgrazing can apparently initiate a positive feedback between the changing surface conditions and the increasing aridity (desertification). A carpet of grass does more than simply hold the soil down; it also conserves, and even replenishes, precious moisture. At night, the grass, which is a poor conductor, cools off rapidly and may collect enough dew to sustain the vegetation through dry periods.

A covering of vegetation probably also causes an amount of convective precipitation greater than that which would occur if the ground were bare. Because the albedo of vegetation is lower than that of most dry soils, more solar energy is absorbed at the surface if it is covered with vegetation. This increased absorption of energy at the surface decreases the stability of the atmosphere, making convection more likely.

Once the grass covering in a semiarid zone is severely reduced by human activities, a number of factors may contribute to a further deterioration. The greater percentage of exposed soil reflects more solar energy, and the moisture loss increases. Increased dust mixed upward into the atmosphere may increase the stability by absorbing solar radiation aloft rather than at the surface. Less dew forms at night, further drying the area.

One major desert of the world that may owe its existence, at least in part, to human activities is the Thar Desert, sometimes called the Indian Desert. Covering portions of western India and Pakistan, it receives up to 40 centimeters of rain a year—enough to support light grazing and a sparse population. This area was not always a desert, and some climatologists believe that desertification was a direct result of large numbers of people overtaxing the fragile land. Similar arguments have been applied to the Sahel region of the Sahara Desert in northern Africa.

Deforestation may also alter the surface energy budget enough to cause important changes in the climate, at least locally. Large forests have low albedos, and thus absorb almost all of the radiation that falls upon them. They also retain moisture and greatly reduce the wind speed at the ground. At night, the forest reduces the amount of heat escaping from the ground, thereby contributing to warmer nocturnal temperatures and longer growing seasons. When forests are removed, the albedo increases, surface evaporation increases, and the diurnal and annual temperature ranges at the surface increase. Especially in winter, the albedo is increased greatly when a tall forest is replaced with a snow-covered field. Tall, thickly planted trees, especially conifers, do not remain snow covered for long after a storm and hence absorb considerable amounts of solar energy (over 50 percent of the insolation). In contrast, a snow-covered field may reflect as much as 95 percent of the incoming radiation. This effect is obvious in satellite pictures of the northeastern United States in wintertime; the dark forested Adirondack Mountains stand out clearly against the white background of the surrounding snow-covered fields.

In some areas, deforestation may so modify the local climate that it is nearly impossible for the forest to grow again. In the late 1800s, for example, the forests in some of the low areas of the Nittany Valley in central Pennsylvania were removed to feed hungry iron and steel mills. The loss of the trees caused such a lowering of temperatures on calm, clear nights that tender, young trees are frequently killed by the extreme cold. Temperatures as low as $-40°C$ ($-40°F$) have been recorded in these barren areas while temperatures in nearby forests were ten degrees warmer.

One of the greatest changes in surface characteristics, of course occurs with urbanization. Highways, parking lots, and rooftops replace vegetation.

Most of these artificial surfaces have higher heat capacities and conductivities than vegetation, and they all have much lower water availabilities. The net results of all of these effects is a general warming of the city compared to the countryside, as we saw in the discussion of the urban heat island in section 11.5.5.

12.8.2 HUMAN VOLCANOES

Human activities that burn fossil fuels are supplying the atmosphere with solar energy that was absorbed from the sun millions of years ago. The total amount of heat added by humans in 1970 was about 10×10^{12} watts, which sounds like a lot of heat until it is compared with the amount of solar radiation received at the earth's surface (7.5×10^{16} watts). Thus, the human contribution to the surface heat flux is about 0.0001 times that of the sun, and most scientists agree that it has a negligible effect on global climate at this time. There have been estimates, however, that one hundred years from now, humans might contribute as much as 1 percent of the solar input. This amount would be significant and would probably increase the global temperature by 1C°.

Although anthropogenic heat sources may be insignificant on a global scale, they are certainly important right now on urban scales. On Manhattan Island, New York, for example, the human heat input is over two times that from the sun in wintertime. In many cities, the contribution by humans is greater than one-third the solar input. These sources of heat are the major factor in producing the urban heat island in the wintertime.

Some of the most intense local sources of heat are the cooling towers that dissipate waste heat produced by nuclear power generators. At present, typical power plants dissipate approximately 4×10^9 watts, enough to raise the temperature of the lowest kilometer of the atmosphere over a square 20 kilometers on a side by about 1C° per day if the heat were confined to this volume. Fortunately, the atmosphere does not allow the heat to accumulate this fast; horizontal winds and vertical convection currents disperse the heat away from the source. However, under marginally stable conditions, the extra heat may be sufficient to generate cumulus clouds and even thunderstorms (see plate 14). And future power parks propose to dissipate as much as 40×10^9 watts, ten times that of the present plants. This amount will almost certainly modify the climate locally. Some of the modifications may have beneficial aspects such as a longer growing season, warmer winters, and increased rainfall.

THE CLIMATE OF THE FUTURE—WHERE DO WE GO FROM HERE? 12.9

We are left with the present climate, which has permitted a sudden increase in population because of the extraordinarily good growing conditions in parts of North America and Asia. What can we expect in the future? Because of the many, and often contradictory, theories concerning climate change, we cannot predict with any confidence the likely changes based on the scientific theories. Probably the

FIGURE 12.25 *Probability of a climatic transition analogous to the changes between maxima and minima in climatic fluctuations of periods (interior numbers) in years as a function of elapsed time after present.*

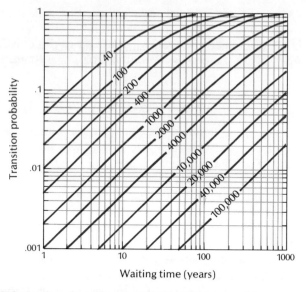

best strategy is to give the probabilities that the climate will return to one of its past states based on the assumption that all of the past changes were purely random events. The statistical likelihoods of returning to some of the past climates are given in figure 12.25.

Figure 12.25 gives the probability (left side of figure) that a transition in climate appropriate to past changes of various periods (given in years) will occur within a certain number of years in the future (the waiting time). An example will clarify the interpretation of the diagram. Suppose we wish to estimate the probability of a change in climate within the next ten years corresponding to the period of the Little Ice Age. Let us assume that the Little Ice Age was associated with climatic fluctuations on a time scale of 400 years. Following the curve labeled "400" in figure 12.25, we see that a change to this type of climate has a probability of only .005 (5 in 1000) of occurring next year. However, the odds increase to .05 (5 in 100) of a change occurring within the next ten years. There is a nearly 50 percent chance that such a change will occur sometime in the next one hundred years.

Actually, a change for the warmer is just as probable when we consider figure 12.25 by itself. However, when we realize that we already live in a relatively optimal climate, any change seems likely to be for the worse. The important points are that (1) important changes in climate are likely to occur within a hundred years and that (2) the rapid increase of population during the past one hundred years has made it harder for humans to adjust in socially acceptable ways to any adverse change in the climate. Technology cannot keep up with the population, there is no place to migrate to, and so a reaction in population appears to be the only answer. We will leave it to the readers to discuss a responsible plan of action.

Questions

1 Why is there so much snow over the Arctic and Greenland when the precipitation there is very light?

2 What mistake do newspaper articles make when they point to several cold winters as being evidence of a new ice age?

3 Discuss possible ways the world will respond to the population explosion.

4 Give at least three methods of inferring past climates.

5 Sketch a mean 700-mb flow pattern that would give California a cool, wet January while giving the central U.S. a warm, dry January.

6 How can temperature extremes occur without a change in mean temperature at a given latitude?

7 Why are warm, dry days far less damaging to snow on ski slopes than warm, moist days?

8 List 5 factors that might have helped produce past ice ages.

9 If a clam shell taken from an ancient sea bed had a large amount of the isotope of oxygen 0^{18}, what can you say about the climate when the clam was alive?

10 What is a danger of inferring global climate change from California Bristlecone Pine tree rings?

11 What are the three variations in the earth's orbital parameters that could cause changes in the climate?

12 To what extent do you think humans were responsible for the starvation in the Sahel during the late 1960s?

Problem

13 Discuss the impact of a change of mean temperature of $-3C°$ on life at your location. Consider the effect on agriculture, snowfall, rainfall, wildlife and human activities (such as outdoor sports). Try to find both positive and negative impacts.

THIRTEEN

Air Pollution Meteorology

smog • sulfur dioxide • carbon monoxide • exit velocity • effective stack height • dispersion • mixing depth • ventilation factor • stagnation • fumigation • background pollution • acid rain • carbon dioxide

Many of the creations of people add undesirable particles or gases to the atmosphere. Some of the most important pollutants include coal dust, incompletely burned rubbish, photochemical "smog," sulfur dioxide, and carbon monoxide. Different cities have different pollution problems; for example, carbon monoxide is the major pollutant in Washington, D.C., rubbish particles and sulfur dioxide plague New York City, and Los Angeles is infamous for its photochemical smog.

In the United States, about half of the total pollution produced in 1974 came from the automobile. Car exhausts spit unburned hydrocarbons, carbon monoxide, and oxides of nitrogen into the atmosphere. In sunny cities such as Los Angeles or Phoenix, Arizona, the oxides of nitrogen combine with the hydrocarbons (left over from the gasoline), first to form ozone, and then complicated organic particles. These particles look like a smoky fog and therefore are called *smog*, although they are really neither smoke nor fog. Smog irritates human and animal eyes and kills vegetation. Carbon monoxide from cars can also be an important health hazard, especially on busy streets.

Stationary sources such as power plants or home space heaters that burn coal are primarily responsible for the surfur dioxide and coal dust in the atmosphere. Coal dust is the easier to control. It used to blacken the skies in London and St. Louis, Missouri; in fact, the famous "London fogs" were caused almost entirely by coal dust from home coal-burning fireplaces. When these were banned, the fogs disappeared. In the following sections, we discuss some of the meteorological aspects of the air pollution problem.

13.1 AIR POLLUTION AND THE WEATHER

There are really two completely different kinds of air pollution problems related to meteorology; one kind deals with the effects of weather on pollution, the other with the effects of pollution on weather and climate. Weather influences pollution in a number of ways: the winds transport pollution from one place to another, and the dilution of pollution depends on various meteorological factors. Therefore, a knowledge of atmospheric conditions is useful in planning, executing, and evaluating air pollution control. On the other hand, weather can be changed by pollution: increased pollution decreases visibility, causes raindrops to become drops of acid, and may increase precipitation (although this possibility is somewhat controversial).

The effects of pollution on climate are even more controversial: increased amounts of carbon dioxide from combustion processes increase the absorption of outgoing radiation from the ground, thereby reducing the overall loss of heat to space and producing a gradual warming trend. The opposite effect is possible when particles emitted by industry into the atmosphere reduce the intensity of sunlight reaching the surface, thereby causing lower temperatures. Much is uncertain about the magnitude of these effects, but they must be studied in detail so that measures can be taken in time to prevent either disastrous warming, which would produce widespread melting of polar ice and consequent flooding, or the equally undesirable extension of ice sheets associated with a reduction in mean temperature.

EFFECTS OF THE ATMOSPHERE ON POLLUTION 13.2

The atmospheric temperature, moisture, and wind structure determine the fate of pollutants from the time they leave the source to the time they reach receptors, such as plants, animals, or people. Thus, a meteorologist is asked where pollution from a given source is going, what the concentration will be at the receptor, and whether pollution levels will be much higher tomorrow than today. Based on the meteorologist's advice, industry may be asked to reduce operations under adverse conditions, or switch to a cleaner fuel.

For convenience, we will discuss the effects of atmospheric variables on three parts of the pollution cycle: the initial rise of the pollutant from the source, the transport downwind from the source, and the dilution or dispersion of the contaminant.

The initial rise from a smokestack is important because the higher the material rises initially, the smaller will be the concentrations near the ground later. The actual amount of rise depends on five variables, three of which are nonmeteorological and may be partially controlled by the polluter. These three are the temperature of the effluent, the initial exit velocity, and the cross-sectional area of the source, e.g., the area of the discharging stack. Then there are two meteorological variables, wind speed and the variation of temperature with height, called the lapse rate, over which the polluter has little control. On a windy day, smoke plumes do not rise very high, as can easily be verified visually. The effect of temperature variation with height is more subtle. Generally, when the air is more unstable, with warm, light air near the surface and colder air aloft, the smoke rises higher. In the morning, when there is frequently a low-level inversion, with cold, dense air underlying warm air, smoke spreads out horizontally soon after leaving the source. Sometimes, under the stable conditions of early morning when relative humidities are high, smoke and water vapor emitted by industry may produce dense fogs which drift slowly downwind.

In order to compute the characteristics of a plume, it is convenient to define an "effective" stack height (figure 13.1), which is the sum of the actual stack height, h, plus the aditional rise, Δh, caused by the exit velocity and buoyancy of the warm plume. Although Δh obviously depends on the five variables

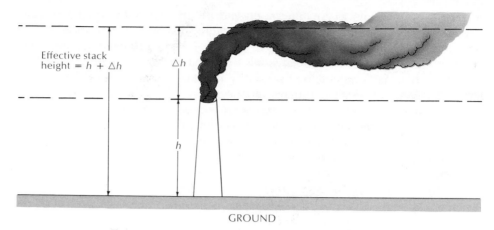

FIGURE 13.1   *Effective stack height.*

previously mentioned, there is still no generally accepted formula for this complicated relationship. Because it is advantageous to make the effective stack height as great as possible in order to reduce concentrations near the ground, modern plants are often equipped with high stacks, and the effluent is emitted at high temperatures. Of course, when there is a strong wind, the effect of the high temperature on raising the effective height is diminished; however, conditions of fast winds usually do not lead to serious pollution problems, as we shall see.

13.2.1 HORIZONTAL TRANSPORT OF POLLUTANTS

The question of horizontal transport of pollution appears at first to be quite simple: The pollutants travel with the wind, so all we need to forecast the trajectory of the pollutants are wind speed and direction. Unfortunately, the wind behaves in a very complicated manner over time and space, especially near the ground. In the first place, wind speed and direction are not steady but vary from minute to minute and hour to hour. There are theoretical reasons why it is convenient to consider hour-average concentrations of pollutants. To estimate such concentrations, we should use hour-average winds. Consistent with this requirement, winds are reported at many weather stations every hour. One difficulty with the hourly wind reports, however, is that these winds are not really hour-average winds but, instead, represent about one-minute averages. This difference may cause serious errors in computing the transport of pollutants.

Within an hour, wind directions fluctuate over perhaps 30 degrees (figure 2.8). The effect of these fluctuations is to spray the pollutants in different directions, very much as water is spread from a waving garden hose. Even though the center of the pollutant "cloud" will move with the hour-average wind, the effect of the fluctuations is to dilute the concentrations and spread the material over a wider area. Thus, wind fluctuations lead to dispersion of pollutant material.

To make use of the fact that the center of the pollutant plume moves with the hour-mean wind, we need to know the wind at the source of the pollution and all along the plume's trajectory. Unfortunately, wind observations are almost

382

never available exactly where they are needed. Hourly winds are usually measured at airports, not near the centers of industrial activity or human habitation. In addition, it is rarely legitimate to assume that the surface wind directions and speeds are uniform across an area as large as a city. Most cities are near large water bodies, or have hills nearby, and some like Los Angeles are close to high mountains. All these topographic features produce local (mesoscale) circulations on a scale much smaller than what we can see on ordinary weather maps.

In some cities where air pollution has been especially serious, mesoscale wind circulations have been studied in detail. Such studies utilize constant-level balloons, numerous radiosonde stations, or helicopters. For example, we know the local circulations in considerable detail for New York, Chicago, the San Francisco Bay area, and Los Angeles. From simple models of typical flow patterns over these cities, if we have measurements of wind at a few locations, we can make a good estimate of winds at other locations in the same cities. However, detailed wind surveys are expensive. An alternate technique is to construct mathematical models of the flow patterns in a city, given various characteristics of the large-scale flow and the detailed topography of the city. This type of modeling is quite difficult, and efforts in this direction have only just started.

One problem with estimating the wind field is that the most serious air pollution problems arise with very light winds. Under these conditions, the mesoscale circulation features tend to dominate the larger-scale circulations, and the transport patterns are quite erratic. We do not know how to deal with such strong local variations of concentrations which can arise under these circumstances. All we can really say is that light winds permit pockets of very strong pollution to develop.

Not only does the wind vary horizontally from place to place and time to time, but it also varies with height. Over flat terrain, the wind speed increases very rapidly in the lowest few meters but then changes quite slowly with increasing height. Also, the direction is nearly constant with elevation up to 100 meters or so, unless the air is quite stable. But more significant changes in direction occur higher up.

Typically, the wind is measured at heights about 10 meters above the ground. These observations are quite good indicators of the general wind in the lowest 100 meters. Above a level of about 100 meters, a slow turning of the wind direction begins. This turning is clockwise (veering) in the northern hemisphere; that is, a south wind near the ground becomes more westerly aloft. Also, the speed increases slowly with height, a fact which should be considered for accurate dispersion calculations. These general rules apply when the air is unstable. When the air is stable and the winds are light, the vertical distribution of wind is erratic, contributing to the difficulty of quantitative computations. In summary, measuring or calculating the transport winds for a given city is usually a very difficult problem.

13.2.2 DILUTION OF POLLUTANTS BY MIXING

After the transport wind has been determined, we must find out how fast the pollutant is diluted. If the dilution, or dispersion, is rapid, pollutant concentrations will generally be low, and no serious problems arise. But

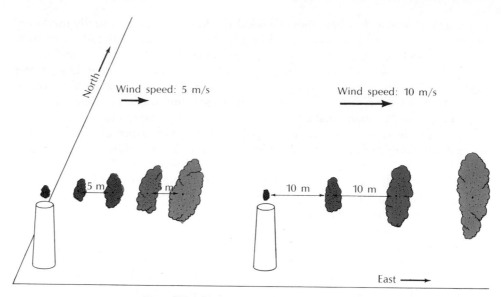

FIGURE 13.2 *Effects of wind speed and horizontal eddies on the dispersion of pollutants.*

on certain days, pollutants are diluted very slowly, and harmful pollutants may reach dangerous levels.

The rate of dispersion is primarily governed by the vertical variation of temperature and by the wind speed. To a lesser degree, it depends on the roughness of the terrain. Also, the topography, such as hills and mountains, can limit pollutant dilution.

The effect of the wind speed on dispersion can be seen in figure 13.2. Suppose that a stack emits a burst of smoke every second. If there is a wind of 10 m/s (about 20 mi/h), the distance between smoke puffs will be 10 meters. If the wind speed is 5 m/s, the distance between puffs will be 5 meters. In general, the faster the wind, the greater the distance between puffs, and the smaller the concentration. Thus, pollution problems are never found with strong winds; instead, so-called air pollution episodes occur only when the wind is weak.

There is another, but less important, effect of strong winds. Wind blowing over any terrain, but particularly over rough terrain, will produce vertical eddies. Such eddies (turbulence) mix polluted air with clear air, so that the clean air becomes less clean, and the dirty air less dirty. In fact, turbulence acts like Robin Hood—it takes pollution from regions that have too much and gives to regions where there is little.

Small-scale horizontal variations, or eddies, in the mean wind direction also act to disperse the pollution across the mean wind, as indicated by the increasing size of the puffs downwind of the source in figure 13.2. These eddies produce a horizontal fanning of the pollution downwind, which reduces the concentration per unit area in the plume.

384

13.2.3 THE MIXED LAYER

More important than wind for producing eddies, however, is thermally produced convection, which depends strongly on the surface heating and the vertical temperature distribution. Whenever air is heated from below, as on a sunny day, convection currents, which are another form of turbulence, are generated. Convection currents mix air with different characteristics, so polluted air becomes cleaner. In terms of the temperature distribution, mixing by eddies occurs when temperature decreases rapidly with height, that is, when air is unstable.

Usually, on sunny days, the heating from the ground is so strong that the well-mixed portion of the atmosphere extends to fairly high levels, to one kilometer or more. Above that mixed layer, an inversion usually exists. Inversions are quite stable and tend to prevent mixing with air above. Hence, the distance from the ground to the bottom of the first inversion is called the *mixing depth*. This depth plays an important role in air pollution prediction. The mixing depth limits the extent to which pollution can spread vertically: the smaller the mixing depth, the greater the concentration.

Figure 13.3 shows a map of the average annual morning mixing depth (in hundreds of meters) over the United States. The greatest mixing depths occur near the coasts, where the warm waters (compared to the land) maintain a relatively unstable layer of air near the ground.

Again, it is quite clear that the daytime mixing depth increases with increased heating of the ground and will be, for example, greater in summer than

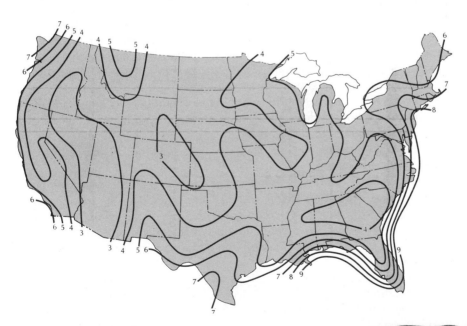

FIGURE 13.3 *Mean annual morning mixing depths (in hundreds of meters).*

385

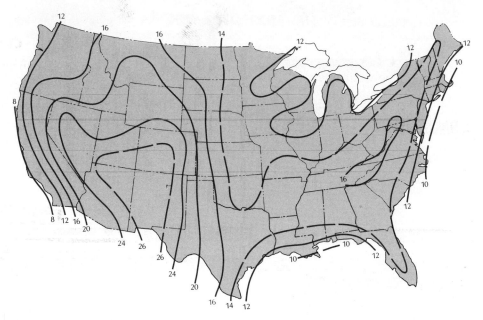

FIGURE 13.4 Mean afternoon mixing depths (in hundreds of meters).

in winter. For example, figure 13.4 shows the annual average mixing depth in the afternoon, which may be compared with the morning values in figure 13.3. In many inland places, the afternoon mixing depth are four times greater than the morning values, owing to solar heating at the ground. Variations near the coasts, where the diurnal variations in water temperature are small, are much less. But the daytime mixing depth depends on other things besides the heat input at the surface. In particular, it depends on the stability of the atmosphere before the heating begins. For example, if the air aloft is slowly sinking (subsiding), it warms as it is compressed by the higher pressure. This warming aloft produces stability, and even strong heating from below under such conditions cannot produce a large mixing depth. General subsidence and stability occur most often in anticyclones, especially near the high-pressure centers. This is where daytime mixing depths are smallest and winds are lightest.

The effects of both the wind speed and the depth of the mixed layer are combined in the ventilation factor, which is the product of the mixing depth and the average wind between the ground and the top of the mixed layer. The concentration varies inversely as the ventilation factor; for example, if the mixing depth and the wind speed are both doubled, the pollution concentration is reduced by a factor of four.

Figure 13.5 shows an analysis of the ventilation factors over the United States on the day when the ventilation was the least over two consecutive days for the five-year period from 1960 to 1964. On this day, the atmosphere's ability to dilute pollution was a minimum. In this figure, the lowest numbers represent the greatest potential for the development of high concentrations of pollution. In general, the western United States had the lowest ventilation factors, and,

386

therefore, the greatest potential for air pollution episodes. Of course, it takes more than low ventilation factors to produce severe air pollution; it also requires sources of pollution. Since many of the sparsely settled regions of the West have few sources of pollutants, the air may be quite clean there even with low ventilation factors.

As we have seen, both mixing depth and wind speed tend to be low in anticyclones; hence, high-pressure areas are often associated with large concentrations of pollutants. Normally, anticyclones move fairly rapidly across a weather map, so conditions at any one point will not get out of hand. But occasionally, most likely in the summer or fall, highs may remain in the same area for a week or more, giving rise to a stagnation situation. (Winter anticyclones seldom stagnate, but when they do, their especially low mixing depths have produced some of the most devastating pollution episodes.) Under such conditions, air pollution meteorologists must alert the public and industry in order to reduce the concentration of harmful effluents to a minimum.

In some parts of the world, highs may stagnate for much longer periods. For example, in summer, west coasts of continents are influenced by anticyclones for months at a time. Not only that, but cold ocean currents increase the stability of the air and reduce the mixing depths. This is why especially strict air pollution regulations are needed for West Coast cities such as Los Angeles.

Another factor that influences the pollution climate is topography; mountains can restrict horizontal spreading. There are mountains both to the east

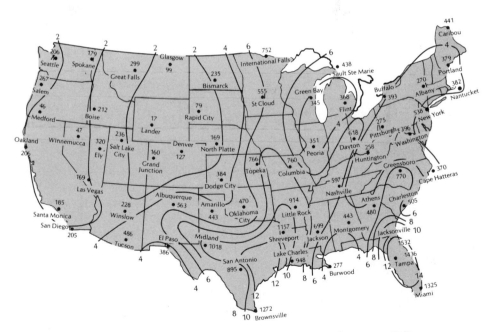

FIGURE 13.5 *Station values of ventilation factors (in m²/s) and lines of equal ventilation factors (in hundreds of m²/s) for slowest dilution episodes lasting two days during the period 1960–1964.*

387

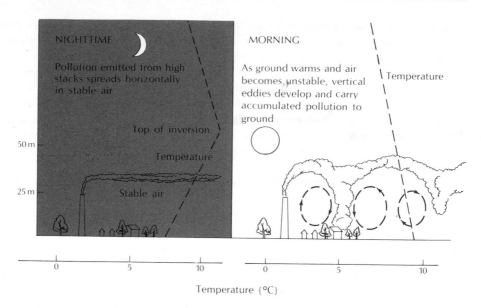

FIGURE 13.6 *Illustration of the fumigation process.*

and north of Los Angeles, adding to the problems there. Topography plays an especially important role in the case of deep, narrow mountain valleys, which are frequently sites of pollution disasters. Such disasters occurred in the Meuse Valley of Belgium in 1930, when sixty-three people died, and in the Monongahela Valley of western Pennsylvania (Donora) in 1948, when twenty-two succumbed. In both places, low-level inversions persisted for several days, trapping industrial pollutants in a stagnant layer of air trapped between the valley walls.

So far, we have considered mixing depths and air pollution potential only for the daytime. At night, wind speeds tend to be less than during the daytime, and, as we will see, the same is true for mixing depths, so concentrations tend to be larger at night than during the daytime.

Concentrations are not always higher at the ground level at night, however. If the air is very stable, with the coldest air at the ground and the temperature increasing upward, the pollutants emitted from high stacks may accumulate in a thin layer a short distance above the ground during the night (figure 13.6). After sunrise, when the inversion is destroyed by solar heating, vertical convective eddies may mix the accumulated pollutants downward to the surface, a process known as *fumigation*.

13.2.4 MIXING DEPTHS OVER CITIES

At night, a mixed layer does form near the ground in cities. However, the mechanism is different from that during the daytime. Before air reaches a city, it is usually quite stable, particularly on clear, dry nights with little wind. Under such conditions, there is almost no vertical mixing, and a plume from an elevated source may not reach the ground for tens of kilometers. Under these conditions, the temperature usually increases vertically to a height of one kilometer or more; we have a *surface inversion* (see figure 13.7). We noted in chapter 11 that cities, because of their high heat capacities and artificial heat sources, are heat islands. Therefore, nighttime surface temperatures are several degrees warmer in the city than in the surrounding countryside.

The effect of the urban heating is to establish an unstable layer in the otherwise stable air close to the ground as illustrated in figure 13.7. Thus, a new mixing depth over the city is established from the surface to the inversion. An example of the urban mixing depth for Cincinnati, Ohio, at sunrise is shown in figure 13.8. A general inversion covers the entire region at an elevation of 366 meters (1200 feet), where the temperature is a warm 25°C (77°F). In the rural areas surrounding downtown Cincinnati, the temperatures are as low as 18°C (65°F). In the city, however, the heat island produces a shallow mixed layer in which the temperature decreases from 22 to 20°C (72 to 68°F) from the surface to a height of about 67 meters (220 feet) above the ground. This mixed layer of warm air is carried downwind from the city and overspreads the cooler air over the rural surface. Such city-induced mixing depths are different in every city, generally increasing with the size of the city. For example, typical mixing depths are 300 meters in New York City, 150 meters in Columbus, Ohio, and 100 meters in Johnstown, Pennsylvania. Although the heat-island effect is subject to intensive research, there is as yet no simple method to estimate accurately nighttime mixing depths, which depend on meteorological as well as city characteristics.

Because nighttime mixing depths in cities are generally smaller than daytime mixing depths, pollution from low-level sources (e.g., apartment house incinerators) is worse at night. However, very high stacks often extend to levels above the mixing depth and will not contribute to the pollution of the city night air at the surface (figure 13.7).

Reasonably good techniques exist for estimating concentrations of pollution from individual continuous sources to a distance of 10 kilometers or so

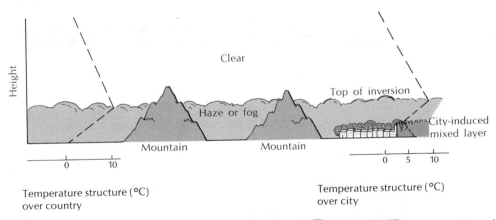

FIGURE 13.7 Nocturnal lapse rates over a city and the surrounding countryside.

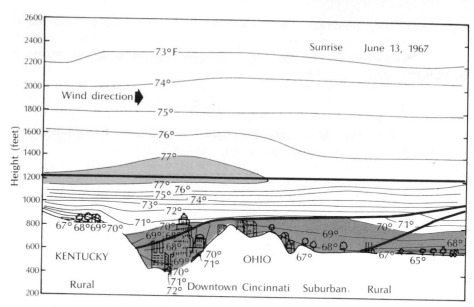

FIGURE 13.8 *Vertical cross section showing temperatures (in °F) and mixing depth over Cincinnati, Ohio, at sunrise. [m = ft (0.305); °C = $\frac{5}{9}$ (°F − 32)]*

away from the source, given the source strength, wind, lapse rate, and mixing depth. In the case of cities, there are numerous sources. A great deal of work has been done on modeling the pollution in individual cities by estimating the total pollution emitted over each square kilometer. However, separate calculations have to be made for different weather situations and for different pollutants. The purpose of these city models is to determine how much of the pollution in a given area is caused by each source so that the effect of curtailing certain sources can be established. Thus, for example, it has been shown that in central New York City almost 95 percent of the sulphur dioxide (SO_2) came from apartment house heating nearby (before recent control measures), and only a very small fraction from power plants, even though the power plants emitted about the same amount of SO_2 as the apartments. The reason was that the power plants had high stacks and therefore, produced little effect nearby, although they had a relatively large effect on the suburbs. It is clear that in this case the use of cleaner fuels in the power plants would have had little effect on the quality of the air in the inner city. It is also noteworthy that even in such examples where the meteorological question of the cause and dispersion of the pollutant is answered, the political question of what to do about it remains, and, in many cases, turns out to be more difficult to solve than the original scientific problem.

13.2.5 EVALUATION OF BACKGROUND POLLUTION

An important problem in air pollution control is an evaluation of background pollution, which is the normal amount of pollution present at a given place. With certain wind directions, air quality may be greatly influ-

390

enced by distant pollution sources, creating both meteorological and political problems. As an example, much of the pollution at Windsor, Ontario, comes from Detroit, and fully half of the pollution at Dusseldorf, West Germany, comes from the Ruhr Valley.

It has become possible to extend some of the techniques designed for estimation of short-range diffusion to distances of 100 kilometers or so. New difficulties arise in such estimates, but, on the other hand, some simplifications are possible. One difficulty is that the large distances involved mean that the pollutants must travel many hours, over which time meteorological variables such as wind and mixing depth undergo important changes. In addition, the pollutants change chemically, interact with the ground, and may be affected by rain and clouds. At present, we know much better how the pollution gets into the atmosphere than how it eventually disappears, because there still are many unknowns in the chemical behavior of the various pollutants.

On the other hand, it is possible to simplify the problem by representing whole cities as point sources some distance upstream of the actual cities, and to arrive at good concentration estimates for considerable distances downstream.

Under unusual weather conditions, pollutants can be traced over more than 1500 kilometers; e.g., particles in Oklahoma have been traced to sources in the northwestern Indiana industrial area; and the acid rains in Scandinavia (which will be discussed later) have been ascribed to sources in West Germany. Such estimates are naturally quite controversial.

EFFECTS OF POLLUTION ON THE ATMOSPHERE 13.3

It is quite likely that pollution modifies the weather now, and that these effects will grow in the future. However, the details of the future influence of pollution on weather are quite debatable and speculative.

The most obvious effects occur on the local scale. For example, the visibility is sharply reduced on smoggy days which occur occasionally in some cities, and frequently in others, such as Los Angeles. Along with visibility, direct sunlight reaching the ground is reduced. In a recent study in St. Louis, Missouri, it was found that direct sunlight was reduced in the city by as much as 40 percent on "dirty" days. However, it is not clear whether even such dramatic reductions of sunlight have significant effects on temperature, because much of the light will reach the ground anyway due to scattering by the polluting aerosols (particles).

13.3.1 EFFECTS OF POLLUTION ON CLOUDS AND RAINFALL

There are many reports of air pollution changing regional weather. Perhaps the most important modification is known as the Laporte effect. As discussed in chapter 11, Laporte, Indiana, is located east of major steel mills and other industries south of Chicago. Rainfall at Laporte has been observed to be systematically higher than at other stations nearby. If this effect does not occur by chance, the increase could be caused by the excess particles, heat, or moisture produced by the industries upwind of Laporte.

An increase of precipitation similar to that observed in Laporte has been reported downwind from some pulp mills in western Washington. Analysis of the raindrops there suggests that some particles introduced into the clouds by the plumes are large enough to capture waterdrops when falling and thus attain sufficient size to reach the ground as precipitation.

In western Pennsylvania, large power plants seem to have increased the precipitation downstream compared to what it had been prior to the construction of the plants; unfortunately, the natural variability of precipitation is so large that this result was not "statistically significant." A dramatic example of the effect of the heat emitted by such large power plants is shown in Plate 14, which depicts growing cumulus clouds situated directly over the smokestacks of the Keystone power plant, located near Indiana, Pennsylvania. On these two days, the atmosphere was very nearly unstable and therefore sensitive to the extra amount of heat and moisture produced by the power plant. This extra heat triggered the release of the natural instability of the atmosphere, and the result was the cumulus clouds.

Another spectacular example of the effect of pollution on local weather is illustrated in Plate 15a, which shows two areas of growing cumulus clouds generated by the smoke and heat from grass fires in the Everglades of Florida. Intense wildfires can also produce strong enough updrafts and convergence of low-level air to spawn whirlwinds of fire—miniature tornadoes that resemble dust devils.

As the previous examples prove, pollution can, at least under just the right conditions, significantly modify the local weather. On a larger scale, we have a number of indications that clouds and precipitation can be increased by pollution emitted tens of kilometers upwind, but none of these observations are conclusive.

A damaging effect of pollution on a regional scale which has important political implications is *acid rain*. Acid rain, which is simply dilute acid (usually sulfuric acid) falling from the sky, is caused by the dissolving of gaseous pollutants such as sulfur dioxide in precipitation. [Sulfur dioxide (SO_2) is a major pollutant produced by the burning of coal and oil.]

Figure 13.9 shows the pH of precipitation falling on the United States during June, 1966. A pH of 7 is neutral; less than 7 is acid. The greatest acidity at this time occurred in the industrial Northeast. Prior to 1930, there was little acidity in this region. The increase from 1930 to the 1960s apparently continued into the 1970s.

Acid rain has damaged forests in Norway and Sweden and has apparently caused many lakes and streams in Norway to become completely barren. It is also causing the rapid dissolving of marble statues in many large cities around the world. Construction of air trajectories and estimates of concentration of SO_2 have indicated that as much as 80 percent of the sulphate in the rain (which causes the acidity) falling in Scandinavia is produced by the emissions of sulphur dioxide in England and West Germany. The acidity of the rain is particularly large only on days when winds blow effluents from one of these two countries over Scandinavia. Similarly, pH values as low as 3.1 (considerable acidity) have been observed

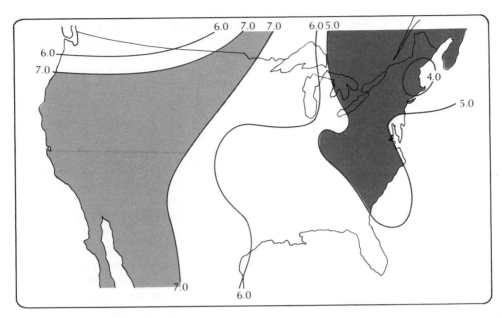

FIGURE 13.9   *The pH of precipitation during June, 1966.*

downwind from the high-stack power plants in western Pennsylvania. This result suggests that although sufficiently high stacks can be built to prevent large concentrations of SO₂ near the ground, the plume can still pollute clouds and rain downwind.

A different form of regional pollution by industry is the increase in number of liquid drops in the atmosphere introduced by huge cooling towers. The hot water in these towers also evaporates, adding large amounts of water vapor to the air. In cold air with high relative humidity, the additional water added to the air can produce widespread fogging, and even highway icing to distances of over 150 kilometers.

13.3.2   CAN POLLUTION CHANGE THE CLIMATE?

Of all the possible effects of pollution on weather, those on climate are the most nebulous but are nevertheless the most important. A mean cooling of the atmosphere by only a few degrees Celsius, according to some estimates, could cause the polar ice to spread over the whole world. A warming by a similar amount could melt the polar ice caps and inundate large low-lying areas of the world, including cities such as New York, Leningrad, and London.

Although some of the effects on a given meteorological variable (e.g., temperature) by individual polluting agents can be estimated fairly well, the ultimate effect on climate is very difficult to estimate, because once one variable changes, others will, too. For example, if the temperature increases, the absolute amount of water vapor is likely to increase; wind circulations will change, and so will the amount and types of clouds. All these changes may feed back to affect the

393

temperature, either magnifying or diminishing the original change. So far, no mathematical models have been able to predict the original effect and all the possible feedbacks. Another basic difficulty is that the causes of natural climatic changes are not known, so we do not know whether a certain observed change is artificial or natural. Of course, the artificial effects may become so large that they override natural climatic changes.

Of all the pollutants likely to change climate, carbon dioxide (CO_2) has been studied the most thoroughly. We know the amount of CO_2 put out by industrial and other combustion, and we have quite a good idea what fraction of this actually stays in the atmosphere (about 50 percent). Most of the remainder presumably is absorbed by the oceans, although a small part goes into vegetation, creating thicker jungles. Figure 13.10 shows that the average annual concentration of CO_2 increased from about 310 to 320 parts per million between 1958 and 1970. The figure also shows an extrapolation for the concentration in the year 2000, when the expected value is 380 parts per million, or almost 20 percent higher than the present concentration. Such an extrapolation is based on certain assumptions, which may be hazardous. One such assumption, for example, is that the ocean will continue to absorb a large fraction of the CO_2 in the atmosphere. Also, this prediction ignores the worldwide energy shortage, which may increase the shift from oil and natural gas to atomic fuel and decrease the emission of CO_2. On the other hand, a shift to coal would make no difference.

Additional CO_2 in the atmosphere does not affect incoming short-wave solar radiation but absorbs more long-wave radiation emitted by the earth and reradiates additional infrared radiation to the ground, thus raising the temper-

FIGURE 13.10 *Increase   of   carbon
dioxide in the atmosphere since 1860.*

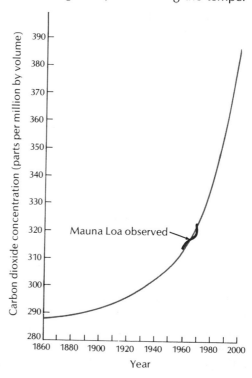

ature of the air near the surface. It also increases radiation emitted into space and therefore cools the upper atmosphere. The warming of the lower layers presumably means that evaporation increases at the surface, thus further increasing the infrared absorption and reradiation of atmospheric radiation. Therefore, the temperature would probably be further increased by this feedback effect.

Taking these factors into account, but not allowing for large-scale atmospheric circulation changes, recent studies have shown that a doubling of CO_2 in the atmosphere would raise temperatures near the ground by about $2C°$. Thus, the anticipated increase of atmospheric CO_2 by the year 2000 should raise the average surface temperature by about $\frac{1}{2}C°$ and the temperatures in the polar regions by about three times as much. Such increases are not believed to be serious, but computations suggest that climate modifications by CO_2 probably will become important late in the twentieth century. Therefore, it is extremely important to monitor worldwide concentrations of CO_2 and to plan measures to counteract the exponentially increasing growth.

Of the other gases, excess water vapor and sulfur dioxide may eventually affect climate, not so much directly, but after sulfate particles have formed from them. Also, there is some evidence that cirrus cloudiness has increased proportionately to increased aircraft traffic. Preliminary computations suggest that such increases have no large effect on surface temperature; however, some measurements have shown that the solar radiation may be reduced significantly. If confirmed, this reduction could be damaging to agriculture.

Sulphur dioxide is converted to sulphate particles, which comprise an important part of the total particle load in the stratosphere. Even in the troposphere, sulphates are the most plentiful artificial particles. Other particles in this category include coal dust and other substances emitted as particles. Also, some particles (e.g., smog) form chemically from oxides of nitrogen and hydrocarbons. The estimated concentrations of all such particles are quite uncertain; however, it seems clear that tropospheric particles from natural sources (volcanic particles, sea salt, dust) are much more plentiful than artificial particles. At present, only about 20 percent of all the particles are artificial. However, some crude estimates suggest that by the year 2000, perhaps half the tropospheric particles will be introduced by human activity.

Increased particle concentrations can affect two aspects of climate—precipitation and radiative transfer. Of these, the second is believed to be the more important. Thus, it seems likely that the reflectivity (albedo) of the atmosphere is changed by the addition of particles. But even the sign of the change is not known. For example, if the particles are bright, the albedo is likely to increase, reflecting more sunlight back into space and cooling the air. However, if dark particles are emitted over an otherwise bright surface (an extreme case being snow), the albedo will decrease, and more radiation will be available to heat the atmosphere.

In any case, the magnitude of the effect of particles on the radiation budget is extremely uncertain. It depends on the optical properties of the particles, which in turn depend on particle composition and size distribution. Furthermore, particles radiate in the infrared and therefore can alter the outgoing long-wave radiation. It is clear, therefore, that our knowledge of the effects of particles in the atmosphere is badly inadequate. We must know more about scattering, absorption, and emission of natural and artificial particles and must predict what kinds of particles are likely to be added to the atmosphere. At the same time, we must monitor changes of particle concentrations, reflected sunlight, and emitted radiation.

So far, we have been concerned entirely with the troposphere. Stratospheric pollution is also important for two reasons. First, the stratosphere has lower densities than the troposphere, so a given amount of pollution has relatively larger effects in the stratosphere. Second, the residence time in the stratosphere is long because of the weak vertical mixing. For example, at 20 kilometers, the residence time of fine particles is about a year. One year after introduction, about one-third of the initial number still remains.

Nature occasionally pollutes the stratosphere with particles from volcanic eruptions. At about 20 kilometers, there exists a rather permanent layer of particles which is attributed to continuing volcanic eruptions. The two most severe eruptions in the nineteenth century may have been assoicated with measurable cooling at ground level; the 1963 eruption of Mt. Agung in Indonesia was followed by a stratospheric warming of about 6C°. Measurement of the quantity of particles introduced and computations based on their optical properties have suggested that a 5C° rise is quite reasonable. However, no definite change in surface weather has been linked to the eruption of Mt. Agung.

Because volcanic particles in the stratosphere may have had important effects on the stratosphere, and in extreme cases, even on the troposphere, the question has been asked whether artificial pollution adds perceptibly to such effects. At present, human activity does not contribute much to the concentration of contaminants in the stratosphere. European supersonic aircraft and military high-altitude planes have had no detectable effect while flying mostly at levels below 20 kilometers, where the residence time is shorter. Atomic testing in the 1950s introduced particles into the stratosphere, but no definite effect on weather has been demonstrated. And although a possible decrease of stratospheric ozone following certain atomic tests has been suggested on the basis of theoretical chemical arguments, it has not been conclusively confirmed by observations.

13.4 STRATOSPHERIC POLLUTION AND THREATS TO THE OZONE LAYER

Much of the concern about stratospheric pollution has centered about the introduction of supersonic transports (SSTs). These are relatively large and fly at an altitude of about 20 kilometers. It is estimated that about 500 SSTs must fly to make using them profitable. Thus, many calculations have been made of possible effects of various pollutants introduced by this number of planes, each flying seven hours per day, over routes most likely to be popular.

The principal emissions of concern are SO_2, H_2O, oxides of nitrogen, and soot. Carbon dioxide is also emitted, but the SSTs will be relatively unimportant sources compared to the private and industrial sources which were discussed earlier. It is estimated that enough water vapor will be emitted to increase the water vapor content of the stratosphere (which is just a few parts per million) by 10 percent. Three potential effects of this increased water vapor would be the following:

1 Increased cloudiness in the very cold regions of the winter pole (about −80°C), where there now exist some thin, iridescent clouds, called *mother-of-pearl* clouds

2 Change in the infrared radiation to space and toward the ground

3 Interaction of the water vapor with stratospheric ozone

Of these three items, quantitative estimates have shown that the effect of the additional long-wave radiation would be negligible. Almost nothing is known about the first item. We know that condensation trails are almost never seen behind stratospheric planes because the relative humidities are so low in most of the stratosphere. But, then, no stratospheric aircraft has been observed in regions around the cold polar night, which is an area where mother-of-pearl clouds sometimes exist. Presumably, not many SSTs will fly through these regions, so the problem of additional cloudiness and radiative changes associated with them may be unimportant. Also, arctic clouds are rare in the stratosphere, are very thin, and are therefore unlikely to affect climate significantly.

The amount of ozone in the stratosphere profoundly affects the climate of the stratosphere and may also influence the climate close to the ground. The upper stratosphere is almost as warm as the air near the ground because the ozone absorbs ultraviolet radiation from the sun (see figure 3.2). Therefore, any reduction in ozone will permit more transmission of ultraviolet light through the stratosphere, producing a cooling of the stratosphere and a warming at the ground. Any interference with ozone is potentially dangerous, for ozone protects us from ultraviolet light. If ozone is removed, ultraviolet light is increased, thus increasing the incidence of skin cancer. It has been estimated that a 20 percent decrease of ozone will produce a 20 percent increase in skin cancer.

Originally it was believed that the oxides of nitrogen emitted by SSTs posed a serious threat to the ozone layer. However, the original calculations underestimated the importance of two chemical reactions. One of these combined oxides of nitrogen with another chemical to form a compound which does not attack ozone. The other speeded the transformation of oxides of nitrogen into nitric acid, which is easily removed from the stratosphere. As a result, the threat of SSTs to the ozone layer is now believed to be small. In fact, relatively low-flying SSTs, such as the British-French *Concorde*, may actually produce ozone in small quantities.

Oxides of nitrogen are not the only substances that interfere with stratospheric ozone; chlorine is potentially worse. Certain chemicals that are used widely as refrigerants and as propellants in spray cans contain chlorine. When these chemicals slowly seep into the stratosphere, the chlorine is released by solar radiation and starts to destroy ozone. It is believed that chlorine has already destroyed about 1 percent of the ozone (1975). Recent estimates suggest that continued use of these chemicals (called *halocarbons*) at the rate used in 1973 will eventually (in approximately one hundred years) reduce the total ozone by an amount somewhere between 5 percent and 30 percent. The best estimate is a 14 percent reduction, which would increase the incidence of skin cancer in the United States by 30 percent. For this reason, it is generally believed that the use of halocarbons should be reduced, and steps in this direction have been taken.

There is another reason for concern over the use of halocarbons. Halocarbons absorb strongly in the atmospheric window in the infrared (between 8- and 12-micrometer wavelengths; see figure 4.9). In this region, radiation emitted at the ground now escapes into space. Halocarbons can block some of this escaping radiation and reemit the radiation back to the surface, causing warming. This effect is not large, but it is systematic and is likely to be largest in polar regions where the warming could cause significant melting of glaciers. Still, by 2050, it is likely to be only about 10 percent of the effect of increasing CO_2.

There may be other ways in which human activities can destroy ozone. A potential problem of uncertain magnitude arises from the use of certain fertilizers. These fertilizers emit nitrous oxide (N_2O), which may combine with excited oxygen atoms in the stratosphere to form two molecules of nitric oxide (NO). As we have seen, this gas can dissociate ozone.

In summary, human activity, particularly industry, can affect climate in various ways. In most cases, we do not know the magnitude of the changes, and in some cases, not even the sign. But because this matter is so important, we must attack these questions by using mathematical models and measurements of the chemical and physical characteristics of the atmosphere, and by carefully monitoring the worldwide increase of a number of contaminants.

Questions

1   What was the major cause of the famous "London fogs"?

2   What five variables control the amount of rise of a gaseous pollutant from a smokestack?

3   Why do the worst air pollution episodes occur during light wind conditions?

4   What are the two most important meteorological variables that affect the rate of dispersion of a pollutant?

5   At what time of day are mixing depths generally the greatest? In what season?

6   If the mixing depth is 100m and the wind speed is 5 m/s, what is the ventilation factor?

7   Are severe air pollution episodes more likely to occur under an anticyclone or a cyclone? Why?

8   Describe the fumigation process.

9   Is the nocturnal mixing depth usually greater or less over a city compared to a nearby rural area?

10   What is acid rain?

11   What was the percent of the increase in carbon dioxide (global/average) from the period 1960–1970? (See figure 13.10) (Answer: 2.5 percent)

12   Compare the probable effects of increased carbon dioxide and particulates in the atmosphere on the climate.

FOURTEEN

Weather and Water

cryosphere • hydrology • precipitable water • hydrologic cycle • transpiration • potential evapotranspiration • flash floods • droughts

Except for a small addition of water now and then from volcanoes, the total amount of vapor and liquid water in the atmosphere, on and under the earth's surface (the *hydrosphere*), and locked up temporarily as ice (the *cryosphere*) is constant. Thus, we do not really consume water; we merely borrow it for a short time. However, even though the global supply remains constant, local supplies vary tremendously; often, human demands exceed the amount of water available locally. The science of water in all its forms in the atmosphere, in the oceans, in rivers, in glaciers, and underground is called *hydrology*.

14.1 AMOUNT AND DISTRIBUTION OF WATER

The total amount of water in the world, a staggering 13×10^{20} liters, almost defies comprehension. With a global population of about 2×10^9 in 1980, there is an average of 6.5×10^{11}, or almost a trillion liters of water for each person on earth.

The average daily consumption of water per person ranges from about 30 liters in developing countries to 6000 liters in the most developed countries. Because water is reusable, there is clearly more than sufficient total water for human needs.

TABLE 14.1 *World's Estimated Water Supply*

	Liters* (10^{15})	Percent of total
Oceans	1,320,000.0	97.210
Ice caps and glaciers	29,147.0	2.150
Subsurface water	8,394.0	0.620
Freshwater lakes	124.9	0.009
Saline lakes and inland seas	104.1	0.008
Atmosphere	12.9	0.001
Rivers, streams	1.2	0.0001
TOTAL	1,357,784.1	100.0

*1 gallon = 3.785 liters

However, the total amount of water is misleading when it comes to gauging supplies readily available for human activities, which are generally dependent on fresh water. Table 14.1 shows how the world's water supply is distributed. Over 97 percent is contained in the oceans and is contaminated by salt (table 14.2). Another 2.15 percent is locked up as ice and is not readily available. All the

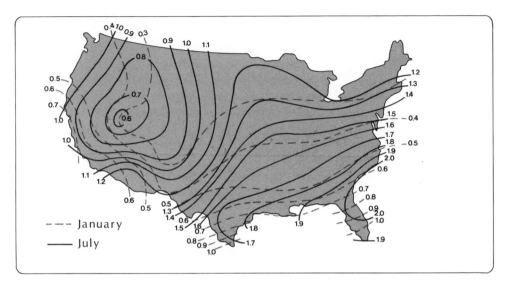

FIGURE 14.1 *Mean January (dashed lines) and July (solid lines) precipitable water (in centimeters).*

freshwater lakes and rivers of the world together hold less than 1 percent of the total supply, and of course this water is concentrated on only a small portion of the land area. For example, the Amazon River discharges enough water to satisfy all the world's demands twice over. And yet, over much of the world, including the western and southwestern United States the demand for water exceeds the supply.

As shown by Table 14.1, the amount of water in the atmosphere represents a small fraction of the total, yet 12.9×10^{15} liters is still a lot of water suspended in the air. If all this water were suddenly precipitated out, it would form a layer about 1 inch (2.54 cm) deep over the entire earth. The amount of precipitation that would result if all the moisture in an atmospheric column were condensed and rained out is called the *precipitable water*. Thus, averaged over the earth, the precipitable water is about 2.54 cm. Figure 14.1 shows the average precipitable water over the United States in January and July. Values are much higher everywhere in the warm season, owing to the increase of saturation vapor pressure with increasing temperatures. The highest values of over 6 cm occur over Florida in July, while in January, the precipitable water is less than 1 cm over many of the northern states.

Water	Salt concentration (ppm)
Distilled	0
Rain	10
Lake Tahoe	70
Lake Michigan	170
Missouri River	360
Pecos River	2,600
Ocean	35,000
Dead Sea	250,000

TABLE 14.2 *Average Concentrations of Dissolved Salts in Water. By definition, water is saline if salt concentration is 1,000 ppm or more.*

14.2 *THE HYDROLOGIC CYCLE*

The water balance presented in the previous section is not a static balance; water is continuously flowing from one category to another. It is almost a certainty that some of the water molecules we are inhaling right now have, in their long and varied past, been part of the Antarctic ice cap, a tropical ocean, a wispy cirrus cloud, and a hailstone hurtling downward toward the surface. The never-ending movement of water and its transformation from one type to another forms the hydrologic cycle, an unending exchange with no beginning and no end. Figure 14.2 gives a simplified view of the hydrologic cycle. Water is evaporated from lakes and streams, oceans, soil, and extracted from the leaves of vegetation (transpiration). As invisible vapor, this water may condense relatively soon after it is evaporated in a cumulus cloud on a sunny day, or be carried for thousands of kilometers over periods of many days before it is cooled sufficiently to produce condensation.

After condensation, much of the water re-evaporates without precipitating, as in the harmless cumulus clouds which are born and die over a period of minutes. Some of the condensed water, however, does precipitate as rain, snow, hail, or even drizzle. After reaching the ground, the water may be stored for a period of time on the surface as snow, or it may begin a variety of journeys de-

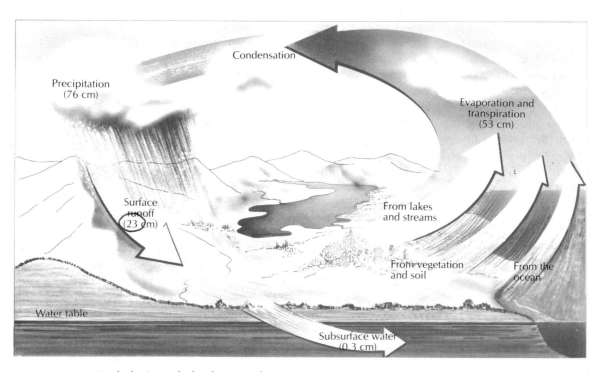

FIGURE 14.2 *Hydrologic cycle for the United States.*

404

pending on the characteristics of the soil and surface temperature. If the soil is relatively dry and pervious to water, most of the rain will soak in. Part of this water is stored in the soil and used later by plants, and another part replenishes the subsurface water table.

If the ground is relatively impervious to water, or if the soil is saturated from previous rains, most of the water will run off into creeks, streams and rivers. Part of this water eventually reaches lakes and the ocean, while another part is evaporated along the way.

If the ground is warmed by the sun soon after light precipitation ends, much of the water will be evaporated. Thus the history of a given volume of water vapor can vary tremendously over even a short time. Over the United States, a yearly average of 53 centimeters (21 inches) of water are added to the atmosphere by evaporation and transpiration. The average annual precipitation, however, is 76 centimeters (30 inches). Thus the United States imports 23 centimeters (9 inches) of water per year. This import is free, and is supplied by the winds transporting water vapor across the continental boundaries.

A more detailed U.S. water budget is presented in figure 14.3. The units are billions of gallons per day. Thus, about 40,000 units of water are carried

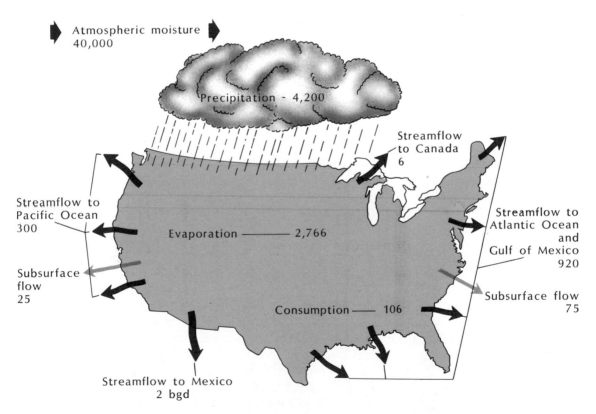

FIGURE 14.3 *Water budget of the conterminous United States. Units in 10^9 gallons per day.*

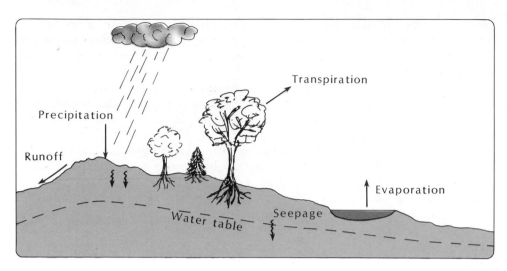

FIGURE 14.4 *Local water balance.*

into the U.S. by the winds. Precipitation of 4,200 units is the only source of water for the surface of the United States. Losses include evaporation from the surface (2,766 units), streamflow to Canada, the Atlantic and Gulf of Mexico, Mexico and the Pacific (1,228 units), subsurface flow (100 units), and evaporation of water consumed by human activities such as irrigation (106 units).

Annual average water budgets over large areas are useful in obtaining an overall view of the hydrologic cycle. However, local water balances are quite different, and for many purposes, such as agriculture, local soil and surface water budgets are quite useful. A local soil water (*SW*) balance, illustrated schematically in figure 14.4, states that the water stored in a region as soil water changes as a result of precipitation (*P*), evaporation (*E*), transpiration (*T*), runoff (*R*) and seepage (*S*). Schematically we may write this balance as

$$\Delta SW = P - E - T - R - S \qquad \qquad \textbf{(14.1)}$$

where ΔSW is the change in soil water. When precipitation exceeds evapotranspiration ($E + T$), the water storage will increase until the soil is saturated. Then surface runoff or seepage into the water table will occur. On the other hand, when evopotranspiration exceeds precipitation, the water stored in the soil decreases.

The rate at which moisture would evaporate and transpire if an unlimited supply of water were available is called *potential evapotranspiration (PE)*. Potential evapotranspiration increases with increasing temperatures and wind speed. A water balance for Vancouver, Washington, is shown in figure 14.5. From January 1 to the middle of April, precipitation exceeds potential evapotransporation, resulting in a water surplus. From the middle of April to about July 1st, *P* is less than *PE*. Soil moisture is extracted during this period so that actual evapotranspiration equals the *PE*. From July 1 to about October 1, there is not enough soil moisture to meet the large *PE,* and a soil water deficit is the result. Around October 1, *P* exceeds *PE* and the soil is recharged until it becomes saturated again around December 1.

By now it is obvious that the amounts of water participating in the hydrologic cycle are enormous, as are the reservoirs of water stored in the oceans, ice caps, lakes, rivers and even the atmosphere itself. Nature does not distribute this water uniformly, however, in space or in time. Thus an arid desert gully which has not had enough water to support life for several years may suddenly become a torrent of water which sweeps away giant boulders when an unusual flow of moist air feeds intense thunderstorms. Or, a subtle shift in the general circulation may cause rainfall to dip below normal for a decade or more, causing drought and wreaking agricultural (and cultural) havoc in the process. In this section we briefly consider two anomalous aspects of the hydrologic cycle, floods and droughts.

14.3.1 FLOODS

Floods occur whenever the soil is unable to absorb heavy precipitation and excessive runoff occurs. If the normal streams and rivers are unable to transport the runoff, they overflow and flooding results. Floods occur on many scales and are associated with quite varied types of weather phenomena. Hurricanes produce heavy rains of 25 centimeters or more over large areas, and have produced some of the most widespread and disastrous floods of history. Almost all of the damage ($2.1 billion) associated with Hurricane Agnes (1972) in the eastern United States was caused by floods. However, even this loss pales by comparison with the loss of over 300,000 lives, mostly by flooding, during the tropical cyclone that hit Bangledesh on November 13, 1970.

More frequent floods occur as a result of heavy rains from thunderstorms, but these floods tend to be localized. Most thunderstorms do not produce

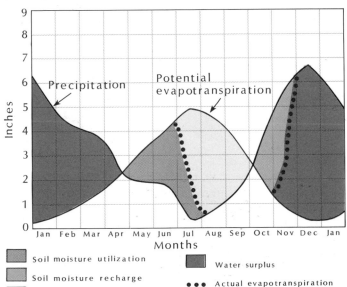

FIGURE 14.5 *Water balance in Vancouver, Washington.*

Soil moisture utilization

Soil moisture recharge

Water deficit

Water surplus

●●● Actual evapotranspiration

significant flooding, because even though rainfall rates associated with each storm may be high (several cm/h), the storms usually cover only a small area and are moving at 10–30 km/h. Occasionally, however, the winds in the upper atmosphere that cause the thunderstorms to move become very light. Under these conditions, individual or groups of thunderstorms can remain virtually stationary, and rainfall can exceed 25 centimeters. Such thunderstorms often occur in mountainous regions, because of the lifting of low-level moist air by the mountains. When heavy rain falls over mountainous areas, the runoff is concentrated in relatively few narrow gullies and enormous buildups of water are possible in a short time. Floods that occur suddenly under such circumstances are called *flash floods*. Some recent flash floods of significance include the Black Hills, South Dakota, flood of June 9, 1972, which drowned 236 people; the Big Thompson, Colorado, flood of July 31, 1976, in which at least 139 people were killed; and the Johnstown, Pennsylvania, flood of July 19–20, 1977, which killed 76 people and cost $200 million in property damage.

For many planning purposes, such as designing sewer systems, it is useful to know the likelihood of extreme rainfalls over given time periods. Figure 14.6 shows a graph of some of the greatest point rainfalls ever observed plotted against the duration of the storm. Records range from 1.23 inches of rain in one minute, which fell in a cloudburst in Unionville, Maryland, to over 1500 inches of rain associated with monsoons of Cherrapunji, India.

From long-term precipitation records, it is possible to estimate the likelihood of receiving a given amount of rain in any period of time. This probability is often expressed as the number of years required to pass before a threshold precipitation rate is likely to be exceeded. Thus a 2-year return period for a 1-hour rainfall of 4 centimeters means that a rainfall of 4 centimeters in 1 hour will occur on the average once every 2 years. Figure 14.7 shows a map of 2-year, 1-hour precipitation (in inches) for the eastern United States. According to this map, a rainfall of 2.8 inches in one hour is likely once every 2 years in southern Florida. Other likely places for extreme hourly rainfalls are the Louisiana coast and the high peaks of the Great Smoky Mountains in North Carolina.

FIGURE 14.6 *Record rainfall amounts over various time periods.*

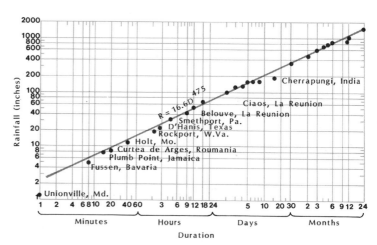

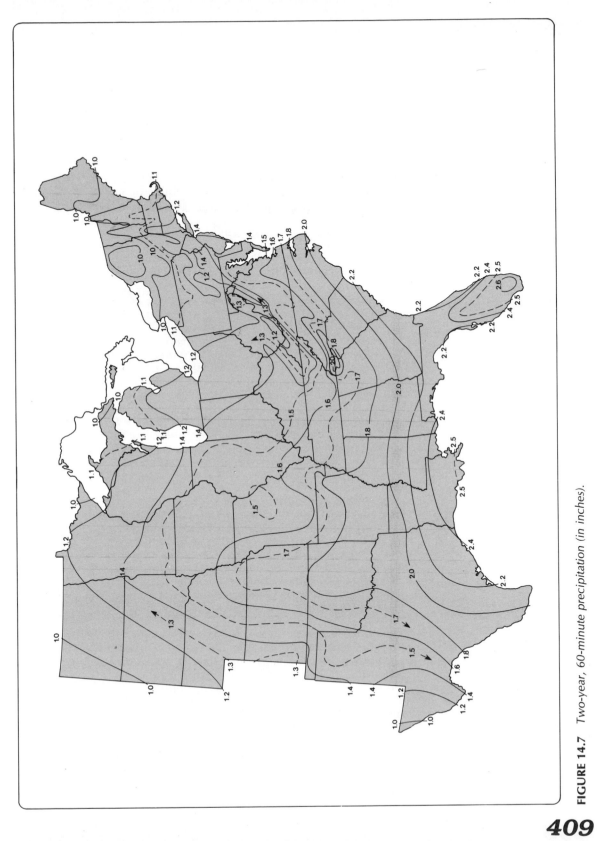

FIGURE 14.7 Two-year, 60-minute precipitation (in inches).

14.3.2 DROUGHTS

Perhaps no other agricultural weather disaster is so frustrating to the farmer as a prolonged drought. Hail, tornadoes, and even floods, do their damage quickly and then depart, but drought lingers day after day, gradually taking its toll—first on vegetation, then on livestock, and finally on the earth itself as hot winds carry away the soil. The several-year drought on the Great Plains in the 1930s which produced the dust-bowl conditions drove thousands of newly settled farmers, lured by the more bountiful rains of the 1920s, from their homes. Nowhere is this tragedy more poetically and poignantly described than in John Steinbeck's Nobel prize-winning novel, *The Grapes of Wrath*:

> To the red country and part of the gray country of Oklahoma, the last rains came gently, and they did not cut the scarred earth. The last rains lifted the corn quickly and scattered weed colonies and grass along the sides of the roads so that the gray country and the dark red country began to disappear under a green cover. In the last part of May the sky grew pale and the clouds that had hung in high puffs for so long in the spring were dissipated. The sun flared down on the growing corn day after day until a line of brown spread along the edge of each green bayonet. The clouds appeared, and went away and in a while they did not try any more . . . The surface of the earth crusted, a thin hard crust, and as the sky became pale, so the earth became pale, pink in the red country and white in the gray country . . .
>
> Then it was June, and the sun shone more fiercely. The brown lines on the corn leaves widened and moved in on the central ribs. The air was thin and the sky more pale; and every day the earth paled . . . the dirt crust broke and the dust formed. Every moving thing lifted the dust into the air: a walking man lifted a thin layer as high as his waist, and a wagon lifted the dust as high as the fence tops, and an automobile boiled a cloud behind it. The dust was long in settling back again.
>
> When June was half gone, the big clouds moved up out of Texas and the Gulf, high heavy clouds, rain-heads. The men in the fields looked up at the clouds and sniffed at them and held wet fingers to sense the wind. And the horses were nervous while the clouds were up. The rain-heads dropped a little spattering and hurried on to some other country. Behind them the sky was pale again and the sun flared. In the dust there were drop craters where the rain had fallen, and there were clean splashes on the corn, and that was all.
>
> A gentle wind followed the rain clouds, driving them on northward, a wind that softly clashed the drying corn. A day went by and the wind increased, steady, unbroken by gusts. The dust from the roads fluffed up and spread out and fell on

the weeds beside the fields, and fell into the fields a little way. Now the wind grew strong and hard and it worked at the rain crust in the corn fields. Little by little the sky darkened by the mixing dust, and the wind felt over the earth, loosened the dust, and carried it away. The wind grew stronger. The rain crust broke and the dust lifted up out of the fields and drove gray plumes into the air like sluggish smoke. The corn threshed the wind and made a dry, rushing sound. The finest dust did not settle back to earth now, but disappeared into the darkening sky.

The wind grew stronger, whisked under stones, carried up straws and old leaves, and even little clods, marking its course as it sailed across the fields. The air and the sky darkened and through them the sun shone redly, and there was a raw sting in the air. During a night the wind raced faster over the land, dug cunningly among the rootlets of the corn, and the corn fought the wind with its weakened leaves until the roots were freed by the prying wind and then each stalk settled wearily sideways toward the earth and pointed the direction of the wind.

The dawn came, but no day. In the gray sky a red sun appeared, a dim red circle that gave a little light, like dusk; and as that day advanced, the dusk slipped back toward darkness, and the wind cried and whimpered over the fallen corn.

Men and women huddled in their houses, and they tied handkerchiefs over their noses when they went out, and wore goggles to protect their eyes.

When the night came again it was black night, for the stars could not pierce the dust to get down, and the window lights could not even spread beyond their own yards . . . In the morning the dust hung like fog, and the sun was as red as ripe new blood. All day the dust sifted down from the sky, and the next day it sifted down. An even blanket covered the earth. It settled on the corn, piled up on the tops of the fence posts, piled up on the wires; it settled on roofs, blanketed the weeds and trees.

The people came out of their houses and smelled the hot stinging air and covered their noses from it. And the children came out of the houses, but they did not run or shout as they would have done after a rain. Men stood by their fences and looked at the ruined corn, drying fast now, only a little green showing through the film of dust. The men were silent and they did not move often. And the women came out of the houses to stand beside their men—to feel whether this time the men would break. The women studied the men's faces secretly, for the corn could go, as long as something else

411

remained. The children stood near by, drawing figures in the dust with bare toes, and the children sent exploring senses out to see whether men and women would break. The children peeked at the faces of the men and women, and then drew careful lines in the dust with their toes. Horses came to the watering troughs and nuzzled the water to clear the surface dust. After a while the faces of the watching men lost their bemused perplexity and became hard and angry and resistant. [From *The Grapes of Wrath*, by John Steinbeck. Copyright 1939, copyright ©1967 by John Steinbeck. Reprinted by permission of the Viking Press, Inc., New York.]

Heat and lack of precipitation are the meteorological factors responsible for most droughts. Both of these conditions occur with large-scale subsidence associated with warm anticyclones that extend from the surface through most of the troposphere. If these anticyclones are temporary, lasting only a few days, the drought is short-lived and the principal damage is confined to nonirrigated crops and annual vegetation. Even streams that depend solely on surface runoff will flow 8–12 days without rain.

The far more serious droughts occur over periods of months or years and are associated with major changes in the general circulation—the upper-level winds. As the wet ascending and dry descending portions of the waves in the westerlies reorient themselves in new positions, some regions experience drier-than-normal periods, while others several thousands of kilometers away have more-than-average rainfall. The drought of 1961–1966 over the northeastern United States is a good example. Table 14.3 shows the percentage of normal rainfall during the four seasons of the years 1962–1965 at New York City and Williston, North Dakota. While the East Coast was receiving only 60–70 percent of its normal rainfall, the Great Plains were blessed with above-average precipitation.

TABLE 14.3 *Percentage of normal rainfall by seasons (1962–1965)*

	Winter	Spring	Summer	Fall
New York	92	58	62	67
Williston, N. Dak.	101	172	125	90

Source: Jerome Namias, "Nature and Possible Causes of the Northeastern U.S. Drought during 1962–1965," *Monthly Weather Review*, Vol. 94, No. 9, pp. 543–54.

The northeast drought was associated with cold temperatures and lower-than-normal pressures off the East Coast during the four-year period. This anomalously deep tropospheric trough, which was especially well developed during the spring, produced a persistent northwesterly flow of dry, cool air over the Northeast. Furthermore, the atmosphere is normally sinking behind troughs and so subsidence contributed to the absence of rainfall. Finally, the location of the trough axis off the coast, and the associated northwesterly winds over the eastern United States, steered the rain-producing cyclones eastward away from the coast, rather than northward along the coast as is typical during wetter times.

During the period of lower-than-average pressure off the East Coast, the pressures were also lower over the Rockies. Therefore, the Great Plains were

412

frequently located on the eastern side of troughs of low pressure, which are favored places for upward motion and, hence, above-normal precipitation.

The reasons for such important, but subtle changes in the general circulation are not well known, but seem to be related to anomalies in ocean temperatures. For example, the Atlantic water temperatures along the coast from Virginia to New England were 2C° lower than normal during the 1961–1966 drought. Note, however, that we said "related to" and not "caused by," for cause-and-effect relationships in meteorology are very difficult to establish. Thus, although it is tempting to say that the cold ocean waters caused the cold atmospheric temperatures and the deep trough, it would be equally plausible to say that the cold atmosphere and the stronger-than-normal northwesterly flow (which would produce upwelling of cold water) caused the cool sea temperatures.

CONSERVATION AND AUGMENTATION OF WATER 14.4

In 1975 the United States withdrew about 339 billion (339×10^9) gallons of water per day (bgd) and returned 232 bgd to the surface or ground water source after being used; thus, the average consumption was 107 bgd (table 14.4). As indicated in table 14.4, most of the water consumed in the United States is used for irrigation (83 percent). Only about 6 percent is used by the public for domestic purposes. The total amount of water used is only about one-fifth of the amount available for use. However, as we have seen, a very irregular distribution of rainfall occurs over the United States, and while the Pacific Northwest and the East have more than enough water, parts of the Great Plains and West require more than is available right now. Their problem is being exacerbated by the flow of population to these warm sunny climates, so much effort is being directed to augmenting and conserving existing water supplies.

As indicated by table 14.4, the greatest potential reduction in water use through conservation is in irrigation. Not only does irrigation eliminate both

TABLE 14.4 *Annual Use of Water in the U.S., excluding Waterpower, in billions of gallons per day (bgd)*

Use	Water withdrawn	Percent of total	Water consumed	Percent of total
Domestic	23.3	7	6.3	6
Business and industry	63.8	19	9.4	9
Agriculture (mostly irrigation)	160.7	47	88.3	83
Steam electric generators	88.9	26	1.4	1
Miscellaneous	1.9	1	1.2	1
Total fresh water use	338.6	100	106.6	100

withdrawal and consumption, it is most prevalent where water supplies are short. One way of reducing the use of irrigation is to reduce evaporative loss of water in canals that deliver the water. As much as 25 percent of the water consumed for irrigation is lost through evaporation before it ever reaches the plants.

Although domestic uses account for only 7 percent of the total water withdrawn, there are savings possible in the home which can be important in regions where water is scarce. Some typical values of the amount of water used for household purposes are given in table 14.5.

TABLE 14.5 *Amount of Water (gallons) Used in Household Activities*

Flush toilet	5
Tub bath	40
Shower bath	20
Dishwasher (1 load)	15
Washing machine (1 load)	25
Waste (faucet dripping at rate of 1 drop per second)	4/day

According to table 14.5, significant domestic savings can be achieved by eliminating waste through leaks in the plumbing, taking showers rather than baths, and running washing machines and dishwashers only when full. Much water is wasted by letting the faucet run while shaving, washing vegetables, or rinsing dishes, when a small amount of water in the sink would do the job equally well.

Augmentation of water supplies may be explored in many ways, including desalination (desalting), enhancement of precipitation through weather modification, and even extracting water from glaciers and icebergs. Water desalination is the most practical method of increasing fresh water supplies, at least near sea coasts. Since 1955, the number of desalination plants has increased from about a dozen plants, producing 2 million gallons per day (mgd), to more than 330 in 1975. These are currently producing about 100 mgd, or about 0.1 percent of the amount of fresh water consumed in the United States. Most of the desalination plants are in California, Texas, and Florida. Desalination plants are still expensive and use large amounts of energy; therefore they are practical only where other sources of water are absent or more costly.

There have been some proposals to obtain fresh water by melting glaciers. Glaciers play an important role in the water balance of some locations, such as the Cascade ranges in Washington. Storing water as snow during the winter, glaciers return the water to streams in the summer when plants are able to use it. The South Fork of the Cascade River, which drains a basin half-covered with perennial snow and ice, peaks in the midsummer when almost no precipitation occurs.

If a thin layer of black carbon dust is spread over a ''clean'' glacier, the decrease in albedo could cause an increase in absorption of solar energy by a factor of two or more, and an increase in runoff by the same factor. Using this technique, the U.S.S.R. has advanced the ice break-up in arctic ports and bays by as much as a month. However, most of the glaciers occur in regions where an adequate water supply already exists, as in the states of Washington and Alaska

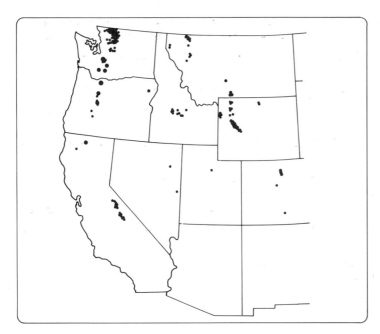

FIGURE 14.8 Location of glaciers in the Western United States.

(figure 14.8). Furthermore, melting a glacier, even if feasible, is a one-shot solution to a water shortage. What do you do after the glacier is melted?

Weather modification projects to increase wintertime snow for summer runoff or to increase rainfall from cumulus clouds have been pursued actively for the past 30 years. Many of these projects have been totally unscientific, in that modification techniques such as cloud seeding have been conducted without adequate controls or measurements to evaluate the result. A few large-scale projects, such as the Florida Area Cumulus Experiment (FACE), have been on much firmer scientific grounds. In FACE, growing cumulus clouds are seeded by aircraft with silver iodide crystals in an attempt to freeze supercooled water and stimulate the growth and merging of the clouds through the release of the latent heat of fusion. A second totally different effort is aimed at seeding orographic clouds in the western mountains during wintertime to increase snowfall. Although there is some evidence for success in these experiments, there is no way of knowing for sure how much precipitation was induced by the seeding. There is also the question whether increasing rain in one area might decrease it in areas downwind. Furthermore, not everyone in an area wants increased precipitation, so there are often political and legal steps taken to prevent weather modification. For all of these reasons weather modification has been a controversial area of research and application, and it is unlikely to have more than a local impact on water supplies in the near future.

Questions

1   If all the water vapor in the atmosphere were suddenly precipitated and spread uniformly over the earth, how deep would the rainfall be?

415

2 Over a typical 2-year period, how much precipitation would you expect in the heaviest 60-minute rainfall event in Miami, Washington, D.C., and St. Louis (figure 14.7)?

3 Explain the general decrease of 2-year, 60-minute precipitation values away from the coasts in figure 14.7.

4 What is the difference between withdrawal and consumption of water?

5 What single activity in the United States consumes over 80 percent of the total water consumption?

6 What is the objective of the FACE program? What is the scientific basis for completing this objective?

7 Compare the average precipitable water over Chicago in January and July. Why does the precipitable water vary more here than in Miami?

8 According to the water balance for Vancouver, Washington (figure 14.5), during what period of the year *could* evaporation be increased if more water were available at the surface?

9 From figure 14.6, estimate the record precipitation over time periods of 1 minute, 10 minutes, 1 hour, 1 day, and 1 month.

Problems

10 Why are flash floods more likely to occur in summer than in winter? Consider temperatures and wind speeds aloft.

11 Discuss the reasons for the maxima in 2-year, 60-minute precipitation values over southern Florida, western North Carolina, and New Orleans (figure 14.7).

12 Estimate the water budget in your house. Does it vary from winter to summer?

FIFTEEN

Meteorological Optics

reflection • refraction • diffraction • scattering • halo • prism • angular distance • sundog • subsun • rainbow • cloudbow • corona • glory • mirage • two-image inferior mirage • superior mirage • Fata Morgana • Green flash • temperature gradient • supernumerary bow

Meteorology is the study of the atmosphere, and optics is the study of light. It is not surprising, therefore, to find that the modification of light by the atmosphere is as capricious as the weather and as variegated as the rainbow. Halos trace lacy geometric patterns across the dome of the sky, a corona provides an ephemeral splash of eye-blinding color, and a mirage turns distant objects into a carnival hall of mirrors. So rich is the range of observable phenomena that even the most casual observer can often be rewarded.

There are many ways that light shining through our atmosphere can be altered by both the air itself and the cloud and haze particles it contains. These include various combinations of four basic mechanisms: *reflection, refraction, diffraction,* and *scattering.* Examples of each of these mechanisms are easy to recognize when they occur in the atmosphere, so from the point of view of basic physics, they would seem to provide a good basis for classifying the optical phenomena we observe. From the point of view of the meteorologist, though, it is not as important that both a halo and a mirage are caused by refraction as it is that a halo is caused by (refraction through) ice crystals and a mirage is caused by (refraction through) temperature gradients in the atmosphere. We will therefore divide the optical phenomena into four broad classes based on whether the light has been modified by ice crystals, water drops, molecules and dust, or temperature gradients.

15.1 ICE-CRYSTAL OPTICS—HALOS, SUN PILLARS, AND SUN DOGS

The four wheels had rims and they had spokes; and their rims were full of eyes round about.[Ezekiel 1:18]

People have long looked with awe at the phenomena that appear in the sky, and they invariably interpret them in terms of their own perceptions of the forces that control their surroundings. Light refracted and reflected by ice crystals in the upper atmosphere can produce a complex of rings, arcs, and bright spots of light whose appearance is remarkably similar to the description that Ezekiel provides us in his vision.

If this interpretation is correct, it is undoubetedly true that Ezekiel was neither the first nor the last person to deduce divine messages from observing the optics of ice crystals. In A.D. 40, the Roman College of Soothsayers boded good from an observation of three suns that appeared in the sky. Nowadays, we could bode nothing but the presence of hexagonal-plate ice crystals, which, with the help of the sun and refraction, can produce two mock suns that sit to either side of the real sun. The Soothsayers, incidentally, missed the mark, as the following year was not a good one for Rome.

The term *halo* is generic for all of the rings, arcs, and spots produced by the reflection and refraction of light by ice crystals in the atmosphere. It might be imagined that this cornucopia of patterns is a result of the almost legendary variability of the forms of the ice crystal. Not so, as only two major types of crystals, the hexagonal plate and the hexagonal column, account for almost all the observed halos. One of the reasons that these two crystals can produce so many different halo types is due to their aerodynamic properties. When the crystals are very small (diameter less than 15—20 micrometers), the constant bombardment by the rapidly moving air molecules keeps the crystals randomly oriented, rather like peanuts in a bag. This penomenon, known as *Brownian motion*, ceases to be important when the crystals grow to the larger sizes where aerodynamic forces dominate: large hexagonal plates fall with their flat faces nearly horizontal, like dinner plates set out for guests, while the large hexagonal columns fall with their long axes horizontal, like wooden pencils scattered on the floor. We have, therefore, three different classes of crystal orientations: randomly oriented crystals of either type, or oriented cloumns, or oriented plates. Each of these crystal groups is

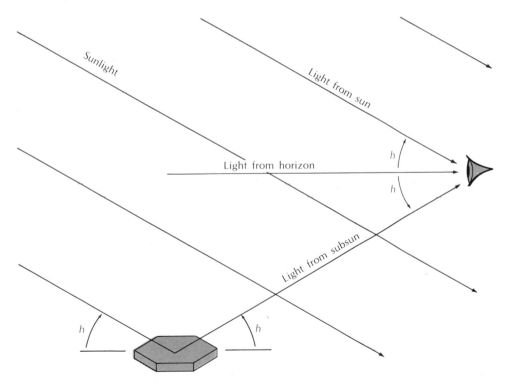

FIGURE 15.1 *Subsun. If the sun is at an angular elevation* h *above the horizon, then its image reflected in a horizontally oriented plate will be seen at an angle* h *below the horizon.*

421

capable of producing its own class of halos. In this book, we will discuss only a few of the over two dozen halo types that can be produced.

Reflection is familiar to everybody as the mechanism that enables us to see our face in a mirror or in a calm lake. When light shines on a smooth surface such as water or ice, it bounces off that surface so that it makes the same angle to the surface when leaving as it did when arriving. When large hexagonal-plate ice crystals are settling through the atmosphere, they behave like thousands of tiny horizontal mirrors which can reflect an image of the sun. This image, the *subsun* (illustrated in figure 15.1), can only be seen from an airplane or a high mountain, for it must appear as far below the horizon as the sun is above the horizon. If instead of plates there are large columns, then reflection off of the columns, which are free to rotate about their horizontal axes, will produce a pillar of light, much like the path of light that extends across a rough ocean surface to meet the setting sun. But in the atmosphere, this *sun pillar* can extend above the sun as well as below it.

When sunlight shines on a water or ice surface, not all of the light will be reflected; some of it passes into the material, and in doing so, is bent so that it travels at a larger angle to the surface. This bending, which is called *refraction* (figure 15.2), is the reason that vertical objects seen under water can appear foreshortened when they are seen from above the surface of the water. (The light rays meet the object at a much smaller angle to the vertical than if the water were removed.) Because the amount of refractive bending of a light ray depends on its color (blue is bent more strongly than red), white light can be separated into its component colors. The familiar instrument for this is the prism. In the atmosphere, tiny hexagonal ice crystals may act as prisms.

A hexagonal ice crystal is really two prisms in one: a 60-degree prism and a 90-degree prism, as illustrated in figure 15.2. Alternate hexagonal faces

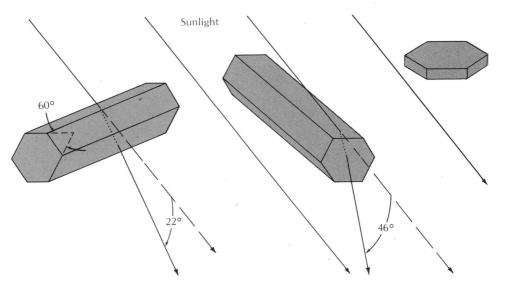

FIGURE 15.2 *Refraction of light by hexagonal ice crystals. Refraction is illustrated for the 60-degree and the 90-degree prisms of a hexagonal column only.*

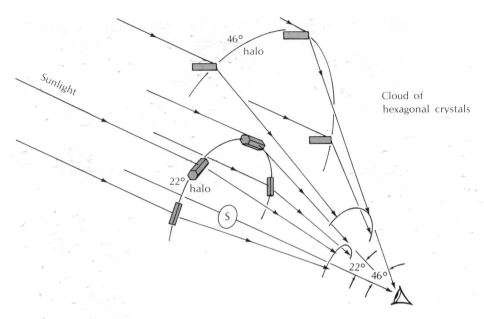

FIGURE 15.3 *Formation of 22-degree and 46-degree halos.*

make an angle of 60 degrees to each other, and the light that shines through this 60-degree prism gets bent from its original direction by an angle that is equal to or greater than 22 degrees. A sky filled with small, and thus randomly oriented, hexagonal ice crystals can produce the *22-degree halo* (figures 15.3 and 15.4), which is a ring of light that surrounds the sun at an angular distance of 22 degrees (about the width of your hand span on the end of your outstretched arm). Inside the ring, it is relatively dark, as light cannot be bent from the direction of the original sunlight by an amount less than 22 degrees, while outside the halo it is bright, because the crystals can bend the light by more than 22 degrees. As red light is bent less by refraction than the other colors, the inner edge of the 22-degree halo is red.

The sides and the ends of the hexagonal crystals make an angle of 90 degrees to each other. This 90-degree ice prism can bend light from its original direction by an angle that is equal to or greater than 46 degrees. Thus, it is possible to form a *46-degree halo* around the sun (it has a radius of about two hand spans). The 46-degree halo (figures 15.3 and 15.4) can be observed, at least in part, only about once or twice a year, whereas the 22-degree halo can be seen upwards to a hundred times a year. An analysis that takes into account the optical properties of small prisms, the aeordynamic properties of falling ice crystals, and the crystallography of growing those crystals in the atmosphere reveals that the 46-degree halo can be produced only by a very narrow size range of hexagonal columns (diameters between about 15 and 25 micrometers) that have been grown very slowly. Neither plates, nor larger or smaller columns, nor rapidly grown columns (the 90-degree prism is not formed well enough) can produce the 46-degree halo. With such restrictions, it is not very surprising that the 46-degree halo is a much rarer sight than the 22-degree halo.

FIGURE 15.4 *Fisheye (180-degree) view of 22-degree and 46-degree halos. The sun has been blocked out to prevent lens flair.*

Although halos of other radii are uncommon, they are not impossible, so there is nothing improbable about Ezekiel's observations of "four wheels." Halos with radii of 8, 17, 32, and 90 degrees have been observed and explained.

When hexagonal crystals are very small, and thus randomly oriented, they will produce a 22-degree halo. What happens when the crystals grow large enough to become oriented by the aerodynamic forces depends on whether we are dealing with a column or a plate. Large hexagonal plates can no longer produce the complete halo but will instead produce two brightly colored spots of light that sit at a distance of about 22 degrees to either side of the sun. Variously called *sun dogs, mock suns,* or *parhelia* (see figure 15.5), these spots are not as common as the halo but can be seen about two dozen times a year over most of the United States and Canada.

Large, oriented, hexagonal colums produce two arcs of light which touch the 22-degree halo at the top and bottom. Called *tangent arcs,* their form changes greatly with the height of the sun (see figure 15.6). When the sun is high in the sky, the upper and lower tangent arcs join to produce the *circumscribed halo* (figure 15.7).

Many other halos are possible, some of which can form only when the sun is low in the sky, and others when the sun is high. One of the most beautiful of all the halos, the *circumhorizontal arc* (plate 15b), cannot form when the sun is less than 58 degrees above the horizon. It is most spectacular when the sun is 68 degrees above the horizon, but in the middle latitudes, the sun gets that high only in June and early July, and then just for a few hours around noon. It is not surprising, therefore, that the circumhorizontal arc is a rare spectacle.

My heart leaps up when I behold
A rainbow in the sky.
            [William Wordsworth, "A Rainbow," 1–2]

Usually seen late in the day after a heavy rainstorm, the rainbow has become a symbol of renewed hope or, in the words of Genesis 9:15, "a covenant between God and every living creature." This optimistic view has not been shared by all cultures. For the ancient Greeks, the rainbow was Iris, the messenger of the gods, who bore news of war and death. Many African and American tribes saw the rainbow as a giant and deadly serpent. This view was extended by the Shoshones to account for the hailstorms that are so common on the high plains of America. While rubbing its back on the icy dome of the sky, the rainbow serpent would chip off small pieces of ice, which would fall to the ground.

The rainbow, undoubtedly the best-known example of meteorological optics short of the blue of the sky, owes its existence to the refraction and reflection of light by raindrops. The light is refracted once as it enters a drop and is then reflected off the inside back of the drop before being refracted again as it exits (figure 15.8). Depending on the angle that the light makes to the drop surface as it enters, the whole process of two refractions and one reflection will bend the light through an angle of anywhere between 138 and 180 degrees, filling the sky between these angular distances from the sun with light. The point 180 degrees from the sun is the head of your shadow, and a point 138 degrees from the sun is

FIGURE 15.5 *Parhelia. Two parhelia and the sun look like three suns in the sky.*

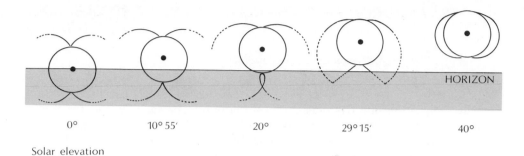

0°   10° 55′   20°   29° 15′   40°

Solar elevation

FIGURE 15.6 *Tangent arcs. The upper and lower tangent arcs to the 22-degree halo change their shape as the sun climbs higher in the sky.*

42 degrees from the head of your shadow, so you will see a disk of light centered on the head of your shadow that has a radius of 42 degrees. The rainbow is the brightly colored edge of this disk of light. Outside the rainbow, the sky becomes perceptibly darker, until at 50 degrees from your shadow another fainter rainbow is formed. This *secondary rainbow* is formed when the light is reflected twice inside the drop before it exits. To distinguish these two rainbows, the rainbow that is formed by only one internal reflection is called the *primary rainbow*. A *tertiary rainbow,* resulting from three internal light reflections, would have to form on the side of the sky near the sun, but it would be very faint. In fact, there is no reliable evidence that one has ever been seen.

FIGURE 15.7 *Circumscribed halo and 22-degree halo when the sun is at elevation of 48 degrees.*

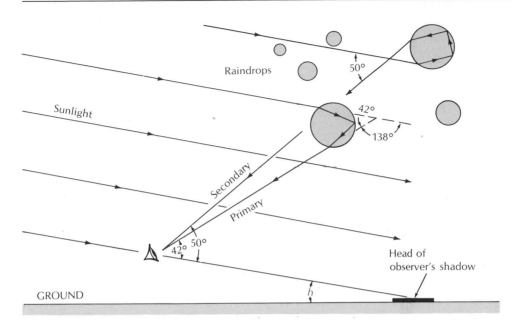

FIGURE 15.8 *Position of primary and secondary rainbows.*

Because both the rainbow and the halo are caused by refraction, we would expect that the order of the colors would be the same in both, and yet we find the red lies on the *inside* of the halo, but on the *outside* of the primary rainbow. The clue to this apparent contradiction is found when we remember that of all the colors, red is bent least by refraction and will therefore lie closest to the sun—on the inside of the halo, which surrounds the sun, and on the outside of the primary rainbow, which is on the other side of the sky. The secondary rainbow is an interesting illustration of this same principle. Here, the red is again on the inside of the bow, but the light that forms this bow has been bent by refraction through an angle of 230 degrees (180 + 50 degrees), thus seemingly turning it inside out. The red is still the closest color to the sun if you were to go around the sky in the other direction, passing the head of your shadow on the way.

There is a cliché that speaks of "all the colors of the rainbow," as though the rainbow exhibited all the colors of the spectrum. However, even casual observation will readily convince you that this is not true. Rainbows differ markedly from one to another, and the colors are not even uniform along a particular bow. This peculiar property will be easier to understand after considering a quite different optical phenomenon, the *corona*, which is also caused by waterdrops.

Often at night, irregularly shifting colors play across the surface of ragged clouds as they drift past a full moon. These colors are seen fairly close to the moon, usually at a distance of between 2 and 10 degrees, or about four to twenty moon diameters. (The sun and moon each has a diameter of about $1/2$ degree, and thus provides a convenient means of estimating angular distances.) On those occasions when the cloud through which the moon is seen appears smooth and uniform rather than ragged, the colors form a series of rings of light around the moon. Called a corona (plate 17a), the colored rings can also be seen around

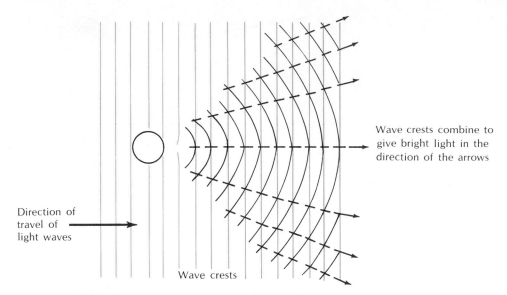

Wave crests combine to give bright light in the direction of the arrows

Direction of travel of light waves

Wave crests

FIGURE 15.9 *Diffraction. Circular waves that originate on either side of the cloud drop combine with the original straight waves to cause light to travel out at an angle to the original direction.*

the sun as it shines through some thin clouds, although it is rarely seen because the light is so dazzlingly brilliant. Newton first mentioned seeing the solar corona by viewing its reflection in still water, and although this method still remains a very good way to cut down on the intensity of the light, it is often easier to view it through a couple of pairs of sunglasses, being careful to block the sun itself from view.

The corona is caused by the diffraction of light around the small drops of water that form the cloud. To explain diffraction, we picture light as a series of waves, much like the waves on the surface of a lake. Imagine some water waves as they pass by a pole sticking out of the water. Each side of the pole causes a disturbance as the waves wash against it and sends out circular patterns of waves. In some places, the crests of the orignal waves and the crests of the circular waves combine to produce very large waves, while at other locations, the crests of one group of waves coincide with the troughs of the other, and they cancel out to give calm water. The same thing happens when light passes by the small water-drops in a cloud (see figure 15.9). Where the two series of light waves combine, we see bright light, which travels at an angle to the original light rays; where they interfere, we see darkness. Because the amount of angular bending of the light by diffraction depends on the color of the light, a red ring may appear in the dark region between two blue rings. The result is a sequence of changing colors repeating over and over and getting fainter as we look farther away from the moon or the sun.

In the corona, the size of the rings of light depends on the size of the cloud drops that produce the diffraction; the smaller the drops, the bigger the rings. As a result, we see a perfectly circular corona when the cloud drops are a uniform size throughout the cloud. Drops that vary in size throughout the cloud will produce an irregular-shaped corona, which sometimes bears scant resem-

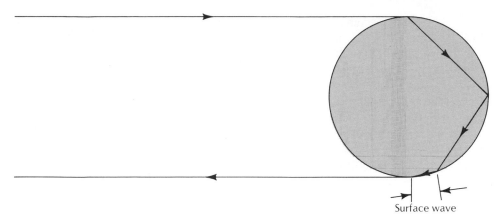

Surface wave

FIGURE 15.10 *Path a ray of light takes through a waterdrop in producing the glory.*

blance to a circle. The irregular patches of color that can result are often called *iridescence*.

In this day of frequent air travel, there is a phenomenon caused by diffraction of light by waterdrops in a cloud that is more familiar to most observers than is the corona. Called the *glory* (plate 18), it is a series of rings of colored light that can be seen around the shadow of the airplane as the shadow passes over a cloud composed of waterdrops. At first glance, the glory just looks like a fainter version of the corona, but on closer examination, it is found to be much more difficult to explain.

Like the corona, the glory is a diffraction pattern, but unlike the corona, the glory diffraction is caused by a ring of light that seems to emanate from the edges of the waterdrop. This light travels back in a direction toward the sun and thus can be seen when you stand with your back to the sun and look at the head of your shadow or the shadow of the airplane in which you are riding. This ring of light arises in a peculiar way; first, the sunlight enters one side of the drop and, like the ray that causes the rainbow, is refracted and then reflected off the back side of the drop. It is then refracted again as it exits the other side of the drop. At the end of this process, however, it has not yet been bent through the complete 180 degrees that is required to send it back toward the sun and, thus, toward your eye. It can, at most, have been bent through about 166 degrees, so in order to make up the additional 14 degrees, the light flows along the surface of the drop in what is known as a *surface wave* (figure 15.10). Having then been bent through the required 180 degrees, the light from all the edges of the drop forms a drop-sized ring that produces the pattern of light which we see as the glory.

Both the glory and the corona can be produced only by cloud drops. Because the size of the rings decreases as the drops get bigger, raindrops would produce very small rings indeed. But the light source for these phenomena is the sun or the moon, which have diameters of about $1/2$ degree. Each ring must therefore be at least $1/2$ degree wide. When the diameter of the rings becomes so small that the spacing between them is only $1/2$ degree, we end up with a continuously bright aureole of light, but no distinct rings. When we see a corona or glory, we can be sure that the drops that caused them are less than 50 micrometers in diameter and thus are well down in the cloud-drop range (recall figure 6.4).

429

We are now in a position to understand why the rainbow does not always appear in the sky just as the simple theory presented earlier would suggest. That theory said that the rainbow arose from light that had been refracted once as it entered the drop, reflected once off the inside of the drop, and then refracted again as it emerged from the drop. If we now consider that the light traveling through the drop is composed of waves, we find that the light exiting the drop has been divided into two different circular wave patterns. Like the two wave patterns that combine to produce the diffraction pattern of the corona, these two wave patterns combine to produce a diffraction pattern at the rainbow. This diffraction pattern produces variations to the simple picture in the rainbow discussed earlier.

At the upper portions of the (primary) rainbow, a number of faint bows can frequently be seen on the inside of the main bow. Called *supernumerary bows,* as if somehow nature had made a mistake by putting them there, these bows are part of the series of diffraction rings, of which the main bow is the brightest. As with all diffraction patterns, when the drops causing the diffraction are very small, the pattern is very broad and the spacing between the successive supernumerary bows increases. The broadness of the pattern has another effect: As the pattern broadens, the different colors of the main bow begin to overlap and cancel each other, so that when the main bow is caused by small cloud drops, it becomes white. This *cloud bow* (see plate 19b and compare with the rainbow in plate 19a), or *white rainbow,* can sometimes be seen from an airplane flying over stratus clouds and can be distinguished from the glory by its very much larger size and lack of color.

The closest approach to "all the colors of the rainbow" is near the foot of the bow, where the largest raindrops can contribute to the brightness of the bow. As a result, the diffraction pattern is very tight, and each color is separated from the others, with no overlap. The large raindrops are flattened by the drag of the air (contrary to the popular myth that raindrops are shaped like teardrops). Near the ground, the orientation of these flattened drops with respect to the observer and the sun is favorable for the production of the rainbow. At higher elevations, however, only spherical drops can produce the proper combination of reflection and refraction that make the rainbow visible. In the upper portion of the rainbow, therefore, the small raindrops (which are constrained to be spherical by surface tension) contribute all the light, and the bow has correspondingly more pallid colors. The small drops also produce a very broad diffraction pattern, which implies a large spacing between the supernumerary bows. Thus the supernumerary bows are most likely to be seen near the top of the rainbow. We see that one of the most beautiful of all natural phenomena, the rainbow, is very complex.

15.3 THE MIRAGE

> As for those who disbelieve, their deeds are as a mirage in the desert. The thirsty one supposeth it to be water till he cometh unto it and findeth it naught, and findeth, in the place thereof, Allah, Who payeth him his due; and Allah is swift at reckoning. [The Koran, Surah 24:39]

The above remarkable simile was written in A.D. 628 (the sixth year of the Hijrah) and suggests that the behavior of the desert *mirage* was common knowledge to the Arabs of that day. Indeed, we even find the mirages of North Africa being described as early as the first century B.C. by the historian Diodorus of Sicily. However, serious scientific investigation into the mirage did not begin until the eighteenth century, when navigators and surveyors found that atmospheric refraction greatly inhibits accurate measurements of position. The nineteenth century looked on the mirage chiefly as a curiosity, and it has only been in this century that its potential as a means of remotely determining the temperature structure of the atmosphere has prompted serious investigation again.

The mirage has had the curious fate of being misdescribed as an "illusion" by dictionaries and encyclopedias alike. This is claimed while simultaneously providing the correct explanation, which ascribes the mirage to the refraction of light by the atmosphere. A mirage is a physical reality and, if thus understood, cannot be illusionary. If you as a car driver see a mirage on the road but imagine that you see water, you are deluded, and the water is an illusion (not the mirage); if you recognize it as a mirage, however, you are not deluded, and there is no illusion. When understood, the images seen by the refraction of light through the atmosphere are no more an illusion than are the images seen by the refraction of light through a microscope, a telescope, or, for that matter, a pair of eyeglasses. While it is unquestionably true that the mirage has been the source of many illusions, so have religion and politics, but that does not make either religion or politics an illusion.

The amount of bending of light by refraction in the atmosphere is very small and so is usually not noticeable. The curved light rays cause an image (the view of a distant boat, for example) to appear slightly displaced from where the object (the boat itself) really is. Whether the image is displaced above or below the position of the object depends on the temperature (strictly speaking, density) structure in the atmosphere. Light rays are always bent so that the colder (denser) air lies on the inside of the curve (figure 15.11). An image will therefore always be displaced toward the warmer (less-dense) air. When the air temperature is greatest near the ground, as it is over a sun-baked road, then the image of a distant object is displaced downward, and the mirage is called an *inferior* (literally, *lower*) *mirage*. When the air temperature increases with height, as is common over lakes on a summer afternoon, the image of a distant object is displaced upward and is thus called a *superior* (literally, *upper*) *mirage*.

The amount of bending of the light rays, and thus the amount of displacement of the image, depends on the rate of the temperature change with height. If the temperature changes very rapidly with height, then the rays are bent strongly; if the temperature varies only a small amount with height, then the rays are hardly bent at all. It is possible, therefore, for the top of a distant object to be seen through a stronger vertical temperature gradient than the bottom of the object. The image of the top will be displaced more than the image of the bottom, and the image will appear either magnified or compressed, depending on whether it was displaced up or down. When the image has been magnified, it is said to be *towering;* when it is compressed, it is said to be *stooping* (see figure 15.11). It is possible for either stooping or towering to occur with either the inferior or the superior mirage. The variety of *imaging* (the formation of images) that can occur

431

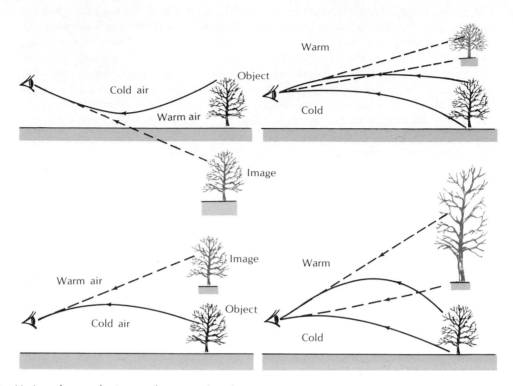

FIGURE 15.11 *Various forms of mirage. The vertical scale is greatly exaggerated.*

in the atmosphere is large, but only a few of the most interesting types will be considered here.

15.3.1 THE INFERIOR MIRAGE

Certainly the most familiar example of a mirage to most people is the appearance of what looks like water. Seen in the distance on a paved road, it vanishes as you approach, only to reappear again farther up the road. For this mirage to be seen, the road must be warmer than the air, so that the road will heat the air immediately above it. A ray of light traveling through this air, which is warmer next to the road than higher up, will be concave up (see figure 15.12). Because the temperature gradients also are stronger next to the road than higher up, the light rays that approach close to the road are more strongly bent than those that pass higher up. They are bent so strongly that a light ray that enters your eye from the direction of the road can actually have originated much higher up, on a tree, for example, or even in the sky. This image of the tree or sky that is seen on the road is reminiscent of the reflections that can be seen on water. Because the image that is seen is inverted, it strengthens the impression that a reflection has

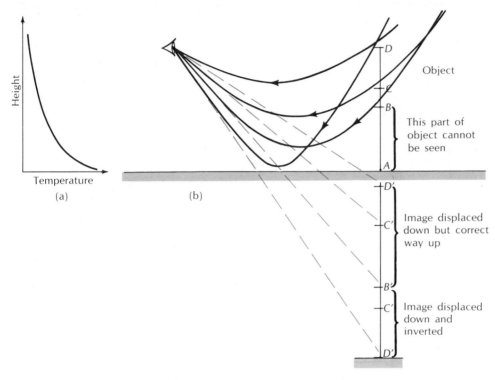

FIGURE 15.12 *(a) Temperature structure that produces a two-image mirage. (b) The rays of light that enter the eye have been bent so that an object* ABCD *will appear to the eye as* D'C'B'C'D', *and the portion between* A *and* B *will have vanished from view.*

occurred, although no reflection has taken place. This, then, is the *two-image inferior mirage:* a distant object is seen once as it would appear ordinarily, and then again inverted below the first image. To produce this effect, the temperature and the temperature gradient must increase near the ground.

There is another fascinating aspect to this type of mirage—it can make distant objects vanish. For example, a small portion of the road can no longer be seen when in its place an image of the sky appears. With increasing distance, very large objects such as people, telephone poles, and even small mountains can disappear, even though in the absence of strong temperature gradients these same objects would be readily visible. The objects cannot be seen because the light rays that ordinarily reach your eye from the object get bent so strongly that they pass up over your head instead. The view you get of a person walking off across a desert is striking. The feet vanish first, only to be replaced with an inverted image of a portion of the legs; then the legs vanish and are replaced by an inverted image of the upper body. Penultimately, the whole body is gone, and the head can be seen both normally and inverted. With a little more distance, the person vanishes completely, although he or she may be a kilometer

FIGURE 15.13 *Two-image inferior mirage seen over water. The base of the hull of the fishing boat (left) has vanished. All of the hull of the sailboat (center) in the middle distance has vanished, whereas the hull, cabin, and base of the sail of the distant sailboat (right) have vanished from view. In each case, the missing image is replaced by an inverted portion of the correct image.*

or less from the observer. The whole scene resembles a person walking farther and farther into a lake, and finally drowning. Another example of this type of mirage is shown in figure 15.13.

Curiously enough, the same appearance of drowning can be seen if a person stands still at a distance of about a kilometer from the observer while the temperature gradient increases with time. Early in the morning, the person is completely visible, but as the day progresses and the sun warms the ground, it looks as if a wall of water (see figure 15.14) flows in over the person from a distant sea. A remarkable description that fits precisely how a mirage would behave on a desert can be found in Exodus 14. In verse 21, the account tells how the waters receded at night (the temperature and the temperature gradient will decrease after the sun sets, so the apparent water will move further away), and verse 22 tells how "the children of Israel went into the midst of the sea upon dry ground, and the waters were a wall unto them on their right hand and on their left" (the apparent water will always remain some distance from the observer on all sides, and as the observer approaches, it will recede). Verse 27 tells us that "the sea returned to his strength when the morning appeared," (the sun heated the ground, the rays were bent more strongly, and the apparent water moved closer). "And the waters returned and covered the chariots, and the horsemen, and all the host of the Pharoah . . . " says verse 28 (as the day gets warmer, the rays bend even more strongly, and the apparent waters flow in over the Egyptians; as they vanish from view, the

434

FIGURE 15.14 *Inferior image. A two-image inferior mirage over the desert gives the impression that there is a wall of water in the distance.*

lower inverted image which looks like a reflection provides convincing evidence that they are indeed being inundated by water). To the Egyptians, it would have been the Israelites who had vanished. They would have given up and gone home. A comparably detailed account of a mirage was not made again for over three millenia. Clearly, the inferior mirage is not qualitatively inferior!

15.3.2 THE SUPERIOR MIRAGE

On most occasions, the *superior mirage* is not as arresting as the inferior mirage. It has one variety, however, that is not only spectacular, but almost legendary: the *Fata Morgana*. The Fata Morgana is named (in Italian) for the fairy Morgan, who was the half sister of King Arthur. Morgan was credited with magical powers which enabled her to build castles out of the air. As a result, the residents of Reggio, Italy, gave her credit for the strange structures that would appear spontaneously out in the straits to the west. Castles, houses, bridges, and, indeed, whole cities would appear, only to vanish again after a short time. Although comparable sightings have been made in many parts of the world, and many different names exist for the phenomenon, the legendary *Fata Morgana* has become generic.

The Fata Morgana is just an extreme form of towering. A distant object, which might be only one point on the surface of the sea, becomes greatly magnified so that it appears like a high wall (figure 15.15). The temperature structure that produces this type of imaging is what is sometimes called a *lifted inversion* (figure 15.16). First, the temperature increases slowly with height, then more rapidly, and above that, more slowly again. We thus have a region of strong tem-

435

FIGURE 15.15 *Fata Morgana. The cliff on the land to the left and the towers that appear in the strait are both the results of refraction. In fact, the land slopes gently into the water and the "towers" are small boats.*

perature gradient between two regions of weaker temperature gradients. Because the amount of magnification depends very critically on the relative strengths of these gradients, minor variations will mean that one point will be greatly magnified

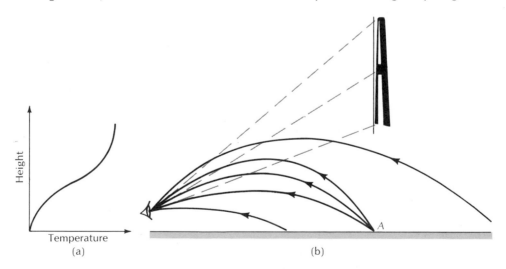

FIGURE 15.16 *(a) Temperature structure that will cause a Fata Morgana. (b) The Fata Morgana occurs when a single point A is seen through a number of different angles and so appears to be drawn out into a long line.*

436

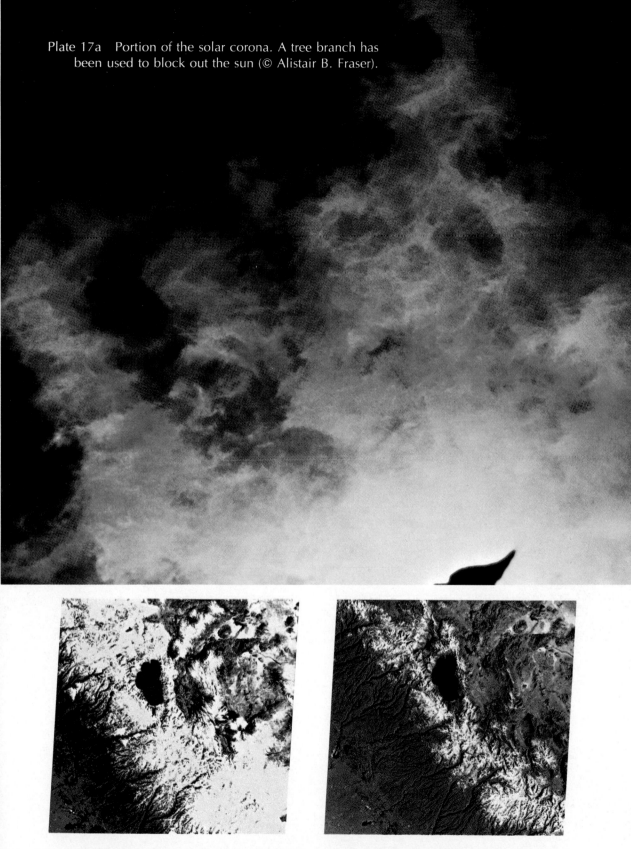

Plate 17a   Portion of the solar corona. A tree branch has been used to block out the sun (© Alistair B. Fraser).

Plate 17b   Landsat views of snow cover over the Sierra Nevadas during (left) a near-normal runoff season (1975) and (right) a drought year (1977) (courtesy of NASA).

Plate 18 Glory around the shadow of an airplane (© Alistair B. Fraser).

while another immediately beside it will, by comparison, appear almost normal. The visual impression, then, can be one of a series of towers, or a wall with gaps in it, although in fact what is being seen is a smooth surface of water. Because the gaps can appear between towers to create the impression of separate buildings, or inside the towers to create the impression of windows, or regularly along the base of a wall to create the impression of a viaduct, a whole city is simulated. As the temperature structure changes, the "city" vanishes as quickly as it came.

15.3.3 THE GREEN FLASH

There is an old Scottish legend that states that those who have seen the *green flash* will never err in matters of love; reason enough, surely, to look for the phenomenon. Occasionally, as the sun is setting or rising, a momentary flash of green light can be seen at the top of the sun (plate 20a). Arriving quite unexpectedly for most people, it seems to contradict normal experience that the low sun is yellow or red.

As the sun's light passes through the atmosphere, it is bent by refraction, so when the bottom of the sun is seen to be just touching the horizon, the sun itself is actually already below the horizon. Its image, however, has been displaced upward. Red light is refracted less than blue light, so when the low sun is examined through a telescope, it can be seen to have a red rim on the bottom and a blue rim on the top. Usually, the blue is lost by scattering as it travels through the atmosphere, and so a green upper rim is seen instead (green is next to blue in the spectrum). Under normal conditions, this green rim is so very thin that it is too small for the unaided eye to detect, but under the appropriate mirage conditions, it can sometimes be so greatly magnified that it becomes arrestingly obvious.

Questions

1 Make three lists of optical phenomena: the first of all those things that only could be seen if you were facing the sun, the second of those phenomena that only could be seen with your back to the sun, and the last of those that might be seen when looking in any direction.

2 Why does red lie on the inside of the halo, the outside of the primary rainbow, and on the inside of the secondary rainbow even though they are all caused by refraction?

3 Why does the cloudbow have more pallid colors than the rainbow?

4 In discussing the two-image mirage, why is it incorrect to speak of it as a view of the object and (an inverted) image rather than two images?

5 Why does the sundog show colors while the subsun does not?

6 If you have seen the green flash, is there any truth to the old Scotish legend?

SIXTEEN

Impact of Weather & Climate on People

Weather and climate have directly affected human evolution since the beginning, influencing the rate and direction of civilization, permeating the spiritual side of life, and engraving their many facets and moods into the various forms of our aesthetic expression. Even in today's plastic world of technology, we find our births and deaths tuned to the rhythms of the seasons, and even to the daily variation in temperatures, clouds, and rain. And now, as the exhaustion of many of our traditional forms of energy looms less than a generation in the future, our interest is renewed in the harnessing of alternative sources of energy intimately associated with the weather, such as sunshine, water, and wind.

16.1 BIOMETEOROLOGY

Whoever wishes to pursue properly the science of medicine must proceed thus. First he ought to consider what effects each season of the year can produce: for the seasons are not alike, but differ widely both in themselves and at their changes.

One should be especially on one's guard against the most violent changes of the seasons, and unless compelled, one should neither purge nor apply cautery or knife to the bowels, until at least ten days have passed. [Hippocrates, *Air, Water, Places*]

The general thrust of Hippocrates' advice concerning the role of weather and climate on human health has been confirmed repeatedly over the centuries in many cultures, and remains valid even in modern, developed countries today. Biometeorology is the study of weather and climate influences on life, including plants, animals, and humans. In this chapter we will scratch the surface of this fascinating subject by presenting some examples of how weather and climate affect people's birth, life, and finally, death.

16.1.1 EFFECT OF WEATHER ON HUMAN HEALTH

Some meteorological phenomena, such as lightning, have immediate and direct effects on human health; other phenomena, such as heat waves or air pollution episodes, cause abnormally high mortality over a period of

440

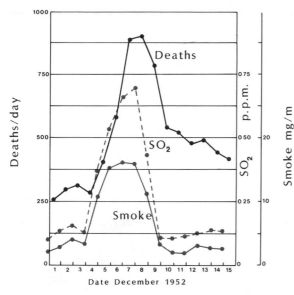

a few days, principally among victims who are already weakened by illness. For example, in December 1952 the number of deaths per day in London increased from about 250 per day to over 800 per day during a smog episode (figure 16.1). Weather also produces a number of subtle psychological and physiological effects which appear to influence many human diseases, as well as sexual behavior and criminal violence. For example, crime shows a seasonal pattern, with murders peaking in summer. Violence in cities is positively correlated with temperature and humidity excesses. Homicides in Los Angeles show an increase during the warm, dry Santa Ana winds, while the general population becomes uncomfortable and irritable. (Raymond Chandler vividly describes the Santa Ana effect in section 17.9). Similar responses are noted by Europeans under foehn* conditions. The reasons for these responses are not well understood, but it is known that changes in the weather affect the body's organs (liver, pancreas, spleen, gall bladder) and glands (pituitary, thyroid, adrenal), which control the flow of hormones into the blood.

Weather influences specific diseases by placing stresses on the body and by affecting the transmission of disease through changes in air temperature and humidity. For example, cold viruses are killed by the ultraviolet radiation in sunlight, so that transmission is favored during the cloudy, dark winter months. Some weather effects on diseases are summarized in table 16.1.

Pollutants in the air produce a variety of ill effects on humans, even in small doses. Table 16.2 lists some effects of pollution and meteorological elements on life.

Through the seasonal variation of diseases and strenuous activities such as shovelling snow (which often induce heart attacks), deaths show an annual cycle in middle latitudes that peaks during winter. In summer, human mortality correlates very highly with periods of abnormally high temperatures. Figure 16.2 shows the daily mortality rates for New York City during two heat waves in July, 1955. The peaks in daily deaths correspond closely with the peaks in temperature.

---

*A *foehn* is a warm dry wind that descends the slopes of the Alps.

TABLE 16.1 *Some short-term effects of weather on diseases*

Disease	Aggravating weather conditions
Tuberculosis	high humidity, fog
Asthma (bronchial)	sudden cooling and falling pressure
Bronchitis	fog
Skin cancer	direct sunlight
Arthritis	sudden cooling
Heart attacks	sudden, strong cooling
Common cold	sudden changes in weather (affects body's thermoregulation mechanism, membrane permeability)
Influenza	Relative humidity below 50 percent favors transmission of influenza virus
Skin diseases	High temperature and humidity

TABLE 16.2 *Effects of pollution and meteorological elements on life*

Agent	Definite effects	Possible effects
Sulfur dioxide and related compounds, such as sulfuric-acid rainwater	(1) Aggravate asthma and chronic bronchitis (2) Impair breathing; lead to lung damage (3) Irritate sensory organs (eyes, nose, throat) (4) Damage vegetation	(5) Damage buildings and works of art, e.g., sulfuric acid dissolves marble statues
Sulfur oxides	(1) Increase mortality (death rate) for short term (2) Increase morbidity (incidence of disease) for short term (3) Aggravate bronchitis and cardiovascular disease (4) Contribute to development of chronic bronchitis, emphysema	(5) Contribute to development of lung cancer
Particulate matter not otherwise specified, such as dust	(1) Restricts visibility (2) Dirties surfaces such as cars, laundry, and buildings (3) Reduces sunlight	(4) Leads to increase in chronic respiratory disease
Oxidants (including ozone)	(1) Aggravate emphysema, asthma, and bronchitis (2) Irritate eyes and respiratory tract; impair athletic performance (3) Increase probability of motor vehicle accidents	(4) Alter oxygen consumption (5) Accelerate aging

442

TABLE 16.2 *(continued)*

Agent	Definite effects	Possible effects
Carbon monoxide	(1) Impairs oxygen transport function	(2) Increases general mortality and coronary mortality rates (3) Impairs central nervous system function (4) Causal factor in arteriosclerosis
Nitrogen dioxide	(1) Discolors atmosphere	(2) Factor in pulmonary emphysema (3) Impairs lung defenses
Lead	(1) Accumulates in body (2) Proves lethal to animals eating contaminated feed	
Fluorides	(1) Damage vegetation; harm animals	(2) Lead to fluorosis of teeth
Ethylene	(1) Damages vegetation; hastens ripening of fruit (used deliberately by some tomato growers to artificially ripen tomatoes)	
Chlorinated hydrocarbon pesticides e.g., DDT	(1) Are stored in body; source usually milk and animal fats (2) Lead to ecological damage	(3) Impair learning and reproduction
Hydrothermal pollutants	(1) Can influence local climate; can interfere with visibility	(2) Have influence on action of hygroscopic pollutants
Airborne microorganisms	(1) Cause airborne infections	
Cold, damp weather	(1) Causes excess mortality from respiratory disease and fatal exposure or frostbite (2) Leads to excess morbidity from respiratory disease and morbidity from frostbite and exposure	(3) Contributes to excess mortality and morbidity from other causes (4) Causes or aggravates rheumatism
Cold, dry weather	(1) Causes mortality from frostbite and exposure (1) Leads to morbidity from frostbite and respiratory disease	(3) Impairs lung function
Hot, dry weather	(1) Causes heat-stroke mortality	

443

TABLE 16.2 *(continued)*

Agent	Definite effects	Possible effects
	(2) Causes excess mortality attributed to other causes	
	(3) Leads to morbidity from heat stroke	
	(4) Impairs function of renal and circulatory tracts; aggravates renal and circulatory diseases	
Hot, damp weather	(1) Increases skin infections	
	(2) Leads to heat-exhaustion mortality	(7) Increases prevalence of infectious agents and vectors
	(3) Causes excess mortality from other causes	
	(4) Causes heat-related morbidity	
	(5) Impairs human performance	
	(6) Aggravates renal and circulatory disease	
Natural sunlight	(1) Leads to fatalities from acute exposure	(5) Increases malignant melanoma
	(2) Causes morbidity due to "burns"	
	(3) Leads to skin cancer	
	(4) Interacts with drugs in susceptible individuals	
	(5) Causes more rapid aging of skin	

Source: Patterns and Perspectives in Environmental Science, report prepared for the National Science Board, National Science Foundation, 1972.

In contrast, the number of deaths per day decreases sharply in the cooler weather immediately following the heat waves. The total variation in death rates between the hot and cool periods is over 50 percent.

Weather not only affects times of death, it affects times of conception as well. Figure 16.3 shows the birth and conception rates by months for 1973 for the entire United States and for Florida. Over the United States, conception rates were considerably higher during the cold months (October–March) than during the warm months (April–September). Florida, which has extremely warm and humid summers, exhibits even a stronger bias in the rate of conception during the comfortable season. For example, the percentage of conception jumps from 7.7 percent in September, to 8.3 percent in October, the month when cold fronts again affect the state. These graphs suggest that weather significantly affects our sexual lives as well as other aspects of our health. There is also statistical evidence

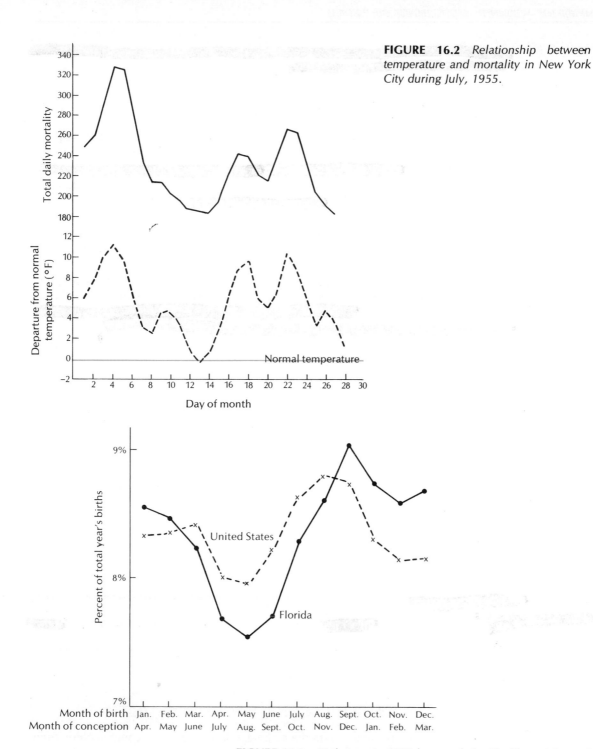

FIGURE 16.2 *Relationship between temperature and mortality in New York City during July, 1955.*

FIGURE 16.3 *Birth rates in 1973 by month for Florida and the entire United States. The data have been normalized to a 31-day month.*

indicating a predominance of males conceived during periods of thermal stress associated with cold and heat waves.

16.1.2 HUMAN HEAT BUDGET

The human body has a number of mechanisms that regulate the inner body temperature to 37°C. Because the body is continuously producing heat, heat loss must occur in order to maintain a constant temperature. A completely insulated human body at rest would experience a temperature increase of 2C° per hour, so that death from overheating would occur in only a few hours.

The body loses heat through conduction and convection, radiation, and evaporation of water from the lungs and skin. The relative contributions of these processes to the total cooling vary greatly depending on air temperature, humidity, clothing, and amount of exercise. A typical budget for a resting, clothed human is given in table 16.3. Heat losses by convection and conduction decrease as the temperature rises. When the air temperature exceeds the body temperature, the heat loss becomes negative and the body absorbs heat from the air.

All humans, regardless of skin color, radiate as black bodies. Radiative heat loss depends on the average temperature of the objects in "view" of the body. Normally this temperature is lower than body temperature and a large loss of heat occurs. Because radiation is a more important component of the human heat balance than convection for a person at rest, it is possible to feel quite warm in a room with air temperatures 10°C if the wall and ceiling temperature are warmer, say 30°C. Conversely, if the air is warm (30°C) and the walls cool (10°C), a person will feel cold.

In hot deserts during the day, both radiation and conduction may add rather than remove heat from the body. Under such conditions, evaporation must supply 100 percent of the required heat loss or death will occur quickly. For this reason it is essential to drink copious amounts of liquids. In extremely hot (shade temperature greater than 36°C) and dry (relative humidity less than 30 percent), the sweat output may be 1−2 liters per hour.

TABLE 16.3 Typical loss of heat by resting, clothed human

Convection and conduction	15%
Radiation	60%
Evaporation	25%

16.1.3 WIND CHILL AND HUMITURE

It is a familiar experience that windy days feel much colder than calm days of the same temperature. Air is such a poor conductor that if no wind is blowing and we are not moving, a thin layer of warm air forms next

446

to the skin and we may feel quite comfortable. Thus, we may sit on the porch of a ski lodge out of the wind and in the sunshine and feel quite comfortable, even though the air temperature may be −5°C.

The effects of wind and air temperature on the cooling rate of the human body have been combined in the wind-chill index, shown in figure 16.4. The wind-chill chart translates the cooling power of the atmosphere with wind to a temperature under calm conditions. Thus, a naked body would lose as much heat in a minute with a temperature of 30°F and a wind of 10mi/h as it would with no wind and a temperature of 16°F.

Wind is not the only variable besides temperature that determines how rapidly the body loses heat. Because evaporation can be a significant part of the human heat budget, the relative humidity is important in determining the rate of body heat loss and therefore the comfort when the temperature is high. For high humidities, evaporation is reduced and discomfort increases; when the humidity is low, evaporation is enhanced, and even a high temperature can be comfortable. An index that combines the effects of temperature and humidity on comfort is the *humiture*. Table 16.4 gives the humiture index for various temperatures and relative humidities. Joggers and athletes may use the humiture to estimate the heat stress placed on the body in summer.

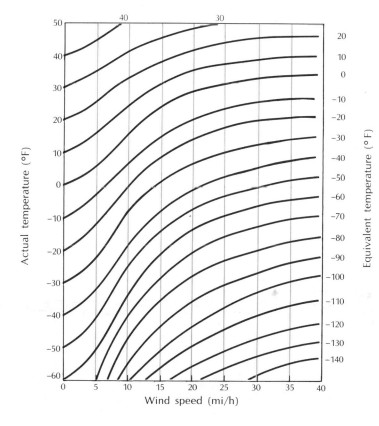

FIGURE 16.4 *Wind-chill chart with equivalent calm-air temperatures as a function of actual air temperature and wind speed.*

TABLE 16.4 *Humiture index—a measure of body heat stress and discomfort*

		Relative humidity									
		10	20	30	40	50	60	70	80	90	100
Temperature (°F)	104	90	98	106	112	120					
	102	87	95	102	108	115	123				
	100	85	92	99	105	111	118				
	98	82	90	96	101	107	114				
	96	79	87	93	98	103	110	116			
	94	77	84	90	95	100	105	111			
	92	75	82	87	91	96	101	106	112		
	90	73	79	84	88	93	97	102	108		
	88	70	76	81	85	89	93	98	102	108	
	86	68	73	78	82	86	90	94	98	103	
	84	66	71	76	79	83	86	90	94	98	103
	82	63	68	73	76	80	83	86	90	93	97
	80	62	66	70	73	77	80	83	86	90	93
	78		63	67	70	74	76	79	82	85	89
	76		60	63	67	70	73	76	79	82	85

16.1.4 LIGHTNING, PLANTS, AND PEOPLE

I flamed amazement; sometimes I did divide,
And burn in many places; on the topmast,
The yards and bowsprit, would I flame distinctly,
Then meet and join.
[*The Tempest*, Act 1, Scene ii]

The form of lightning taken by Ariel as he boarded the King's ship in *The Tempest* was St. Elmo's Fire, a harmless, visible electrical discharge which occurs around the tops of high pointed objects such as ship's masts. Other types of lightning, however, are not so harmless. In an average year, lightning kills more people (200–300) in the United States than any other weather phenomenon, including tornadoes or hurricanes. In addition, lightning from the over 1800 thunderstorms occurring at any time over the earth causes many livestock casualties and starts over seven thousand forest fires each year in the western United States alone.

A generation raised under the friendly, admonishing eyes of Smokey the Bear tends to believe that forest fires are invariably damaging and are to be prevented at all costs. However, forest fires are frequently very beneficial to the ecology of the region. They may control undesirable woody plants, allowing grasses to grow on grazing land, in turn reducing soil erosion and runoff. Occasional small fires will lower the chances of a major conflagration by periodically removing vegetative debris that might otherwise accumulate to dangerous levels. In addition, small fires produce soil fertilization from the ashes, control ticks, poisonous snakes, and other undesirable animals, destroy fungi, and create a better habitat for game animals.

There is one direct beneficial aspect of lightning to vegetation. Lightning is a natural producer of fertilizer. The lightning discharge in air produces ozone, ammonia, and oxides of nitrogen, compounds which react with the rainwater to produce ammonia hydroxide and dilute nitric and nitrous acids which serve as soluble fertilizers. Thus, strikes of lightning near (but not too near!) plants are quite desirable.

Lightning has always had a devastating effect on people, as evidenced by the following amazing story, which occurred over a hundred years ago (see figure 16.5):

> *Mr. Cardan relates that eight harvesters, taking their noonday repast under a maple tree during a thunderstorm, were killed by one stroke of lightning. When approached by their companions, after the storm had cleared away, they seemed to be still at their repast. One was raising a glass to drink, another was in the act of taking a piece of bread, and a third was reaching out his hand to a plate. There they sat as if petrified, in the exact position in which death surprised them.* [Nicholas Camille Flammarion, *The Atmosphere*. Translated from the French; edited by James Glaisher (New York: Harper Brothers, 1874), pp. 439, 440).]

FIGURE 16.5 *"Harvesters killed by lightning."*

FIGURE 16.6 *Fulgarite—sand fused by heat from lightning.*

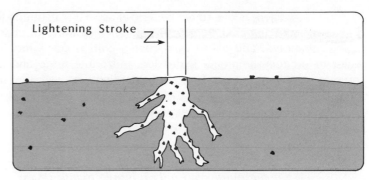

When lightning strikes sandy soils, the extremely high temperature (30,000°K, or 5 times the temperature of the sun) may melt the quartzite sand along the path of the stroke. The resulting glassy object, called a *fulgurite*, resembles petrified tree roots (figure 16.6). Fulgurites are found in all parts of the United States. The best time and place to search for fulgurites are after a thunderstorm on a beach.

When lightning strikes water, the current spreads rapidly in all directions, stunning or killing fish (or people) in the water. Because water conducts electricity so much better than air, it is a dangerous place to be in a thunderstorm. Other dangerous places are on high points of land or near solitary trees. On the other hand, automobiles are one of the safest places to be. Lightning that hits a car will travel around the outside of the car and jump harmlessly to the ground.

Unprotected houses are not safe from lightning. Lightning may cause chimneys to explode, ignite fires, or travel along wires, walls or ceilings inside the house or kill victims inside. Unsafe locations are near plumbing fixtures, telephones, televisions, or electrical appliances.

In 1753, Benjamin Franklin published a short article, "How to Secure House, etc. from Lightning," in *Poor Richard's Almanack*. Franklin's lightning rod, which he refused to patent, attracts lightning from other potential targets and conducts the current safely to the ground.

It is not necessary that it be raining to be in danger of lightning—in fact, it is not even necessary to have a thunderstorm overhead. The median horizontal distance from the source of lightning in the cloud to the strike location on the ground is about 4 kilometers, and ground points as far as 20 kilometers from a thunderstorm have been struck. A good rule of thumb is that if you can hear thunder, you are in danger of being hit by lightning. Since about two-thirds of all lightning deaths occur while the victim was outdoors, it is important to get inside some shelter when thunderstorms are nearby.

Your chance of being hit by lightning are about a million to one. If you are hit, you will either be killed instantly without pain or you will likely recover fully, not recalling the strike at all. People are rarely hit directly by lightning; such a hit would always be fatal. Instead, they are usually struck by small secondary currents radiating outward from the main stroke. Minor strikes often knock the victim unconscious, and the electric current may stop the involuntary heart beats and breathing. Artificial respiration and cardio-pulmonary resuscitation are often helpful.

Freakish lightning stories abound. While many are true, some appear downright incredible. According to one account, in Pittsburgh on July 30, 1892, two people were killed by lightning. Burned on the chest of one victim was a perfect photographic image of the tree which had failed to provide protection to the individual. An equally impossible story originated from Boston on July 10, 1891, when a sleeping cat was supposedly electroplated when lightning stripped silver from a sword hanging on an overhanging wall.

WEATHER AND THE ECONOMY 16.2

Few aspects of the economy are independent of weather. Violent winds, floods, hail, and lightning cause hundreds of millions of dollars worth of damage to property and agriculture every year in the United States. Freezes in Florida periodically destroy large citrus crops, raising prices for oranges in New York. A cloudy, cool Fourth of July can drive thousands from the beaches into the movie theaters. Weather even affects politics: the unusually severe winter and heavy snows in Chicago during the winter of 1978–1979 probably played an important role in the defeat of the incumbent mayor, who was blamed for the slow removal of snow.*

Agriculture is perhaps the most sensitive area of the economy to weather variations. Too much or too little rain, or even moderate rainfall at the wrong time, can severely reduce the yield of a particular crop. Weather affects crop growth and vitality, meat, wool, and dairy production, and the incidence of disease. Temperature and humidity variations affect the life cycle of insects, and peculiar combinations of these variables can result in plagues which darken skies and turn lush fields into bare ground.

The role of weather in business and industry has been studied less systematically than in agriculture. However, it is easy to find examples where weather plays a significant, even dominant role. The construction industry is perhaps most dependent on weather and climate. Not only must day-to-day variation in weather be considered before starting such activities as pouring concrete, but climatic factors which place wind, thermal, and moisture stresses on construction must be allowed for in the design of the structure. The destruction by wind of the first Tacoma Narrows bridge at Tacoma, Washington, in 1940 is an example of an engineering design problem. The necessity of designing sewers, dams, canals, and bridges for the maximum likely rainfall is obvious.

Transportation is an industry which can be paralyzed by weather. Air, water, and land travel can be stopped over large areas for days at a time, costing unrecoverable amounts of money in lost revenues and wasted time. On the other hand, a few industries, such as the motels and restaurants, may actually benefit from such disruptions.

---

*To some extent the blame was unjustified. The heavy snow which caused the problem was likely to occur about once every 100 years. It makes little economic sense to have sufficient machines and personnel available every winter for the one year in a hundred when they are needed.

Weather affects the utilities primarily through consumption of energy. Prolonged cold spells in winter or hot spells in summer can severely tax the capability of power companies to meet demand. Hydroelectric power, of course, is dependent upon precipitation. In the future, an increasing reliance upon solar and wind power will make the energy industry even more sensitive to weather.

Some industries are able to protect themselves from inclement weather to a significant degree: store-front windows may be boarded up as a hurricane approaches, and fruit growers may burn oil or run huge fans to prevent the accumulation of cold air near the surface of orchards on a night when a freeze is forecast. Because such protective measures cost money and time, the cost of taking the action must be weighed against the risk of the event and the loss if the event occurs and no action is taken. A simple cost–benefit analysis can often be used to decide whether or not to take a protective action. If C is the cost of taking action, L is the cost of damage if no action is taken and the adverse weather occurs, and P is the probability of the adverse weather occurring, the cost–benefit analysis says

If P is greater than $\dfrac{C}{L}$, it will pay to take action.

If P is less than $\dfrac{C}{L}$, it will not pay to take action.

If P equals $\dfrac{C}{L}$, it does not matter whether action is taken.

For example, if the potential loss of a strawberry crop due to frost is $500, and the cost of covering the crop is $100, it will pay to take action if the probability of frost is higher than 20 percent.

FIGURE 16.7 *The relationship between average temperatures in September and October and the relative contribution of those months to the total September-to-December sales of women's winter coats.*

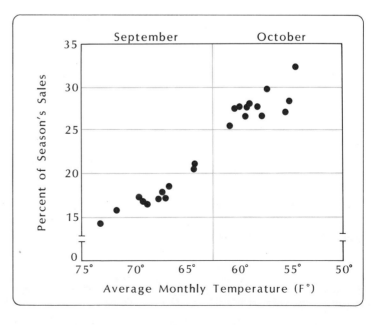

Finally, weather affects the sales of many products. Beer consumption in Rhode Island increases by about 1 percent for every 1F° temperature departure above normal. The percentage of winter coats sold early in the season is inversely proportional to the temperature (figure 16.7). Sales of weather-related clothing, raincoats, and umbrellas are highly correlated with occurrence of rain.

METEOROLOGICAL ASPECTS OF THE ENERGY SHORTAGE 16.3

Not only are the number of energy consumers increasing worldwide, the consumption of energy per person is increasing as people continue to demand higher standards of living through energy-consuming technology. Within the past decade, however, it has become clear to more and more people that the shortage of conventional fossil fuels is real and that drastic changes in the life styles of industrial countries must occur unless alternative sources of energy are developed. In the wake of scares over nuclear power, once promised as the major source of energy in the 21st century, there has been much interest in so-called *soft energy sources,* including sunshine, wind, and water. Of these sources, only hydroelectric power makes a significant contribution to total energy production in the United States (table 16.5). While these sources have been overly advertised as realistic solutions to the energy problem for this generation, they can help to conserve the remaining fossil fuel and buy time for the development of new energy technologies.

Before considering the impact of meteorological sources of energy, we need to distinguish between energy and power, which, although often confused, are really quite different. Energy can be thought of as the total amount of fuel or work needed to perform a certain job or raise the temperature of a substance by a certain amount. Thus 1 calorie of energy is required to raise the temperature of a gram of water 1C°. Or, the amount of energy required to lift a 1 kilogram weight 1 meter is 9.8 Joules (41 calories).

Energy is not associated with a time scale: it doesn't matter whether the heat is applied rapidly or slowly or whether the object is lifted 1 meter in 10 seconds or 10 minutes—the amout of energy required is the same. In contrast, power is the rate of energy consumption or the rate of doing work. To raise the temperature of a gram of water 1C° in 1 second requires 10 times the *power* to produce the same temperature rise in 10 seconds, even though the total *energy*

Type	1978	2000
Coal	18%	16%
Petroleum	49	37
Natural gas	25	18
Nuclear power	4	26
Hydropower	4	3

Total energy consumption in 1978 was 2.3×10^{13} kWh.

TABLE 16.5 *Percentage of total energy consumed in the United States in 1978 and projected percentages for the year 2000*

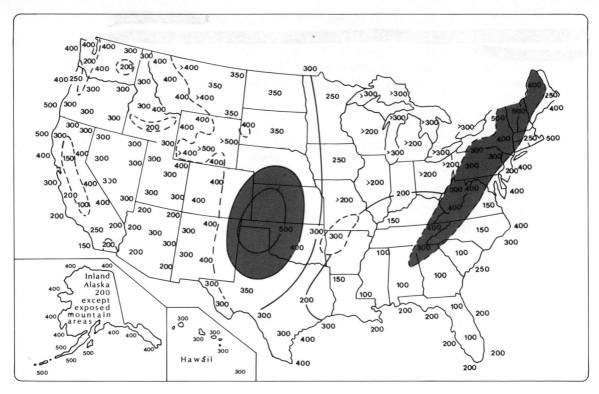

FIGURE 16.8 *Annual mean wind power (W/m²) estimated at 50 m above exposed areas.*

required is the same. In general, two actions may require identical amounts of energy but quite different amounts of power.

There is enough energy associated with solar radiation and wind over the earth every day to supply the projected needs of everyone in the world for the foreseeable future. The problem is that these energy sources are diffuse; they are associated with relatively low power. Figure 16.8 shows the estimated annual mean wind power at a height of 50 meters above the surface in exposed areas. Even over the Great Plains where the winds are relatively steady and fast, the average power is only about 0.5 kW per square meter. The average solar radiation power reaching the surface is the same order of magnitude, several hundred watts per square meter. Thus, without some mechanism of concentrating these energies, they are suitable only for jobs requiring relatively low power.

Figure 16.9 depicts the types of energy consumption in the United States as a percentage of the total energy use. The various categories may be classified as requiring relatively low or high power. Space heating and hot water heating are uses which do not require a lot of power. Solar and wind energy systems are well suited to these needs. On the other hand, transportation requires great amounts of power, and soft energy sources are very inadequate for these purposes. Try to imagine a solar-powered jet aircraft! Because of the greatly different power requirements of the various energy uses, the obvious strategy is to try to convert as

454

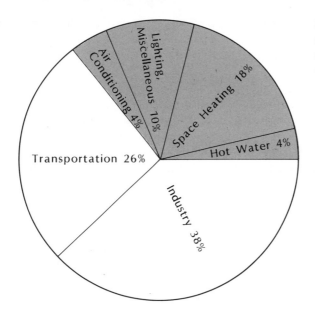

FIGURE 16.9 *Percentage of total U.S. energy demand in 1975. Shaded area denotes residential and commercial uses.*

much of the low-power energy requirements to solar systems and save concentrated energy supplies such as petroleum for the uses requiring high power.

16.3.1 WIND POWER

The wind, an indirect form of solar energy, has been used by people to power boats and pump water for centuries. The windmill has been used to generate small amounts of electricity since 1890, and in the rural United States, thousands of windmill-driven generators have been used to supplement the electrical needs of farms. Most of these small windmills generate about 1 kilowatt of power. Because *each* person in the United States requires about 1 kilowatt of power, and the projected U.S. requirement in the year 2000 is 1.6×10^9 kilowatts, it would clearly be impractical to rely on windmills of this size.

However, bigger, more powerful windmills are possible and in fact have already been tested. The most famous example is the "Grandpa's Knob" windmill, erected in 1941 on a 600-meter (2000-foot) mountain in Vermont known as *Grandpa's Knob*. This windmill consisted of two propeller blades 53 meters (175 feet) in diameter, each weighing 8 tons. The windmill produced energy at an average rate of 1250 kilowatts before flying apart in 1945, hurling one blade 230 meters (750 feet) into the air.

The Grandpa's Knob windmill did, indeed, prove that significant amounts of energy could be produced by the wind, but a number of problems remain, some meteorological, and some economic. First, large windmills can theoretically convert about 35 percent of available wind energy to electricity, but to reach this maximum, steady wind speeds of over 30 km/h are required. However, few populated regions of the United States have low-level wind speeds of this magnitude often enough to produce a dependable supply of wind-driven electric

455

energy. Therefore, windmills would have to be located in selected regions where strong, steady winds occur, such as the Great Plains, along high mountain ridges, or on the oceans. Unfortunately, in these places, storage and transport of the electricity would be a problem.

Even in normally windy areas, some mechanisms for storing the energy are necessary for the inevitable calm periods. Suggested storage mechanisms include batteries, pumping water back into dams when the wind is blowing for later hydroelectric generation, and electrolyzing water into hydrogen and oxygen, which could later be burned as a clean fuel.

The maximum available wind power P (in kilowatts) increases as the cube of the wind speed V (in km/h)

$$P = 1.4 \times 10^{-5} \, A \, V^3 \qquad \qquad \textbf{(16.1)}$$

where A is the area (in m^2) through which the wind passes. Of course, not all of this energy can be captured. A typical windmill with a propeller diameter of D meters and an overall efficiency ε of converting wind to electric energy produces power (in kilowatts) according to

$$P = 1.1 \times 10^{-5} \, \varepsilon \, D^2 V^3 \qquad \qquad \textbf{(16.2)}$$

Thus a windmill with a typical efficiency of 0.35, a blade diameter of 3m, and a wind speed of 30 km/h will produce 0.9 kW of power. Figure 16.10 shows a graph of power versus wind speed for windmills with propellers of various diameters.

Before rushing out to purchase a wind generation system for the household, a number of factors should be considered. First is a careful study of the average steady wind speed at the site. Because of large variations in wind speed and direction over short distances, especially over hilly or forested terrain, the wind climatology at airports may not be representative of your particular site. According to figure 16.10, little energy can be produced until the wind exceeds

FIGURE 16.10 *Wind generator output versus wind speed (efficiency = 0.35).*

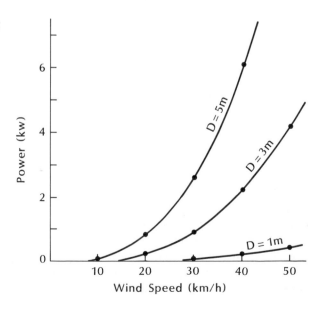

about 30 km/h. The average household requires about 800 kilowatt-hours per month. To produce all of this energy with a windmill with a 3-meter diameter propeller would require a wind slightly greater than 30 km/h blowing all the time, a situation which never occurs. More typically, such speeds might blow only 20 percent of the time, and this provides only 20 percent of the required energy.

If you determine that significant wind speeds do exist, the cost of the wind and energy storage system must be considered. Small wind-generating systems, up to a maximum rated capacity of 1000 watts, cost about $4.00 per maximum rated watt in 1980. Thus a windmill rated at 1000 W would cost about $4,000. After this initial installation cost, the energy is free except for maintenance and repairs.

Perhaps the major problem with wind energy generation on a large scale is the demand for space that windmill "farms" would require. To generate significant amounts of energy, some have envisioned vast forests of giant windmills, each 70 stories high covering thousands of square kilometers of land. One specific proposal consists of 300,000 wind turbines in the Great Plains spaced as close as one per each square mile. Each 250-meter tower would carry 20 turbines powered by 2 blades 15 meters in diameter. This network would provide about 2 $\times$ 10^5 megawatts of electricity, which is slightly over half of the total U.S. electrical generation capacity in 1970 (3.6×10^5 megawatts). However, such an array would pose many safety and environmental problems. Will people tolerate such mammoth structures over such a wide area? What will the safety record be? (Recall the demise of the Grandpa Knob windmill.) How much noise will be generated? How much useful agricultural or recreational land will have to be sacrificed? How often will these giant machines require maintenance and repair? Clearly wind power, like most other forms of energy production, has its serious side effects.

16.3.2 SOLAR POWER

In these days of dwindling supplies of oil, coal, and gas, who has not gazed at the fiery sun on a cold winter day and dreamed of using that energy to heat our homes or power our cars? The sun is indeed a potential source of safe, clean, and abundant energy. The earth intercepts more solar energy than we could ever conceivably use; in only two weeks, it receives more energy than the entire known global supply of fossil fuels.

The solar constant of 2 calories per square centimeter per minute is equivalent to 1400 watts per square meter, or 1400 megawatts per square kilometer. Of course, the amount of radiation reaching the ground is much less, being reduced by clouds and dust in the atmosphere and by low solar elevation angles. Nevertheless, the United States on an annual average receives about 13 percent of this figure, which means that about 180 watts per square meter are available. This rate of heating produces, on the average, about 4 kilowatt-hours (kWh) of energy per square meter each day, which is twice the amount needed to heat and cool the average house. Thus, it is obvious that solar power could make an important contribution to the world's need for power.

Although plenty of solar energy falls on our backyards every day, the technological problems of utilizing this energy on a large scale are substantial. First, no energy arrives at night, which means that supplemental sources or expensive solar energy storage devices are necessary. Variations in the weather, especially in the amount of cloudiness, make the solar energy supply quite variable from day to day and from season to season. Furthermore, some means of concentrating the diffuse solar radiation must be used. Also, the cost of converting solar energy to electricity by large power plants is not yet competitive with those for conventional power sources, although the costs are projected to become about equal by 1990.

Several means of collecting and concentrating solar energy have been proposed. Arrays of reflectors might be set up over large fields to reflect the radiation to a collector on top of a solar tower in the middle of the array. The collector would absorb the concentrated energy, which would later be used in generating electricity or in heating buildings.

Another way of harvesting solar energy avoids the problems of the vagaries of weather by utilizing solar energy collectors on satellites (figure 16.11). Here, with the collector always facing the sun, the full solar constant can be intercepted and converted to electricity on the satellite. The electricity can then be converted to microwave energy and transmitted to antennas on earth, where the microwave energy can be safely and efficiently converted back to electricity. It is estimated that one such satellite system could produce 3000–15,000 megawatts.

The preceding power-generating techniques can produce enough electricity to provide important supplements to other power sources, especially

FIGURE 16.11 *Collection of solar energy by satellite and relay to earth. This system could produce 10,000 mW.*

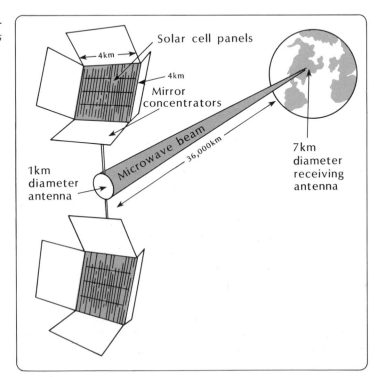

Solar cell panels

4km

4km

Mirror concentrators

Microwave beam

36,000km

1km diameter antenna

7km diameter receiving antenna

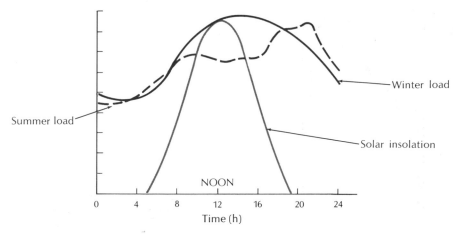

FIGURE 16.12 *Relative hourly variations in solar insolation and winter and summer demands for electricity.*

during periods of heavy electrical use. Fortunately for the earthbound solar energy systems, the peak in solar radiation received at the ground coincides more or less with the peak in electrical demand, as shown in figure 16.12.

For many years, solar energy has been used on a very limited basis to heat water for homes or to provide heat for growing plants. Individual solar power systems for air and water heating and air conditioning are feasible and conserve significant amounts of energy. It should be possible to provide economical home and office solar heating and cooling systems that would generate 60–90 percent of the total heating and cooling requirements for about 75 percent of the United States.

One simple solar heating system for homes, which is available commercially in Australia, Israel, Japan, the Soviet Union, and the United States, is the solar water heater, which utilizes the high heat capacity of water to store the solar energy. A simplified schematic diagram of such a system is shown in figure 16.13. The sun heats the water in collectors which are mounted on the roof. Then the hot water is pumped into a storage tank where it may be tapped by the hot-water heating system in the house. After being used, the cool water is pumped back into the solar collector on the roof.

Solar energy may be used to generate electricity in two ways. *Solar thermal systems* create temperature differences which can then be utilized to create mechanical energy in a turbine. Alternatively, *photovoltaic systems* utilize photocells to convert sunlight directly into electricity. The difficulty with both of these techniques is their high cost, which is projected to be 3 to 7 times more expensive than conventional systems through the year 2000 (figure 16.14).

Although generation of electricity by solar energy is not likely to be competitive with other methods in the next few decades, the use of solar systems for household heating is not a dream of the future but a reality of the present over much of the United States. Data from passive solar systems* in buildings in New

---

*Passive systems are ones in which the flow of heat is by natural means (convection, conduction, radiation) rather than the compressors, pumps, fans, etc., which are used in active systems.

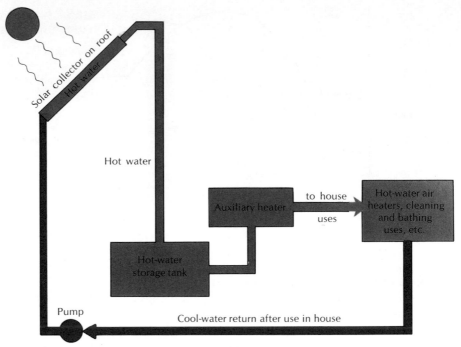

FIGURE 16.13 *Solar water-heating system.*

Mexico, New Jersey, and California show fuel bills under $100 for the entire heating season, which are well below those of neighbors. Besides their low operating cost, passive solar system often add nothing to the original cost of a house, have little chance of malfunction, and pose no serious environmental problems. Figure 16.15 compares the costs of 100 kWh of residential heat from various sources of

FIGURE 16.14 *Estimated relative costs of electricity at plant for various energy systems relative to coal (1.0) for the period 1980–2000.*

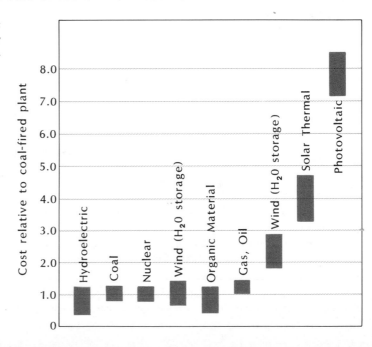

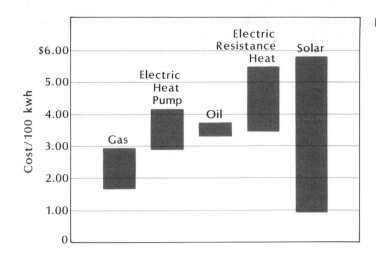

FIGURE 16.15 *Cost in 1979 dollars of 100 kWh of residential heat.*

energy. The great variability in solar costs stems from the variations in complexity and efficiency of active and passive systems.

16.3.3 HYDROELECTRIC POWER

Hydroelectric power generation is a well-tested, proven technology which is close to ideal in terms of economically producing electricity with a minimum of environmental effects. However, the locations in the world where hydroelectric power systems are feasible are limited: the ideal locations are mountainous regions with heavy precipitation. Therefore the total hydroelectric generation in the U.S. was only 4 percent of the total energy requirements in 1978 and is projected to decrease to about 3 percent in the year 2000 (table 16.5). However, the total average annual generation of 271×10^9 kWh in 1978 is still important, offsetting 445×10^6 barrels of oil per year. It has been estimated that a potential exists in the United States for an additional 522×10^9 kWh if all existing dams and new sites were developed to their maximum capacity.

WEATHER AND CULTURE 16.4

Because of its extreme importance to the physical aspects of human existence, it is not surprising that weather has affected the spiritual and mental side of human life as well. The influence of weather and climate can be found in the religious and artistic customs and expressions of all cultures.

16.4.1 WEATHER AND RELIGION

Ellsworth Huntington, in his classic book, *Mainsprings of Civilization,* argues persuasively that the physical environment has affected the evolution of religions in various parts of the world. Religious imagery, taken out of context from the region of its roots, may have little meaning. For example,

461

in Palestine the analogy of God's protection to "the shadow of a great rock in a weary land" (Isaiah 32:2) may be quite comforting to weary desert nomads, but seem rather meaningless to Eskimoes on a frozen tundra. Likewise, representation of heaven as a place where "the sun shall not smite by day" (Psalms 121:6) may not seem too inviting to a Finland shepherd or Siberian farmer.

Hell, as well as heaven, is viewed differently by people from hot and cold climates. Buddhism, which originated in northern India where the late spring and early summer are extremely hot, views hell as a hot spot. A hot hell is no coincidence, for Allahabad, the birthplace of Gautama, endures maximum temperatures in May of 100–110°F every day. In contrast, the hell envisioned by shepherds of high, cold Tibet is one of continuous torture by cold.

16.4.2 WEATHER AND MUSIC

Music is another facet of culture where one would expect to find significant influences by weather. Anyone can think of numerous song titles which consist primarily of weather or climate references (e.g., "The Rain in Spain," "Summertime," "Somewhere Over the Rainbow," and "Raindrops Keep Falling on My Head"). In addition, many classical compositions employ musical imitations of weather elements. James Wagner, in his delightful essay "Music to Watch Weather By," summarizes some of the better-known classical compositions that feature meteorology. For example, *The Four Seasons*, written in the eighteenth century by Antonio Vivaldi, describes the weather and its effect on human activities in each season of the year. *The Four Seasons* expresses the relief felt on a warm spring day at the end of winter, the violence of a summer thunderstorm, the joy of an invigorating autumn day, and the excitement of the bracing outdoor sports of winter.

Toward the end of the eighteenth century, Franz Joseph Haydn composed *The Seasons*, which depicts a progression of similar weather events to those that Vivaldi described. In his orchestral introduction to winter, Haydn depicts musically the fog and mist which are so prevalent in northern Europe in November and December.

Some of the primal awakenings that accompany the return of spring are present in Igor Stravinsky's *Rites of Spring*, which includes sections describing the pagan orgies conducted by primitive tribes in the spring. This section provoked violent reactions from the audience in its Paris premiere in 1913.

An example of the life cycle of the thunderstorm is found in Beethoven's Sixth Symphony (*Pastoral Symphony*). First, the clouds darken and lower, announced by the soft rumble of distant thunder. After the violence of the storm subsides, the soft harmony of a rainbow appears.

Other familiar classical music references to weather include *The Grand Canyon Suite* by Ferde Grofé, which describes the sunrise, sunset, and a midday thunderstorm; Johann Strauss' "Thunder and Lightning Polka;" and Moussorgsky's *Night on Bald Mountain*, which describes a raging mountain top storm.

16.4.3   WEATHER AND LITERATURE

Meteorological imagery abounds in literature and language. The authors of the Bible, drawn from a stock of weather-conscious farmers, shepherds, and sailors, spoke often of the weather in the Near East. Observant of the preculiarities of the climate in this region of the world, the different writers produced a remarkably consistent collection of weather proverbs. For example, the east wind in Palestine blows off the hot, arid Arabian Desert, and according to the Biblical prophets,

> *The east wind dried up her fruit.* [Ezekiel 19:12]

> *An east wind shall come, the wind of the Lord shall*
> *come up from the wilderness, and his spring shall*
> *become dry and his fountain shall be dried up.*
> [Hosea 13:15]

> *When the east wind toucheth it, it shall wither.*
> [Ezekiel 17:10]

> *And behold, seven thin ears, blasted with the east*
> *wind, came up.* [Genesis 12:6]

> *God prepared a vehement east wind* [Jonah 4:8]

If east winds produced drought and heat in biblical lands, the west wind carrying moisture from the Mediterranean Sea brought showers, and, according to Luke 12:54,

> *When ye see a cloud rise out of the west, straightway*
> *ye say there cometh a shower; and so it is.*

The south wind, on the other hand, produced stormy, unstable weather, and the Bible speaks of whirlwinds coming from this direction:

> *As whirlwinds in the south* [Isaiah 21:1]

> *And shall go with whirlwinds of the south* [Zechariah 9:14]

> *Out of the south cometh the whirlwind* [Job 37:9]

Shakespeare often includes the imagery of meteorological storms and weather. In England, southerly winds in advance of approaching Atlantic storms bring mild, moist air over the British Isles. Shakespeare, who was very observant of the British climate, writes in *As You Like It,*

> *Like foggy south, puffing with wind and rain* [3.5.50]

and again, in *King Henry the Fourth, Part II,* he writes

> *When tempests of commotion like the south,*
> *Born with black vapour, doth begin to melt,*
> *And drop upon our bare, unarmed heads.* [2.4.339–41]

Shakespeare also described human struggle through the imagery of meteorological storms. In *Hamlet,* Shakespeare sets the ominous mood of the calm before the storm:

> *But as we often see, against some storm,*
> *A silence in the heavens, the rack stands still,*
> *The bold wind speechless and the orb below*
> *As hush as death: anon the dreadful thunder*
> *Doth rend the region.* [2.2.471—75]

And, speaking of human affairs as well as the weather in *King John,* Shakespeare writes

> *So foul a sky clears not without a storm.* [4.2.108]

16.4.4   WEATHER AND ART

In general, artists at any period of history tend to paint subjects from their environment, including buildings, people, animals, and plants. It should not be surprising, therefore, that artists tend to reflect the weather and climate of their environment. To test the hypothesis that painters living in regions or times of different climates will reflect the differences in their paintings, Hans Neuberger examined more than 12,000 paintings in 41 art museums in the United States and eight European countries. More than 53 percent of the paintings contained some meteorological information. Neuberger found significant differences in cloud cover and visibility in the paintings of artists from America, Great Britain, Holland and Belgium, and Italy and Spain. These differences correlate well with the different climates in these areas; for example, visibility was highest in Spain and Italy, lowest in Great Britain and the Netherlands.

Cloud amounts and types in the paintings also showed regional differences. Showery, convective clouds were most frequent in the Netherlands, least frequent in America. Table 16.6 shows the relative frequencies of cloud amounts in various regions. It is interesting that there was not a single British painting with clear sky!

The painters not only revealed regional differences in climate, they showed climatic variations in time as well. During the Little Ice Age of approximately 1550—1849, the frequency of low and convective clouds and the total

TABLE 16.6   *Relative frequencies (in percent) of cloudiness categories in paintings from various regions (Neuberger,* Weather, *25, 1970).*

Region	Clear	Scattered	Broken	Overcast
America	5	16	47	32
Britain	0	4	48	48
Netherlands	7	11	44	38
France	9	10	49	32
Germany	16	16	42	26
Italy	12	24	42	22
Spain	13	13	38	36
AVERAGE	10	15	43	32

cloud cover were considerably higher than in the periods before and after. Apparently the artists of this cool, dark, and gloomy epoch reflected the change in climate in their paintings.

Questions

1 List some ways that weather conditions affect diseases.

2 Why are colds more common in winter than summer?

3 One likely consequence of shifting from oil or nuclear power to coal would be an increase of sulfur dioxide (SO_2). List some negative aspects of SO_2 on the environment.

4 Why are suntans harmful?

5 Why can you eat more in winter than summer and not gain weight?

6 Why do humidifiers reduce the indoor temperature required for comfort in the winter?

7 Under which condition would you lose more heat: temperature 10°F and wind speed 0, or temperature 30°F and wind speed 25 mi/h?

8 Which kills more people in the U.S. during the average year—tornadoes, hurricanes, floods, or lightning?

9 How does the temperature of lightning compare to that of the sun?

10 What are some safe and dangerous places to be when lightning threatens?

11 How far from a thunderstorm do you have to be in order to be safe from lightning?

12 If 50 percent of all the space heating and hot water heating requirements could be provided by solar energy, what percentage of the total U.S. energy demand would the sun provide (use figure 16.9)?

13 If the wind speed doubles, by how much could the power produced by a windmill increase?

14 How much power would a windmill with propeller diameter of 3m produce with a wind speed of 30 km/h (use figure 16.10)?

Problems

15 List some industries that are affected by weather but are not mentioned in this chapter.

16 You are the manager of an airport, and a tropical cyclone is 1 day away. There is a 25 percent chance of hurricane force winds at your airport. It would cost $100,000 to move all the planes to a safe location. If hurricane force winds occur and the planes are not moved, the damage would be $1,000,000. Should you move the planes or not?

17 Estimate the number of windmills that would be required to replace all the nuclear energy produced in 1978. Use table 16.5 and assume a propeller diameter of 3m, with each windmill producing 1 kW of power 20 percent of the year. (Answer: 525 million)

465

SEVENTEEN

A Year's Weather in the United States

chinook · January thaw · dog days · radiation fog · marine fog · Santa Ana · Indian summer · Great Lakes snowstorms

In the previous chapters, we touched on some of the characters in the weather drama and have tried to make the everyday, as well as the extraordinary, weather episodes more understandable to the appreciative observer. The emphasis has been on how the weather affects people, and on how people influence the weather and climate, both now and in the future. In keeping with these themes, it is appropriate to summarize the concepts in the book by a month-by-month review of a year's weather pageant over the United States in an effort to illustrate how the physics of the atmosphere unite to produce the dazzling progression of meteorological performances in varied settings.

17.1 JANUARY

Jack Frost in Janiveer
Nips the nose of the nascent year*

The emphasis in January over the eastern two-thirds of the country is on intense winter storms followed by savage winds and piercing cold. The development of these storms is favored by the very strong north-south temperature contrasts and the resulting fast jet streams that are present during January. Winter storms typically begin as weak low-pressure centers that form along an old polar front in Texas or along the Gulf Coast. As a trough in the fast upper-level westerlies sweeps across the Rockies, a favorable upper-level divergence pattern is produced for the deepening of the surface low. Mild and moist air from the Gulf of Mexico is carried northward and overruns the cold, denser air mass near the ground. As the air is lifted, rain and, farther north, snow, are produced.

The low moves east-northeast as it intensifies, spreading snow as far north as Illinois and Michigan, and drawing warm, unstable air ahead of its path into the south Atlantic states, where thunderstorms break out ahead of the advancing cold front. Behind the developing low, northerly winds carry cold, dry air well south of the U.S. border into Mexico and Central America.

The low continues to intensify as it moves up the East Coast, now drawing moist air from the Atlantic as well as from the Gulf of Mexico to feed the heavy rain and snowfall. Paralyzing snows—$^1/_3$ meter or more—occur to the west

---

*Unless noted otherwise, the weather proverbs in this chapter are taken from Richard Inwards, *Weather Lore* (London: Elliot Stock, 1893).

Plate 19a (above)   Rainbow taken with the same lens as the cloud bow in Plate 19b (© Alistair B. Fraser).

Plate 19b (below)   Cloud bow; the cloud bow appears nearly white (© Alistair B. Fraser).

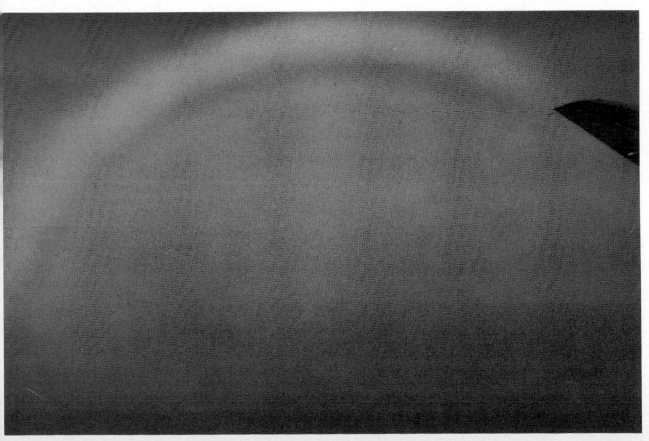

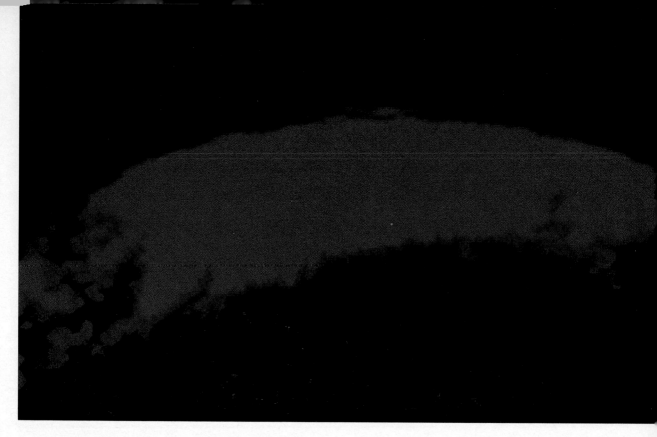

Plate 20a (above)   The green flash. As the sun rises or sets, the upper rim occasionally will grow and momentarily produce a brilliant flash of green light (© Alistair B. Fraser).

Plate 20b (below)   Fog in the Bald Eagle Valley of Pennsylvania (courtesy of Richard A. Anthes).

of the storm track in the mountains of North Carolina, Virginia, West Virginia, and Pennsylvania. Northeast winds of near-hurricane force peak the Atlantic waves into 3-meter crests which crash on lonely New Jersey beaches. Snow spreads into New England, while the winds at Mt. Washington in New Hampshire howl at 240 km/h with a temperature of −30°C (−22°F).

Behind the storm, cold-wave warnings in the upper Midwest signal a fresh blast of arctic air which only recently covered the icy Yukon. Piercing cold air sweeps southward out of the dark, frozen interior of western Canada. In the North, the snow cover turns dry and hard, squeaking under the feet of the few who must venture outside.

As the polar anticyclone passes southward over Minnesota and Wisconsin and turns eastward to cover Illinois, Indiana, and Kentucky, winds diminish in the East. At night, the quiet of the northern forests is disturbed by trees cracking with cold under skies filled with stars that blaze with a passionate intensity never seen through summer hazes. The earth's surface radiates freely throughout the long night to the cold void of space, unhampered by any water-vapor blanket.

The coldest weather of the years occurs in the calm, nighttime centers of these intense anticyclones. The morning radio calls out the toll:

International Falls, Minn.	−41°C	−41°F
Bismarck, N. Dak.	−38	−36
Minneapolis, Minn.	−30	−21
Madison, Wis.	−26	−15
Chicago, Ill.	−23	−9
Indianapolis, Ind.	−18	−1
Louisville, Ky.	−14	7
Atlanta, Ga.	−10	14
Tallahassee, Fla.	−9	16
Orlando, Fla.	−2	28
Miami, Fla.	3	37

The emphasis in the East is cold, verifying again one of the truest proverbs:

As the day lengthens, so the cold strengthens.

Only in the Deep South, where the weather is more directly influenced by the solar radiation, do the coldest temperatures occur in December near the solstice.

While the East suffers through the cold waves, the eastern slopes of the Rockies may experience temporary relief from winter by the amazing *chinook,* a strong, warm wind named after the Chinook Indians of the lower Columbia River. The chinook pours down the sides of the Rockies, warming and drying as it is compressed by the higher pressure. It is impossible to describe this wind better than the following account, published in 1896:

> *Picture to yourself a wild waste of snow, wind-beaten and blizzard-furrowed until the vast expanse resembles a billowy white sea. The frigid air, blowing half a gale, is filled with needle-like snow and ice crystals which sting the flesh like the bites of poisonous insects, and sift through the finest crevices. . . . Great herds of range cattle,*

469

which roam at will and thrive on the nutritious grasses indigenous to the northern slope, wander aimlessly here and there, or more frequently drift with the wind in vain attempts to find food and shelter; moaning in distress from cold and hunger, their noses hung with bloody icicles, their legs galled and bleeding from breaking the hard snow crust as they travel. . . . Would the chinook never come? The wind veered and backed, now howling as if in derision, and anon becoming calm, as if in contemplation of the desolation on the face of nature, while the poor dumb animals continued their ceaseless tramp, crying with pain and starvation. At last . . . at about the hour of sunset, there was a change which experienced plainsmen interpreted as favorable to the coming of the warm southwest wind. At sunset the temperature was only −13°, the air scarcely in motion, but occasionally seemed to descend from overhead. Over the mountains in the southwest a great bank of black clouds hung, dark and awesome, whose wide expanse was unbroken by line or break; only at the upper edge, the curled and serrated cloud, blown into tatters by wind, was seen to be the advance courier of the long-prayed-for chinook. How eagerly we watched its approach! How we strained our hearing for the first welcome sigh of the gentle breath! But it was not until 11:35 P.M. that the first influence was felt. First, a puff of heat, summer-like in comparison with what had existed for two weeks, and we run to our instrument shelter to observe the temperature. Up goes the mercury, 34° in seven minutes. Now the wind has come with a 25-mile velocity. Now the cattle stop traveling, and with muzzles turned toward the wind, low with satisfaction. Weary with two weeks standing on their feet they lie down in the snow, for they know that their salvation has come: that now their bodies will not freeze to the ground.

The wind increases in strength and warmth; it blows now in one steady roar; the temperature has risen to 38°; the great expanse of snow 30 inches deep on a level is becoming damp and honeycombed by the hot wind.

Twelve hours afterward there are bare, brown hills everywhere; the plains are covered with floods of water. In a few days the wind will evaporate the moisture, and the roads will be dry and hard. Were it not for the chinook winds the northern slope country would not be habitable, nor could domestic animals survive the winters. [A. B. Coe, "How the Chinook Came in 1896," *Monthly Weather Review*, Vol. 24, 1896, p. 413]

Some of the most rapid changes in temperature occur with the onset of the chinook. For example, on January 22, 1943, in Spearfish, South Dakota, the temperature rose from −20°C to 7°C (−4°F to 45°F) in two minutes! The effects of the chinook are not all beneficial, however. Hurricane-force winds can cause extensive damage in the lee of the Rockies, as the residents of Boulder, Colorado, know.

In the Far West, January is a cloudy, stormy month in the north and cool and dry in the south. The precipitation that falls in the Pacific Northwest is caused by a frequent progression of intense Pacific cyclones, whose centers of low pressure usually cross the Canadian coast north of the United States. The moist Pacific origin of these frequent storms and the mountain barriers blocking the onshore winds produce a favorable situation for intense precipitation events—snow in the mountains and cold, dreary rain along the coast. Annual snowfall on the western flanks of the Sierra Nevada and the Cascade mountains exceeds 1000 centimeters in some places.

The southern portion of the Pacific Coast is much drier than the north. Moisture-bearing storms, even in January, usually pass well to the north. The land is also warmer than the cool Pacific, so air from the Pacific warms and the relative humidity decreases, producing the famous southern California sunshine day after day. Occasionally, however, slow-moving Pacific storms move inland far enough south to produce heavy rains and mud slides along the steep slopes of the southern California coast.

One final idiosyncrasy about January weather—in most of the eastern United States, including Washington, D.C., New York, and Boston, around the middle of the month, it is common to experience a temporary respite from the severe cold. After days of subfreezing temperatures, northern residents are awakened by the unfamiliar sound of water dripping off snow-covered roofs. The "January thaw" has arrived. The January thaw is a climatic anomaly in which unseasonably mild weather tends to recur at nearly the same time every year, usually during the period of January 20–23. Although the annual occurence of the January thaw is not a certainty, temperatures at least 5C° above the seasonal average are more than six times likely to occur in the New York area between January 19 and 24 as between December 4 and 12 or February 7 and 12. We really do not understand why the January thaw occurs, but we appreciate its temporary relief nevertheless.

FEBRUARY 17.2

There is always one fine week in February.

A February spring is not worth a pin. [Proverbs of Cornwall, England]

The records show that the groundhog month, February, is nearly as cold as January, but there are usually some hints in most places of warmer days ahead. Toward the end of the month, the sun is above the horizon at Latitude 40°N for 11 hours and 7 minutes, compared to a stay of only 9 hours and 20 minutes on December 25. The nearly two extra hours of daylight provide more of a psychological lift than a sensible warming to the winter weary. However, in the South, on days when low-pressure systems to the west bring semitropical air from the Gulf of Mexico, daytime temperatures can be quite warm. Along the Gulf Coast, days of 22°C (72°F) are not uncommon, and early spring flowers such as daffodils appear during this month.

But in the United States, do not be misled by a rare, fine February day, for February can offer as intense cold and deep snowfalls as January, although they usually do not last as long. And even though February is not the coldest month of the year in the South on the average, some severely cold temperatures have occurred during this month. For example, the cold wave of February, 1899, brought subzero temperatures (°F) into Florida.

The ice storm is a particularly damaging, although frequently very beautiful, visitor to the United States during February. Like snowstorms, ice storms are associated with extratropical cyclones and the overrunning precipitation of warm fronts. The difference is in the temperature of the air aloft. In the snowstorm, the temperatures are below freezing through the entire depth of the storm. In ice storms, however, the warm air overrunning the cold surface air remains above freezing through a thick enough layer to melt the snowflakes falling from still higher (and colder) regions. The cold, dense air next to the ground, however, is reluctant to move, especially in mountainous regions, where it remains trapped in the valleys. The rain falls into the subfreezing layer near the ground and freezes upon contact with any surface—trees, telephone wires, roads, or cars. The weight of the accumulated water can be immense: 11 tons of ice have been supported by telegraph poles before snapping; individual wires may accumulate 450 kilograms of ice. Birds are found with their feet frozen to the branches of trees, and deer in the forests have their feet and legs slashed by the edges of ice-covered snow as they break through the sharp crust of ice. In the United States, only Florida, New Mexico, Arizona, and the southern part of California normally escape ice storms. The region most visited by these quiet, but devastating storms extends from Texas into Kansas, then eastward across the Ohio Valley into the Middle Atlantic states.

February is perhaps the best month for skiing in the United States. Most ski resorts have had two or three months of snowstorms and so have developed a deep base of snow. In the Northeast, the depth may have reached $1^1/_2$ meters; in the West, 6 meters is more likely. Also, the days are noticeably longer and the sun correspondingly brighter in February than in either December or January, making skiing more comfortable.

17.3 MARCH

> *March was so angry with an old woman for thinking*
> *he was a summer month, that he borrowed a day*
> *from his brother February, and froze her and her*
> *flocks to death.* [From island of Kythmos, Greece]

In March, spring gains a firm foothold in the South and occasionally offers tantalizing appetizers of 20°C (68°F) temperatures to the North. The sun crosses the equator and rises at the North Pole this month, and the days lengthen rapidly with each passing week, providing roughly 15 minutes per day more sunshine after a week. In the North, March temperatures are 5–10C° higher than in February.

The strong March sun in the South and the lingering bitter cold in Canada produce very strong north-south temperature gradients across the United States in March, creating a favorable environment for intense storms. March storms may present a tremendous variety of weather, with blizzard conditions in the cold air over the Dakotas, and near-hot, humid conditions with thunderstorms in the warm sector. In the Great Plains, the greatest amount of snow falls in March as the intense lows pass to the south. For example, Rapid City, South Dakota, normally receives 56 centimeters in March, compared with 38 in February and 36 in January.

March is known as the windy month. High surface winds are common in March for two reasons, the first being the frequent passage of the intense cyclones just noted. The second reason is that the atmosphere in March is becoming more unstable. The instability (steep lapse rate) is caused by the strengthening solar radiation warming the ground and lower atmosphere while the atmosphere aloft retains its wintertime cold temperatures. This instability favors vertical mixing, which transports fast-moving air to the surface from aloft.

Along the Pacific Coast, March is a relatively wet month, although not so wet as January or February. Tempered by the cool Pacific, the temperatures along the West Coast respond slowly to the increasing power of the sun, and temperatures in March are only slightly warmer than those in February.

17.4   **APRIL**

April weather
Rain and sunshine, both together

The reputation April has for producing showers with sunny interludes is well deserved, for by April, the lower part of the atmosphere is becoming quite warm over most of the United States, while the air at higher levels remains cold, thus producing the instability conducive to shower formation. This instability, coupled with the occasional intense cyclones that persist into April, favors the development of tornadoes, those monsters of nature that are nowhere else so prevalent and intense as they are in the United States.

The atmosphere begins its preparation for tornadoes in a mild enough manner, when a strong anticyclone anchored over the southeastern United States begins swirling hot, humid air across the Texas Gulf Coast and northward into Kansas and Oklahoma, ahead of a developing cyclone in the lee of the Rockies. At first, the southerly winds bring comfortably warm temperatures and pleasant spring weather to the Midwest, but as the jet stream dips southward across the Rockies, funneling cold, dry air over the warm, moist air at the surface, the synoptic stage is set for the sudden release of instability through the mechanisms of the squall line, the thunderstorm, and all too often, the tornado.

First, the southwestern sky darkens; lightning illuminates black clouds, and static fills the radio. The warm, moist wind blows gustily from the southeast, carrying tropical moisture into the developing squall. The line of darkness approaches, and ahead of the rain curtain, the characteristic funnel protrudes from the cloud base, twisting and hissing like an angry serpent. Now, the visible funnel reaches the ground, scouring it clean of trees, buildings, and everything else rising above the ground. Then, a few minutes after its arrival, it is gone, continuing its malevolent tract to the northeast. Cold rain washes the air and drenches the devastation as the line of thunderstorms pass. Then, the bright April sunshine returns, and quiet descends over the wasteland.

17.5 MAY

A swarm of bees in May is worth a load of hay.

By May, the month so rhapsodized by poets, the last frosts of winter have been banished to the extreme northern United States and higher mountains. In most places, May is a superb month, with mild days and cool nights. Warmed by two months of sunlight, the cold of the Canadian prairies has begun to slacken, and with it, the north-south temperature gradient across the United States. With the demise of the huge north-south thermal contrast, the intense winter cyclones move northward into Canada, and in their absence, the likelihood of solid overcast days with steady precipitation diminishes. Rainfall begins to occur in showers and thundershowers, for the air is relatively unstable. But the showers are fleeting, and the potent sun, now only a month from its zenith, goes about its business of warming the land and the atmosphere.

May is a growing and flowering month, as indicated by the above proverb extolling the pollenating virtues of bees. Fruit trees bloom in the North, and gardens in the South.

On the West Coast, the precipitation decreases markedly in May, even in Seattle, where the average cloudiness dips below 70 percent for the first time since the previous September. The semipermanent anticyclone which resides in the Pacific builds northward, and the cyclonic storms are shunted off to the north, away from the U.S. coast.

However, May is not all flowers and light, for the forces of the North do not give up easily. Occasionally, cold pools of air in the upper atmosphere break away from their polar origins and drift slowly over portions of the United States. When these upper-air cold pools anchor themselves over New England, cool, cloudy, windy, and sometimes drizzly weather may persist for three or four consecutive days. If the cold air aloft comes down across the Rockies, severe weather can ensure, including snow in the mountains and tornadoes and hailstorms in the Great Plains along the leading edge of the cold air. But these temporary setbacks do not mar the affection most of us feel for the merry month of May.

It never clouds up in a June night for a rain.

As the proverb states, over most of the United States, rainfall in June is normally of the convective thundershowery variety and, therefore, likely to occur in the afternoon, when surface temperatures are warmest. Even when weak extratropical cyclones and their attendant warm and cold fronts struggle sluggishly across the continent in the weakened flow aloft (small horizontal temperature differences; therefore, slow upper-level winds), the low-level convergence and associated lifting are more likely to produce transient showers than steady rains. In addition, the proverb has the best chance of being right in June, which has the shortest nights of the year—only eight hours in the extreme northern United States. Thus, there simply is not much time for clouding up at night.

June in the South is the first month of tropical weather; even weak polar fronts rarely penetrate those states south of 35°N. Daytime temperatures are hot, nighttime temperatures warm, and humidities always high. Thunderstorms break out nearly every afternoon in the tropical air mass.

Along the coastal regions of the United States, June is a good month for the development of the sea breeze. The land, under the influence of the strongest solar radiation of the year, heats up more than the surrounding waters, which are still cool from the previous winter. We have seen effects of this differential heating before, on a global scale. Here, during the day, it produces welcome onshore winds which extend 15−80 kilometers inland. Along the leading edge of this sea breeze (called the *sea-breeze front*), convergence frequently generates a line of showers and even thunderstorms, and many places near the southeast coast can expect a daily shower more or less at the same time each day as the sea-breeze front passes. An example of these sea-breeze thundershowers located a few kilometers inland from the Florida coast is shown in plate 9.

JULY *17.7*

Whatever July and August do not boil, September cannot fry.

In most of the United States, the hottest weather of the year occurs in July, with August running a close second. Even though the sun has started its retreat toward southern latitudes, the net energy budget over the United States is still positive; that is, the atmosphere receives more heat than it loses during most of July.

By coincidence, a typical heat wave might begin as Sirius, the Dog star, rises and sets with the sun, an event which occurs sometime in mid-July and is the reason for the term *dog days*. The heat wave begins slowly as an old polar high ceases its southward drift and settles down for a lengthy visit over the south Atlantic states. At first, the air is warm, comfortably dry, and clear, but as the atmosphere aloft slowly subsides and warms by compression, extreme stability is produced, even with hot surface temperatures. As the warm high aloft increases in intensity, the winds through a deep layer of the troposphere become light and variable and finally die away altogether. Pollution, both natural (pollen and dust) and artificial (gases and particles), is neither carried away horizontally by the winds nor mixed vertically in the stable air and so accumulates in the lower kilometer of the atmosphere. A yellowish haze deepens with each hour over the eastern United States. The sun turns from brilliant white to yellow, then a brassy orange, and finally, a burnt crimson as the heat wave and the haze reach a peak. The warm high-pressure system is now firmly entrenched, maintaining in part its foothold on the United States by shunting cooler air and cyclonic disturbances clockwise around the edge of its huge circulation into Canada.

Daytime temperatures reach 35°C(95°F) as far north as Maine, exceeding even Miami's 32°C(90°F) reading, from which sea breezes provide a little relief. Nighttime temperatures fail to break the 27°C(80°F) level; brown-outs occur in the cities as air conditioners labor twenty-four hours a day in the torrid heat. Absolute humidities reach their maximum values of the year as the heat wave enters its second week.

Finally, an imbalance in the upper-level circulation permits some cool Canadian air to start southward, driven by the northerly winds associated with a vigorous wave in the upper-level westerlies over Canada. As this wave intensifies, the stagnant high over the United States weakens slightly, then moves grudgingly off the coast, permitting cool, clear Canadian air to replace the muggy air of the heat wave. Ahead of the advancing relief, thunderstorms erupt, providing welcome moisture for parched lands. Behind the Canadian front, the sun sparkles in a cool, deep-blue sky. The heat wave is broken.

17.8 AUGUST

August sunshine and bright nights ripen the grapes.

August weather over most of the United States is similar to July's, although the days are noticeably shorter and the temperature drops a degree or so from the July maximum. The exception to this slight cooling is along the Pacific coast, where water temperatures are still rising. Hence, in Los Angeles and San Francisco, August is slightly warmer than July.

In the East, August days continue hot and humid. During the longer nights, local fogs become more common. These *radiation fogs* form under conditions of weak pressure gradients (light winds), clear skies (strong cooling by radiation), and long nights (long time for cooling to occur). They are concentrated in the valleys where cold air drains from the higher surrounding land and collects into stagnant pools of cool, moist air, as shown by the early morning fog in the Bald Eagle Valley of Pennsylvania (plate 20b). Local differences in the early morning temperatures under such conditions can be quite large and are strongly correlated with topography. Thus, at night, the temperature on top of a 100-meter hill under clear skies might be 15°C(59°F) while in the valley there is a dense fog and a temperature of 10°C(50°F). These valley fogs are not very thick and tend to evaporate soon after sunrise.

It is noteworthy that fogs "burn off" from below, not from above. Very little of the rising sun's heat is absorbed by the top of the fog deck. Part is reflected from the top of the fog; the rest penetrates through the fog and warms the ground and the air in contact with the ground. The lowest fog droplets evaporate in this warmer air. Furthermore, as time-lapse photography shows, local hot spots develop underneath the fog and generate thermals which rise through the fog deck and penetrate the top, mixing drier air downward and helping dissipate the fog. Because such radiation fogs are associated with calm anticyclonic conditions, they are strong indicators of fair weather; hence the proverb

When the fog falls, fair weather follows.

In contrast to the inland radiation fogs, the *marine fogs* of the West Coast are caused by the cooling of low-level air by cold water. San Francisco is perhaps most famous for its fogs, great banks which in the summer pour inward through the Golden Gate in response to the sea breeze. These fogs do not survive more than a few kilometers inland during the day, as they are eroded from below by surface heating. They are beneficial near the coast because they frequently contribute moisture—as much as 0.1 centimeter—to vegetation during this otherwise dry season. For example, the needles of the redwood tree strain the tiny droplets from the air and combine them into larger droplets, which then drip off the trees, providing necessary water. It has also been shown that redwood needles absorb directly the equivalent of 0.1 centimeter of rain on foggy nights. These redwoods grow only on the coast. Sixty-five kilometers inland, where the fogs rarely reach, redwoods cannot survive.

SEPTEMBER 17.9

June—too soon;
July—stand by;
August—look out you must
September—remember
October—all over [Captain Nares]

Then up and spake an old Sailor,
Had sailed to the Spanish Main,
I pray thee, put into yonder port,
For I fear a hurricane.

Last night, the moon had a golden ring,
And to-night no moon we see!'
The skipper, he blew a whiff from his pipe
And a scornful laugh laughed he.
[Longfellow, "The Wreck of the Hesperus"]

The unwary visitor from the north would hardly suspect anything amiss as he says farewell to the golden red sun on a hot, still September evening in the Florida

Keys. Not a cloud disturbs the brassy sky, not even the usual evening clouds that form over the Gulf Stream. And even the usual southeast breeze has died away completely, leaving a strangely tranquil twilight.

The night begins uncomfortably warm; no wind moves the humid air. Walking outside, searching in vain for relief, he sees the gibbous face of the waxing moon shrouded by fine, powdery clouds which cast a faint orange ring in the night sky. Left unsatisfied by even whisper of cool air, he sinks restlessly back on his bed.

Later at night, he is conscious of an unusual sound—soft at first, but gradually increasing in volume. The gentle lapping of small wavelets on the beach is drowned out by a rhythmic cadence of swells breaking over the sandy shoals far out in the bay.

The morning dawns without a sun; thickening and lowering cirrostratus clouds infect the eastern sky. At last, the wind has returned, but this time from the northwest, an unusual direction for September, but welcome nevertheless. The needle of an old barometer on the wall suddenly becomes unstuck from its rusted position at 1012 millibars and drops to 1001 millibars.

Later in the afternoon, high and middle clouds cover the entire sky, and low, dark clouds appear on the northeast horizon. The wind is blowing noisily in from the northwest now, with brief gusts reaching 50 km/h. Whitecaps cover the longer, high swells that roll relentlessly toward the coast from the northeast.

The first rain falls an hour later as an outer band of the hurricane sweeps across the islands. Two centimeters fall in fifteen minutes; then, the rain abruptly ceases as the dark clouds race on. But other nimbus clouds scud in across the frenzied sea, bringing more rain and ever-increasing winds. The top of a thirty-year-old coconut palm snaps under a brief gust reaching 130 km/h.

The gray sea is now completely covered with spindrift, which merges imperceptibly with the raindrops streaking almost horizontally in the howling wind. Quickly now, the wind and water rise to the climax. The western coast and the landscape are rearranged under the fury of the wind and tons of water; new islands are created, old ones covered forever. Salt spray is driven everywhere, even tens of kilometers inland, where it damages salt-sensitive plants.

Then, as if by supernatural intervention, the 150-km/h winds subside within a few minutes, the low clouds open, and a dim sun appears through a thin overcast. Towering white clouds, illuminated by the sun, wall the eye of the hurricane. The wind blows fitfully, first one way, then another. The barometer, now at its ebb, reads 950 millibars.

But the other side of the wall is approaching, and quickly the winds resume their battering, this time from the southeast. Now, the east coast of the island is disfigured, bearing the full intensity of the wind-driven storm tide. Trees which had barely managed to withstand the northwest winds by leaning toward the southeast are snapped back by blows from the opposite directions.

The hurricane leaves as gradually as it came—the squalls become less frequent and less violent, occasional breaks in the clouds uncover the moon, and the winds subside little by little. The sea slowly returns to normal, adjusting gradually to the lighter winds. The hurricane has crossed the Keys and moved into the Gulf of Mexico.

Hurricanes, dramatic as they can be, are rare refugees from the tropics, even in September along the Florida coast, where they are most likely to occur. For most of the United States, September is a superb outdoor recreation month for those who like sunshine. It is remarkable that in nearly every part of the country, September is one of the clearest months of the year. (Note the annual variation of cloudiness in the climatic graphs in appendix B.) Also, water temperatures reach a maximum in September, making this month ideal for beach vacations almost everywhere along both coasts.

The reason for the minimum in cloudiness is related to the stability of the air and the infrequent appearance of extratropical cyclones. The autumn atmosphere is generally stable because, in contrast to spring, the air aloft is warm after the summer months, and the ground is beginning to cool under the lengthening nights. These cooler surface temperatures and warm air aloft are also responsible for the frequent formations of fog during September evenings. Thunderstorms become less frequent in September for the same reason—increasing stability of the air.

Although the arctic region begins to cool substantially in September as it bids farewell to the sun, the temperatures in southern Canada are still mild. Therefore, the north-south temperature contrast over the United States remains small during September, and extratropical storm systems remain weak.

An unwelcome September visitor to California, a bad cousin of the beloved Great Plains chinook, is the dessicating *norther,* or *Santa Ana.* This hot northeast wind sweeps downward out of the Sierra Nevada, producing temperatures over 40°C(104°F) and humidities as low as 5 percent, burning and shriveling any unprotected vegetation (and people) in its path, and frequently causing an extreme fire hazard. The hot, dry wind causes nervousness, depression, and even suicides in people. As Raymond Chandler wrote about the Santa Ana: "On nights like that, every booze party ends in a fight. Meek little wives feel the edge of the carving knife and study their husbands' necks. Anything can happen."*

The cause of the Santa Ana is similar to that of the chinook—air descending steep slopes, warming and drying as it is compressed. Santa Ana winds are favored when high pressure builds over the Pacific Northwest states with low pressure to the south over Mexico, producing a strong north-south pressure gradient and associated geostrophic east winds.

OCTOBER *17.10*

Dry your barley in October, or you'll always be sober.
(Because if this is not done, there will be no malt.)
[The Reverend C. Swainson]

By October, the forces of the North clearly mean business. Snow covers the dark arctic once again, and hour after hour, the earth radiates the hard-earned warmth of summer to space. The polar front intensifies and thrusts farther and farther south,

---

The Midnight Raymond Chandler (Boston: Houghton Mifflin Co., 1971), p.7.

pushing the warm air back toward the tropics with each passing wave in the strengthening westerlies. Smoke curls from the chimneys again in houses across the land as the first frosts of autumn march southward. The first cold front of the season sweeps past Miami, dropping the dew point into the 40s(°F) and ending five months of tropical reign.

October along the Pacific coast, particularly the northwest coast, is the transition month between dry, sunny summer weather and the cold, wet winter season. The anticyclonic nose of the Pacific High, which has extended over the Northwest for the past four months, begins to weaken, and vigorous Pacific storms sweep inland on the now unprotected coast. Seattle's precipitation, for example, doubles from about 4 centimeters in September to over 8 centimeters in October, while the clouds once again cover the sky over 70 percent of the time.

The first widespread snowfalls of the season whiten the aspens of the Rockies in October, each day extending to lower elevations. These early-season snowfalls occur when cold upper-level troughs of low pressure dip southward from Canada along the Rockies, and easterly flow near the surface carries moisture from the Gulf of Mexico up the eastern slopes of the mountains.

October is a changeable month, and it is possible in the East to have frost and 25°C temperatures in the same week. Frequently, an intense polar anticyclone carries early morning frosty weather as far south as Virginia, but then slows down over the Carolinas, and finally stalls over Georgia. The first morning is decidedly cold, and even during the day, sweaters are comfortable in the 10°C temperatures. However, on the second night, the warm air in the south begins its long return northward around the western perimeter of the high, and temperatures fall only into the 40s (°F).

On the second day, the sun and the increasing southerly flow combine to produce 15°C weather as far north as New York. Indian summer has arrived.

The synoptic conditions for Indian summer are very similar to those which produce the August heat waves, only now, the mean temperature is 10C° lower, and the extra warmth associated with the sunshine, sinking of dry air, and light winds is welcome. High pressure aloft and at the surface produces light winds throughout the troposphere, and cool, foggy mornings give way to warm, hazy afternoons. Smoke from burning leaves adds to the natural haze and reddens the evening sky.

As we noted previously, the configuration of a warm high-pressure system at the surface and aloft tends to be slowly changing, with cold air isolated in Canada and cyclone families traveling around the perimeter of the anticyclone. Thus, the pleasant Indian-summer weather may linger for a week or so, allowing the last of the grapes to ripen and the farmers to complete the autumn harvest.

Indian summer—the name itself suggests the full-bodied flavor of a mellow wine. Indian summer—nature's last banquet before grimly settling down to the austere business of winter. Indian summer—the time of the year when the last of the autumn sun's rays strike gold and red in the forests. Indian summer—frost on the pumpkins and fresh apple cider along the roadways. Drink deeply, but slowly; savor every drop, and remember.

No warmth, no cheerfulness, no healthful ease,
No comfortable feel in any member, no shade, no
shine, no butterflies, no bees,
No fruits, no flowers, no leaves, no birds. No-Vember
[From *The Works of Thomas Hood*, Epes Sargent, ed.
(New York: Putnam, 1865), p. 332]

November slashes most of the country like a cold blade of steel. There are no more reprieves like Indian summer, no more pleasant nights of sitting on the porch, no more running barefoot down dusty paths, and no more late-season bonuses from the garden. The colorful leaves of last week are ripped away in an angry north wind, and the golden harvest moon turns white with cold.

In no other month is there such a change of weather. Nearly everywhere the cloudiness increases dramatically to near-wintertime levels. Look at the climatic charts (appendix B) of Chicago, for example, where the cloudiness jumps nearly 20 percentage points in November. While the clouds increase, the mean temperatures tumble precipitously—from 13 to 4°C (55 to 39°F) at Chicago, from 15 to 7°C (59 to 44°F) in St. Louis, and from 9 to 0°C (48 to 32°F) in Minneapolis.

Significant snows reappear in St. Louis, Boston, Washington, D.C., Minneapolis, Minnesota, and Rapid City, South Dakota. Only in the extreme south does the weather improve or remain hospitable. In Miami, for example, the precipitation decreases from 21 centimeters in October to 7 centimeters in November, and the mean temperature drops to a comfortable 23°C (73°F).

In November, the famous Great Lakes snowstorms begin to assault the leeward shores of Lakes Superior, Huron, Michigan, Erie, and Ontario and turn the lakes themselves into nightmares for those commercial ships seeking to make one more run. In contrast to the large-scale heavy snows associated with extratropical cyclones, the Great Lakes storms produce their maximum snowfall after the low and cold front have passed. An ideal synoptic situation occurs after a deep cyclone (very low pressure) moves northward along the eastern seaboard, with a strong Canadian high situated over the Dakotas. The large west-east pressure gradient accelerates frigid, dry air along a trajectory from the Canadian ice fields across the Great Lakes. Pouring over the relatively warm waters, the air in the lowest kilometer quickly becomes saturated. As the moistened air reaches the opposite shore, amazing (to those who have never seen a Great Lakes snow squall) amounts of snow can be dropped. The efficiency of the lakes as snow producers can be seen in figure 17.1, which shows the mean seasonal snowfall over the northeastern United States. Maxima in snowfall exist along the southern and eastern shores of the lakes, which are downwind of the prevailing west and northwest flow. For a specific example of the downwind Great Lake effect, consider Milwaukee, Wisconsin, and Muskegon, Michigan, which are located only 96 kilometers apart on opposite shores of Lake Michigan. Milwaukee receives 109 centimeters (43 in), of snow mostly from extratropical storm systems, but Muskegon, with its extra bounty from the lake, receives double that amount, 220 centimeters (87 in).

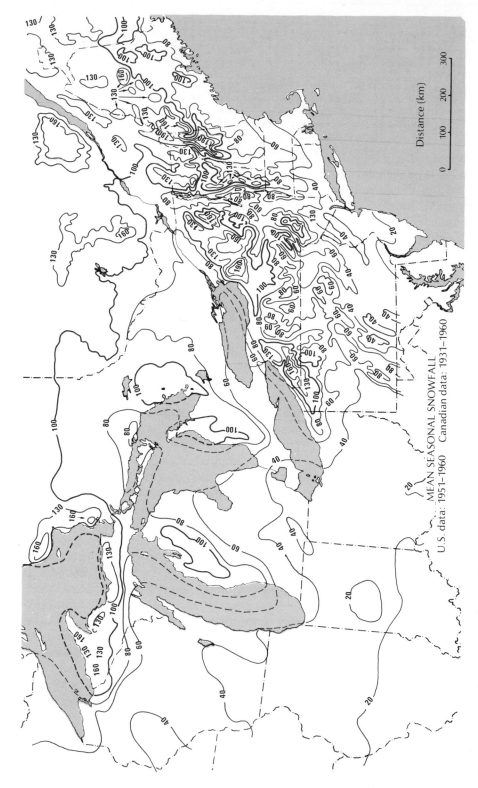

FIGURE 17.1 *Mean seasonal snowfall (in inches) over northeastern United States.* [cm = in (2.54)]

The heat added from the Great Lakes to the atmosphere also contributes to the heavy lake snowstorms. Because warm air is less dense than cold air, a trough of lower pressure tends to form over the Great Lakes whenever the water is significantly warmer than the surrounding air. This trough of low pressure produces convergence of surface air, which then rises and contributes to the precipitation process.

The Great Lakes snowstorms continue through the winter until the lakes freeze and ice stops their role as a heat and moisture source. Although most of the Great Lakes do not freeze until late December or January, it is interesting to mention here how the freezing occurs. One of the water's peculiar properties is that it reaches a maximum density at a temperature of 4°C (39°F). The water near the bottom of deep freshwater lakes in the northern United States (such as Lake Superior) has a temperature of 4°C year around. During the summer, the surface waters reach a temperature of 15–20 °C (59–68°F) and therefore float on the denser water below. As the subfreezing blasts of the November winds blow across the lakes, the water at the surface is chilled, becomes dense, and sinks, very much like convection currents in the atmosphere. This vertical mixing through cooling from above is efficient until the entire lake temperature equals 4°C. Then, further cooling produces lighter water, vertical mixing is inhibited, and freezing of the uppermost layer may readily occur.

DECEMBER 17.12

December cold with snow, good for rye.
A green Christmas makes a fat churchyard.

December is the darkest month of the year everywhere in terms of the time the sun spends above the horizon. In Fairbanks, Alaska, the sun appears for only $3^1/_2$ hours on December 21, and then only if the skies are clear. Adding to the dreariness of a $20^1/_2$-hour night are the frequent ice fogs in Fairbanks. In most parts of Alaska, the winter air is dry enough that even in the coldest weather, only a few ice crystals precipitate, and visibility remains good. In cities, however, significant amounts of water vapor are emitted by cars, aircraft engines, and the combustion of fuels. Here, under the calm anticyclonic conditions of December, where radiative cooling frequently drops the surface temperature to -45°C (-49°F), dense fogs of ice crystals hover over the city. The inversion during these episodes, which may last for a week or more, is extremely well developed. Temperatures may rise by 15C° in only 300 meters, making it desirable to live on the mountain slopes rather than in the valleys.

The cold polar atmosphere can sometimes play strange tricks on the senses. Imagine being in the remote outpost of Barrow, Alaska, which is located on the shore of the Arctic Ocean at a latitude of 71°N. At noon on November 21, the sun sets for an advertised two months, until its scheduled reappearance on January 21. But two days later, the southern horizon brightens again, and a distorted, yet unmistakable image of the sun reappears, casting a feeble yellow glow

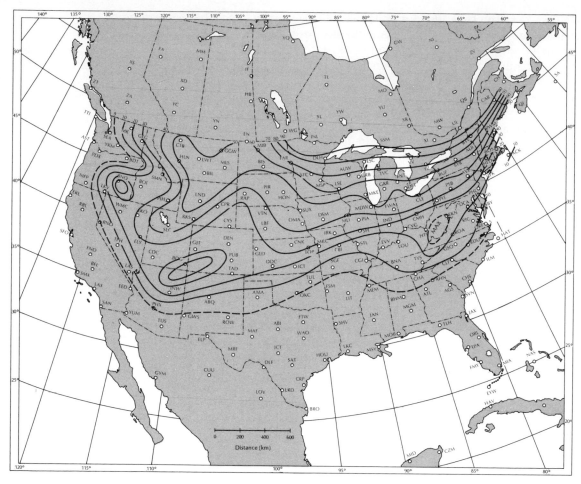

FIGURE 17.2 *Probability of a white Christmas, defined as at least 2.5 centimeters of snow on ground.*

over the icy sea. A wild hope springs to the heart. Are the astonomers wrong? Has the tilt of the earth's axis suddenly changed? Will the sun steadily climb in the sky, erasing the clouds and snow like a bad dream? But even as we eagerly watch, our hopes are shattered as the sun scintillates, bends, wobbles, and plunges a final time below the frozen horizon. The whole scene has been a cruel hoax, caused by refraction. The cold, dense atmosphere has bent the sun's rays over the horizon for one brief moment.

Over the continental United States, December is dark for meteorological as well as astronomical reasons, the brightest times being indoors during the holiday season. December is one of the cloudiest months of the year, particularly in the Pacific Northwest and along a band extending from the mountainous areas of Pennslyvania northeastward into New England. For example, cloudiness reaches 81 percent in Seattle, 90 percent in Portland, Oregon, and 74 percent in

484

Caribou, Maine. Thus, it must have been in December when the frustrated poet from Maine wrote the following (note that *dirty* was formerly a synonym for *cloudy*):

> *Dirty days hath September,*
> *April, June and November;*
> *From January up to May,*
> *The rain it raineth every day.*
> *All the rest have thirty-one,*
> *Without a blessed gleam of sun;*
> *And if any of them had two-and-thirty,*
> *They'd be just as wet and twice as dirty.*

December begins the snowfall season over much of the United States as cyclones sweep across the middle and southern sections of the country with increasing frequency. December is the month when many people, not yet tired of cold weather, appreciate a snowstorm, especially if the snowfall occurs during the Christmas season. Still, the chances of snow cover on Christmas day are small, as shown in figure 17.2, except in the northern third of the country and in the mountainous regions. Thus, most locations wait through green Christmas after green Christmas, dreaming in vain for that elusive magic morning of white.

As a final note, it is interesting to see throughout the centuries of weather wisdom how people come to expect and trust "normal," or what they consider "proper," weather for the season, such as hot weather in July or snow at Christmas. Exceptions to these ideas of what is "right" for the season are commonly greeted with suspicion, and even as portenders of evil, as the proverb relating green Christmases to fat churchyards (many graves), or the following odes to January indicate:

"Normal" (to be trusted)	*When oak trees bend with snow in January, good crops may be expected.*
"Abnormal" (to be regarded with suspicion)	*In January should sun appear, March and April pay full dear.*
	A January spring is worth naething. [Scottish proverb]

These forebodings are, for the most part, misapprehensions of what is really quite characteristic of the atmosphere, which rarely behaves normally or follows the neat climate tables or the smoothly drawn curves on the annual graphs of temperature and rainfall. These "normals" are really made up of an average of many "abnormal" days. As an extreme example, consider a mythical climate in which half of the days have temperatures of 35°C, and the other half have temperatures of 15°C. The average temperature of 25°C would never occur! Thus, the "normal" weather may be actually less likely than the so-called atmospheric *freaks* such as April snows or January thaws. As the Norwegian proverb tells it,

> *There are many weathers in five days, and more in*
> *a month.*

485

Questions

1 Explain the proverb, "As the day lengthens, so the cold strengthens."

2 Why is the chinook, which blows out onto the Great Plains from high, cold mountains, warm and dry?

3 Why is March a windy month over most of the United States?

4 Why is tornado formation most common in April over much of the United States?

5 Explain the proverb, "It never clouds up in a June night for a rain."

6 Why are heat waves in July sometimes called *dog days?*

7 Why are radiation fogs more common in August and September than in March?

8 How do radiation fogs evaporate (from the top, bottom, or uniformly)?

9 Why do California redwoods grow best near the coast?

10 What synoptic weather patterns favor Indian summer? Sketch a typical surface isobar pattern for an Indian summer over the Ohio Valley.

11 What are the probabilities of a White Christmas in the following cities: Washington, D.C., Chattanooga, Madison, Denver, Salt Lake City, Seattle?

Appendices

$\mathcal{A}$ CLOUD TYPES & GENERA

Clouds are classified on the basis of their appearance and the approximate height at which they occur. Figure A.1 illustrates the basic cloud types. Their names are combinations of these root words:

> *Cirrus*—feathery or fibrous
> *Stratus*—stratified or layered
> *Cumulus*—heaped up
> *Alto*—middle
> *Nimbus*—rain

Adjectives which are often used to further describe the basic clouds are

> *Castellanus*—turreted (cirrocumulus and altocumulus)
> *Congestus*—crowded together (cumulus)
> *Fractus*—broken (stratus, cumulus)
> *Humilis*—lowly, weakly developed in the vertical (cumulus)
> *Lenticularis*—lens-shaped (cirrostratus, altocumulus, stratocumulus)
> *Uncinus*—hook-shaped (cirrus)

The formation of clouds is discussed in chapter 6.

The official World Meteorological Organization (WMO) classification of clouds includes ten genera. The WMO definition and description of these genera give the essential characteristics of the major cloud types that people are likely to encounter. Differences in clouds within a particular genus have led to a subdivision of most of the genera into species. There are fourteen species applied to different genera. The subdivision into species is beyond the requirements of this book and is not included. The reader may find additional information in the *International Cloud Atlas Volume I*, published by the World Meteorological Organization, 1956. The following descriptions were taken from this atlas.

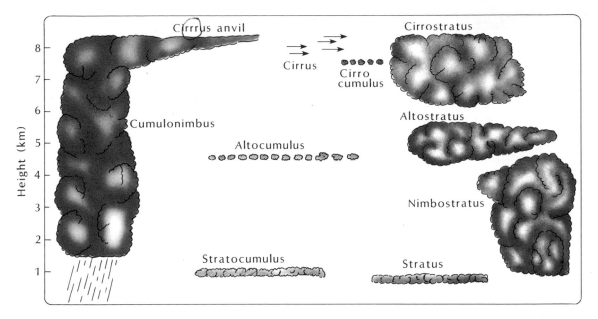

FIGURE A.1   *Illustration of cloud types and typical elevations.*

CIRRUS

Detached clouds in the form of white, delicate filaments or white or mostly white patches or narrow bands. These clouds have a fibrous (hairlike) appearance, or a silky sheen, or both.

CIRROCUMULUS

Thin, white patch, sheet, or layer of cloud without shading, composed of very small elements in the form of grains, ripples, etc., merged or separate, and more or less regularly arranged; most of the elements have an apparent width of less than one degree.

CIRROSTRATUS

Transparent, whitish cloud veil of fibrous (hairlike) or smooth appearance, totally or partly covering the sky, and generally producing halo phenomena.

489

ALTOCUMULUS

White or gray, or both white and gray, patch, sheet, or layer of cloud, generally with shading, composed of laminae, rounded masses, rolls, etc., which are sometimes partly fibrous or diffuse and which may or may not be merged; most of the regularly arranged small elements usually have an apparent width of between one and five degrees.

ALTOSTRATUS

Grayish or bluish cloud sheet or layer of striated, fibrous, or uniform appearance, totally or partly covering the sky, and having parts thin enough to reveal the sun at least vaguely, as through ground glass. Altostratus does not show halo phenomena.

NIMBOSTRATUS

Gray cloud layer, often dark, the appearance of which is rendered diffuse by more or less continuously falling rain or snow, which in most cases reaches the ground. It is thick enough throughout to blot out the sun.

Low, ragged clouds frequently occur below the layer, with which they may or may not merge.

STRATOCUMULUS

Gray or whitish, or both gray and whitish, patch, sheet, or layer of cloud which almost always has dark parts, composed of tessellations, rounded masses, rolls, etc., which are nonfibrous (except for virga) and which may or may not be merged; most of the regularly arranged small elements have an apparent width of more than five degrees.

STRATUS

Generally gray cloud layer with a fairly uniform base, which may give drizzle, ice prisms, or snow grains. When the sun is visible through the cloud, its outline is clearly discernible. Stratus does not produce halo phenomena except, possibly, at very low temperatures.

Sometimes stratus appears in the form of ragged patches.

CUMULUS

Detached clouds, generally dense and with sharp outlines, developing vertically in the form of rising mounds, domes, or towers, of which the bulging upper part often resembles a cauliflower. The sunlit parts of these clouds are mostly brilliant white; their base is relatively dark and nearly horizontal.

Sometimes cumulus is ragged.

CUMULONIMBUS

Heavy and dense cloud, with a considerable vertical extent, in the form of a mountain or huge towers. At least part of its upper portion is usually smooth, or fibrous or striated, and nearly always flattened; this part often spreads out in the shape of an anvil or vast plume.

Under the base of this cloud, which is often very dark, there are frequently low ragged clouds either merged with it or not, and precipitation sometimes in the form of virga.

491

B CLIMATIC SUMMARIES OF SELECTED U.S. CITIES

The following graphical summaries of the climate at selected U.S. cities were prepared from the data presented in *World Survey of Climatology: Climates of North America,* Vol. 11, Reid Bryson and Kenneth Hare, eds. (Amsterdam: Elsevier Scientific Publishing Company, 1974).

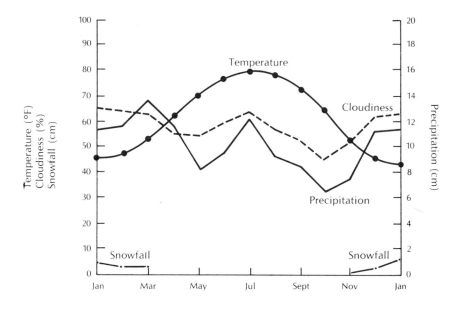

Atlanta, Georgia: 34°N 84°W;
elev. 308 meters

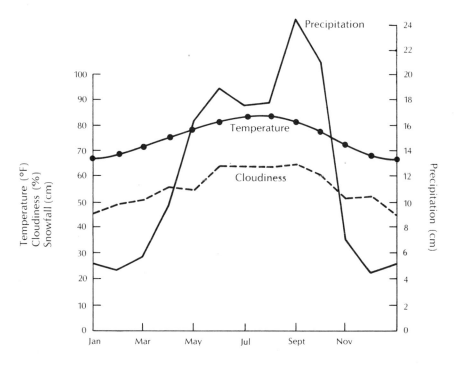

Miami, Florida: 26°N 80°W;
elev. 2 meters

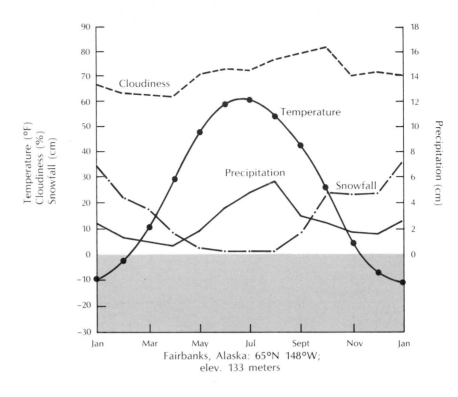

Fairbanks, Alaska: 65°N 148°W;
elev. 133 meters

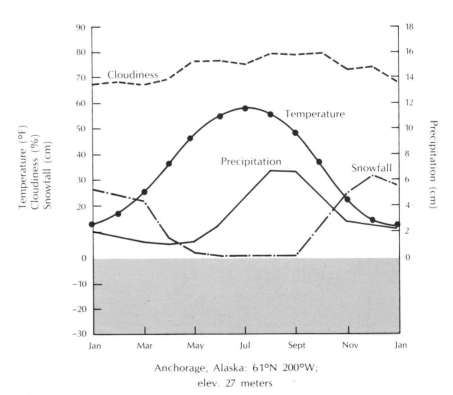

Anchorage, Alaska: 61°N 200°W;
elev. 27 meters

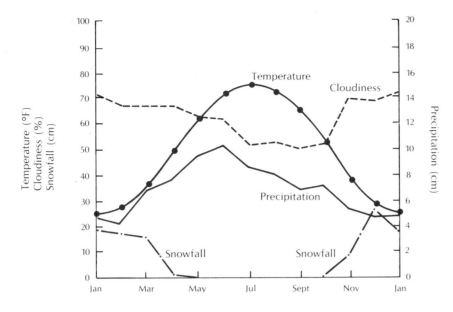

Chicago, Illinois: 42°N 88°W;
elev. 185 meters

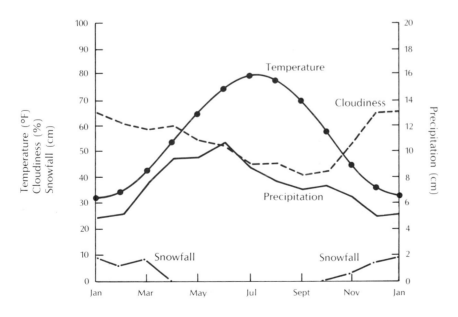

St. Louis, Missouri: 39°N 90°W;
elev. 142 meters

495

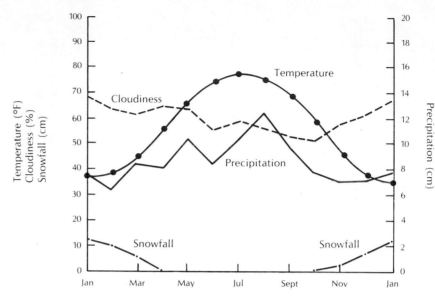

Washington, D.C.: 39°N 77°W;
elev. 4 meters

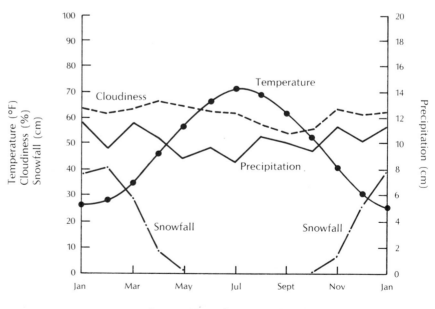

Boston, Massachusetts: 42°N 71°W;
elev. 192 meters

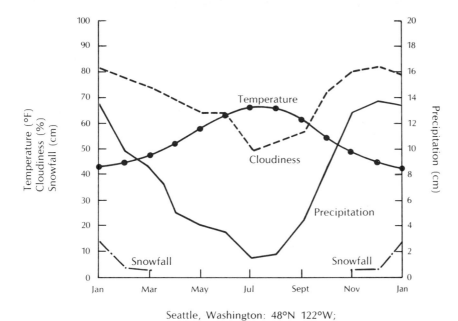

Seattle, Washington: 48°N 122°W;
elev. 4 meters

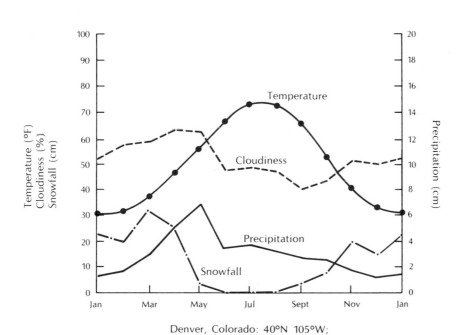

Denver, Colorado: 40°N 105°W;
elev. 1610 meters

497

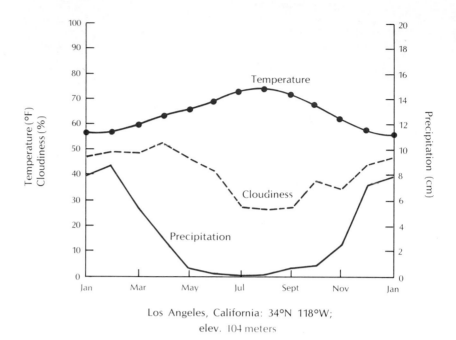

Los Angeles, California: 34°N 118°W;
elev. 104 meters

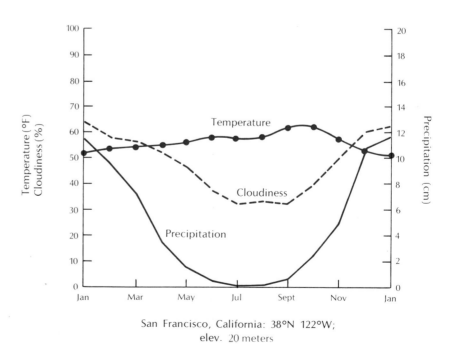

San Francisco, California: 38°N 122°W;
elev. 20 meters

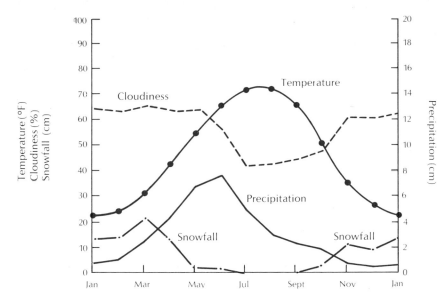

Rapid City, South Dakota: 44°N 103°W;
elev. 993 meters

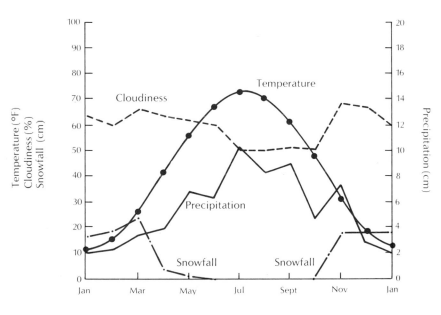

Minneapolis-St. Paul, Minnesota: 45°N 93°W;
elev. 254 meters

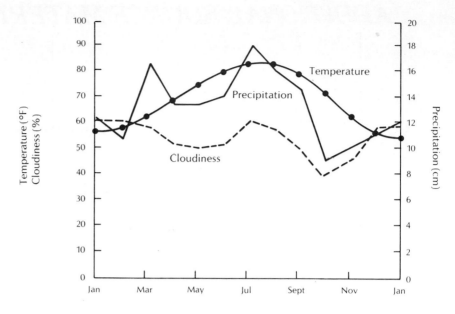

New Orleans, Louisiana: 30°N 90°W;
elev. 3 meters

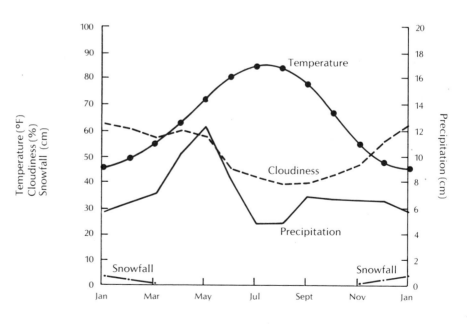

Dallas, Texas: 33°N 97°W;
elev. 146 meters

ADDITIONAL SOURCE MATERIALS

This appendix contains a sample selected from literally hundreds of additional sources for information on meteorology and related fields.

ADDRESSES

For air pollution information:

> Air Pollution Control Association
> 4400 Fifth Avenue
> Pittsburgh, Pennsylvania 15213

The National Center for Atmospheric Research, sponsored by the National Science Foundation, and the National Weather Service are good sources for meteorological information:

> National Center for Atmospheric Research
> Boulder, Colorado 80303
> Attn: Public Information Office

> Public Information Publications
> National Weather Service
> National Oceanic and Atmospheric Administration
> Rockville, Maryland 20852

Also, much climatology data can be obtained from local National Weather Service offices or the

> National Climatic Center
> Asheville, North Carolina 28801

Another useful address for World Meteorological Organization publications on meteorology is

> WMO Publication Center
> UNIPUB, Inc.
> P.O. Box 433
> New York, New York 10016

The following books may be obtained from the publishers. Prices and publishers' or distributors' addresses can be found in *Books in Print,* available in larger libraries and book stores. In the few cases where this information is not listed, the data are provided here. Material published by the American Meteorological Society (AMS) is available directly from the AMS, 45 Beacon Street, Boston, MA 02108. A publications list with prices is provided on request. Material published by the Government Printing Office is available from the Superintendent of Documents, U.S. Government Printing Office, Washington, D.C. 20402. Prices are provided in the subject bibliography on weather, SB234, available on request. Additional books are described in the December 1979 issue of *Weatherwise.*

Battan, Louis J.: *Radar Meteorology,* Chicago: The University of Chicago Press, 1959. A nontechnical description of the use of radar in studying the atmosphere.

_____: *Nature of Violent Storms,* Garden City, New York: Doubleday and Co., Inc., 1961. Discussion of thunderstorms, hurricanes, and tornadoes.

_____: *Harvesting the Clouds,* Garden City, New York: Anchor Books, Doubleday and Co., Inc., 1969. A lively, descriptive book on the various aspects of weather modification, including the LaPorte anomaly, cloud seeding, and hurricane modification.

Bentley, W. A. and W. J. Humphreys: *Snow Crystals,* 180 Varich Street, New York, New York: Dover Publications, 1962. A remarkable and beautiful book first published in 1931, containing 2453 illustrations, most of them striking photographs of the many varieties of ice and snow crystals.

Byers, H. R.: *General Meteorology,* New York: McGraw-Hill Book Company, 1974. A comprehensive, elementary meteorology textbook.

Calder, Nigel: *The Weather Machine,* New York: Viking Press, 1974. An interesting and a sometimes controversial look at the atmosphere with emphasis on climate change theories.

Dunn, G. E. and B. I. Miller: *Atlantic Hurricanes,* Baton Rouge, Louisiana: Louisiana State University Press, 1964.. A very readable book on the many aspects of tropical storms.

Edinger, J. G.: *Watching for the Wind,* Garden City, New York: Anchor Books, Doubleday and Co., Inc., 1967. Enjoyable style with emphasis on local meteorology.

Flora, S. D.: *Hailstorms of the United States,* Norman, Oklahoma: University of Oklahoma Press, 1956. A description of these damaging storms in the United States.

Hodges, L.: *Environmental Pollution,* New York: Holt, Rinehart and Winston, 1973. Complete treatment of all aspects of pollution, including air, water, noise, thermal, pesticide, etc. pollution.

Hays, James D.: *Our Changing Climate,* New York: Atheneum, 1977. Nontechnical book with emphasis on ice ages, past and future.

Heuer, Kenneth: *Rainbows, Halos, and Other Wonders: Light and*

Color in the Atmosphere, New York: Dodd, Mead, 1978. An introductory book describing optical phenomena produced when light interacts with the atmosphere.

Hughes, Patrick: *American Weather Stories,* Washington, D.C.: U.S. Government Printing Office, 1976. Weather as it has affected American history. Seven stories trace the American weather experience from the hurricanes that threatened Columbus to the foul weather that has plagued American presidents on their inauguration days.

————: *A Century of Weather Service,* New York: Gordon and Breach, 1970. A history of meteorological service in the United States.

Inadvertent Climate Modification, Report of the Study of Man's Impact on Climate (SMIC), Cambridge, Massachusetts: The MIT Press, 1971. Fascinating review of the many aspects of the effect of human activities on climate.

LaChapelle, E. R.: *Field Guide to Snow Crystals,* Seattle, Washington: University of Washington Press, 1969. Discussion of the meteorological factors producing different types of snow crystals; 57 photographs.

Landsberg, H. E.: *Weather and Health,* Garden City, New York: Doubleday and Co., Inc., 1969. An introduction to the fascinating link between weather and human life.

Lowry, W. D.: *Weather and Life: An Introduction to Biometeorology,* Corvalis, Oregon: Oregon State University Book Store, Inc., 1967. A look at the many effects of weather on life.

Ludlum, David M.: *Early American Hurricanes 1492–1870.* Boston, American Meteorological Society, 1963. *Early American Winters 1604–1820,* Boston: AMS, 1966. *Early American Winters II 1821–1870,* Boston :AMS, 1968. *Early American Tornadoes 1586–1870,* Boston, AMS, 1970. This four-volume "History of American Weather" provides a thoroughly documented look at the weather events of the country's past. Based on records, diaries, letters, newspapers, and other primary sources.

————: *Weather Record Book,* Princeton, N. J.: Weatherwise, Inc., 1971. Describes major weather events in the United States from 1871 to 1979. Provides various "records"—coldest, hottest, longest dry spell, etc. Good sections on all kinds of storms.

Mather, J. R.: *Climatology: Fundamentals and Applications,* New York: McGraw-Hill Book Company, 1974. Includes many interesting applications of climatology, such as those in health, architecture, and industry.

Neuberger, H. and J. Cahir: *Principles of Climatology,* New York: Holt, Rinehart and Winston, 1969. Brief treatment of factors behind the climate; numerous examples.

Ruffner, James A., and Bair, Frank E., eds.: *The Weather Almanac,* 2nd ed., Detroit: Gale, 1977. (New York: Avon, in paper, 1979). Includes weather and climate data for the United States, information on air quality, data for key cities around the world, information on storms and other natural disasters, discussions of weather fundamentals, an introduction to forecasting, and a glossary.

Scorer, Richard: *Clouds of the World: A Complete Color Encyclopedia,* Great Britain: David and Charles, 1972. A complete reference book on clouds. Photographs from all over the world, many of them in color, illustrate and explain the physics and mechanics of clouds.

Sloane, Eric: *Folklore of American Weather,* New York: Meredith Press, 1963. A collection of weather proverbs with a brief explanation regarding the scientific basis (if any) of each saying.

————: *For Spacious Skies,* New York: Crowell, 1978. This book is just one example of Eric Sloane's many works which combine the artist's skill with his long-standing interest in both Americana and weather. Includes some of authors paintings in color.

Stanford, John: *Tornado: Accounts of Tornadoes in Iowa,* A popular book that explains tornadoes and provides safety education. Focus is on Iowa, but concepts apply wherever tornadoes occur.

Stewart, G. R.: *Storm,* New York: Modern Library, Inc., 1947. A well-written novel of the birth, life, and death of a Pacific storm and its dramatic effect on the personal lives of people in the western United States.

Thompson, Philip D. and O'Brien, Robert: *Weather,* New York: Time-Life 1968. Distributed by Silver-Burdett. This handsomely illustrated, oversize volume provides an overview of the topic of weather by grouping a number of high-interest, interrelated topics.

Trewartha, G. T.: *An Introduction to Climate,* 4th ed., New York: McGraw-Hill Book Company, 1968. A general, thorough treatment of regional climatology.

U.S. Department of Commerce, NOAA: *Hurricane, the Greatest Storm on Earth,* Washington, D.C.: U.S. Government Printing Office, 1977. Illustrated brochure on hurricanes and hurricane research. Good introduction and overview. Includes a bibliography.

Tornado, Washington, D.C.: U.S. Government Printing Office, rev. 1978. This booklet tells what tornadoes are and what they can do, when and where they are most likely to occur, what they look like, and most importantly, what can be done to minimize their destructive, possible lethal effects.

Uman, Martin A.: *Understanding Lightning,* Pittsburgh, Pa.: Bek Industries, 1971. This popular level work groups into sections frequently asked questions about lightning. Includes a chapter on ball lightning. Information on how to take lightning photographs is included.

CLOUD CHARTS

Cloud charts can be obtained from PI Publications, National Oceanic and Atmospheric Administration, Rockville, Maryland, or from

Science Associates
Nassau Street
Princeton, New Jersey 08540

Many films may be purchased, rented for a modest price, or borrowed without cost. Information is available from the suppliers. Additional sources are given in the October 1979 issue of *Weatherwise*.

Aetna Life and Casualty
Film Librarian
Public Relations and Advertising Department
151 Farmington Ave
Hartford, CT 06115
  Have one film, ''Hurricane'' available for loan. Need 6 months advance notice.

American Meteorological Society
45 Beacon St.
Boston, MA 02108
  Films available for loan for chapter meetings.

Bell Telephone
Local Area Business Offices
  Have one film, ''Unchained Goddess,'' from Bell Science Series. Very entertaining introductory film suitable for secondary schools or introductory courses. Available for free loan. Need advance notice.

The Walter A. Bohan Company
2026 Oakton St.
Park Ridge, IL 60068
  Large selection of movies dealing with satellite meteorology. Range from introductory level to advanced synoptic. Brochures available. For sale only.

The California Institute of Earth, Planetary, and Life Sciences
12208 Northeast 137th Pl.
Kirkland, WA 98033
  A number of films on hurricane development as viewed by satellite. Brochures available. For sale or rent.

Encyclopedia Britannica Educational Corporation
425 No. Michigan Ave.
Chicago, IL 60611
  A large selection of films for sale or rent. Mostly at the introductory level. Catalog available.

Indiana University Audio-Visual Center
Bloomington, IN 47401
  Some films including ''The Weather Machine.'' Rental only.

McGraw-Hill Films
1221 Avenue of the Americas
New York, NY 10020
  Several introductory level films. For sale or rent. Brochures available. Some examples are *The Inconstant Air* (excellent general film) and *The Flaming Sky.*

MESOMET, Inc.
Suite 502, 190 North State Street
Chicago, IL 60601
(312) 263-5921
  A number of slide sets dealing with meteorology and climatology. Suitable for teaching or public presentations. Brochure available

Motion Picture Service
Department of Commerce—NOAA
12231 Wilkins Ave.
Rockville, MD 20852
  Selection of films from introductory level to advanced. Short term loan at no charge. Brochure available.

NASA Films
Local NASA Office
  Several firms on satellite meteorology. Available for loan or sale. Brochure available.

National Audio Visual Center
National Archives and Records Service
General Services Administration
Washington, DC 20409
  Very large selection of films ranging from introductory level to advanced. Topics include basic meteorology, severe storms, and air pollution. For sale or rent. Complete catalog for sale.

NCAR Films
National Center for Atmospheric Research
P.O. Box 3000
Boulder, CO 80307
  Several films for sale or rent. Deal with advanced topics. Brochures available.

Pyramid Films
Box 1048
Santa Monica, CA 90406
  Handles "In Search Of—" series. Very entertaining films for general audience. For sale or rent.

Time Life Multimedia
100 Eisenhower Drive
Paramus, NJ 07852
  Handles "NOVA"series films. For sale or rent.

Wards Modern Learning Aids Division
P.O. Box 1712
Rochester, NY 14603

Universal Education and Visual Arts
100 Universal City Plaza
Universal City, CA 91608
  A number of films ranging from introductory level to more ad-
  vanced. For sale or rent. Separate catalog of films available on
  request. Examples include:
  Above the Horizon (general meteorology)
  Formation of Raindrops
  Solar Radiation I (the best one)
  Solar Radiation II
  Atmospheric Electricity (interesting equipment)
  Convective Clouds
  Sea Surface Meteorology (electrical phenomena emphasized)
  Planetary Circulation of the Atmosphere (may be a bit specialized
  for the nonmeteorologist)
  It's an Ill Wind (air pollution)

Also, a number of films on air pollution are available from

Audio-visual Facility
Public Health Service
U.S. Department of Health, Education, and Welfare
Atlanta, Georgia 30333

Two excellent films from you local Weather Service Office:

  Tornado: Approaching the Unapproachable
  Hurricane!

INSTRUMENTS

You can get an instrument catalog from Science Associates (see address under
Cloud Charts). Weather instruments tend to be expensive, so it pays to shop. Two
other places are

    Taylor Instruments
    Rochester, New York

    Weather Measure Corporation
    San Jose, California

507

LIBRARIES

University, college, and public libraries have a wealth of popular books on meteorology. Many older libraries contain interesting (and sometimes amazing) early books on weather.

PERIODICALS

NONTECHNICAL

Bulletin of the American Meteorological Society: published monthly by the American Meteorological Society, 45 Beacon Street, Boston, Massachusetts 02108.

Weather: published monthly by the Royal Meteorological Society, Cromwell House, High Street, Bracknell, Berks, United Kingdom.

NOAA: published quarterly by the National Oceanic and Atmospheric Administration, Public Information Publications, Rockville, Maryland 20852.

Meteorological Magazine: published monthly by the British Meteorological Office, British Information Services, 845 Third Avenue, New York, New York.

Weatherwise: published bimonthly by the American Meteorological Society, 45 Beacon Street, Boston, Massachusetts 02108.

TECHNICAL

Ecology Today
Journal of Applied Ecology
Journal of Applied Meteorology
Journal of the Atmospheric Sciences
Monthly Weather Review
Quarterly Journal of the Royal Meteorological Society
Tellus (geophysics)
Water and Pollution Control
Water Resources Research

MULTIMEDIA PROGRAM

Weather Around Us, by Richard Anthes. Coordinated text and audiovisual program that serves as a complete introductory course in meteorology. Published by Charles E. Merrill Publishing Co., 1300 Alum Creek Drive, Columbus, Ohio 43216.

WEATHER MAPS AND SATELLITE PICTURES

The *Daily Weather Map* is published in weekly sets by the Public Documents Department, Government Printing Office, Washington, D.C. 20402. The series contains a good readable surface chart, upper-air chart, and temperature and precipitation maps for each day. The 1980 price was $30.00 per year. Make checks payable to "Superintendent of Documents." Composite global satellite pictures are published monthly by the National Environmental Satellite Service (NESS) under the title "Environmental Satellite Imagery." These are available from the National Technical Information Service (NTIS), Sills Building, 5285 Port Royal Road, Springfield, Virginia 22151.

In addition, glossy prints of images obtained from both polar-orbiting and stationary equatorial satellites are available from the National Climatic Center, Asheville, North Carolina 28801. A typical price for a glossy print is $2.75. However, the National Weather Service Office in a major city could be a veritable silver mine of prints and fascimiles for the asking. Finally, the National Aeronautics and Space Administration has Educational Offices at Greenbelt, Maryland; Hampton, Virginia; and Cleveland, Ohio, and can be helpful.

D CONVERSION OF S.I. TO ENGLISH UNITS

NAMES OF INTERNATIONAL UNITS

Physical quantity	Name of unit	Symbol
	Basic units	
length	meter	m
mass	kilogram	kg
time	second	s
temperature	kelvin	K
	Derived units	
area	square meter	m^2
volume	cubic meter	m^3
frequency	hertz	hz (s^{-1})
density	kilogram/cubic meter	kg/m^3
velocity	meter/second	m/s
angular velocity	radian/second	rad/s or s^{-1}
force	newton	N ($kg \cdot m/s^2$)
pressure	newton/square meter	N/m^2
work, energy, heat	joule	J ($N \cdot m$)
power	watt	W (J/s)

CONVERSION FACTORS

Meter: 3.28 feet, 1.09 yards, 39.37 inches
Kilogram: 2.205 pounds, 1.10×10^{-3} tons (short, 2000 pounds)
Square meter: 10.76 square feet, 2.471×10^{-4} acres, 3.861×10^{-7} square miles
Cubic meter: 35.31 cubic feet, 28.38 bushels, 219.97 gallons (U.S. liquid)
Meter per second: 3.60 km/h, 1.94 knots, 2.237 mi/h
Meter per second squared: 3.28 feet/s^2
Newton: 1.0×10^5 dynes ($g \cdot cm^2/s^2$)
Newton per square meter: 1.0×10^{-5} bars, 10 dynes/cm^2, 2.953×10^{-4} inches of mercury at 0°C, 0.01 millibar, 1 pascal
Joule: 9.471×10^{-4} British Thermal Units, 0.2387 calorie, 1.0×10^{-7} ergs, 2.777×10^{-7} kilowatt hours
Watt: 0.239 calorie, 9.485×10^{-4} Btu(s)

Glossary

Absolute temperature   Also known as the *Kelvin temperature,* this temperature is zero at the coldest possible temperature that can exist in the universe, absolute zero. The absolute temperature is obtained by adding 273.15 to the Celsius temperature. The absolute temperature is measured in Kelvins (K) rather than degrees.

Absorption   The process by which radiation transfers energy to matter (such as to the air or ground), thus causing it to warm.

Absorptivity   The fraction of the radiation falling on an object that is absorbed.

Acceleration   The rate of change of velocity with time.

Accretion   See *coalescence.*

Adiabatic process   Any process that does not involve an exchange of heat with the surroundings. The adiabatic process of greatest interest in meteorology involves the vertical motion of air. As air rises, its temperature decreases because it expands at the lower pressures encountered aloft. This happens in spite of the fact that there was no loss of heat from the air parcel to its surroundings.

Adiabatic lapse rate   The rate at which the temperature decreases in a parcel as it is lifted adiabatically. If a cloud is not forming within the parcel, then the temperature decreases at a rate of 10 C° per kilometer (5.5 F°/1000 ft); this is called *dry adiabatic lapse rate.* If a cloud is forming in the parcel, then we have the *wet adiabatic lapse rate* whose value depends upon the actual temperature but is typically about 6 C°/km.

Advection   The transport of some quantity, such as temperature or moisture, by the winds. Thus *warm air advection* means that warm air is being brought into that region by the wind.

Aerosol   A solid or liquid particle in the air.

Air mass   An air mass is a body of air covering a wide area and characterized at any constant level by relatively uniform temperatures and humidities.

Albedo   The ratio of the radiation reflected by an object to that received by it.

Anabatic   A wind blowing up a slope, usually warm; the opposite of katabatic.

Anemometer   An instrument used to measure the strength of the wind.

Aneroid   Not wet; used to describe a barometer that contains no liquid.

Anticyclone   A region of the atmosphere in which the winds rotate in the opposite direction to the earth's rotation. For this to happen, the surface pressure must be higher than the surroundings. The air in such high pressure areas (anticyclones) is usually subsiding, causing clear skies.

Anticyclonic   Rotating about a vertical axis in the opposite direction to the earth's rotation. When viewed from above, anticyclonic rotation is clockwise in the Northern Hemisphere and counterclockwise in the Southern Hemisphere.

Anvil   The term given to the portion of a convective cloud (cumulus or thunderstorm) that spreads out horizontally at the base of a stable layer, such as the stratosphere.

Aphelion   That point in the orbit of the earth that is farthest from the sun.

Atmospheric effect   The atmospheric effect describes the fact that the earth's surface is warmer than it would be in the absence of an atmosphere, because the atmosphere is largely transparent to solar radiation but nearly opaque to terrestrial radiation. As a result the surface of the earth receives not only solar radiation but also the longer wave radiation absorbed by the atmosphere and reradiated.

Backing A counterclockwise turning of the wind, either with height or over a period of time; the opposite of veering.

Barograph A recording barometer.

Barometer The instrument used to measure air pressure.

Bergeron-Findeisen process One of the processes by which cloud drops grow into raindrops. The process involves the freezing of a small fraction of cloud drops and the rapid growth of the resulting ice crystals in the water-saturated environment (which is supersaturated for the ice).

Billow clouds Clouds oriented in broad parallel lines at right angles to the wind.

Black body A hypothetical object that absorbs all of the radiation that falls on it. By Kirchhoff's law, a black body also emits the maximum possible radiation for its temperature.

Boyle's law If the temperature of the air is fixed, then the pressure varies inversely with the volume.

Brownian motion The incessant, random motion of very small particles in the air caused by random collisions with air molecules themselves.

Buoyancy A buoyant parcel of air is one which is less dense than its surroundings so that there is no longer a balance between the pressure gradient force (in the vertical) and the force of gravity; the net force causes the parcel to rise. Often the buoyant parcel is less dense than its surroundings because it is warmer.

Buys-Ballot's law In the Northern Hemisphere, if one stands with his back to the wind then the pressure is lower to the left than to the right; in the Southern Hemisphere the relation is reversed. The law is a qualitative statement of the geostrophic wind relation.

Calorie A unit of heat defined as the amount of heat required to raise the temperature of one gram of water one degree Celsius.

Cap cloud A smooth lenticular-shaped cloud that sits over the top of an isolated mountain. It is formed in a stable atmosphere when moist air is forced up over the peak.

Carbon dioxide A minor constituent gas of the atmosphere (0.03 percent) which is important because it readily absorbs and emits the long-wave radiation (infrared radiation) that so strongly influences the temperature of the earth and the atmosphere.

Ceiling The height of the lowest cloud deck covering more than half the sky.

Centrifugal force An apparent (fictitious) force experienced by a body under rotation, directed outward from the center of curvature of the path.

Centripetal force Any force which causes the air (or any other object) to travel in a curved path. The centripetal force is directed towards the center of curvature of the path.

Charles' law If the pressure of the air is held fixed, then the temperature varies inversely with the density.

Chinook A warm dry wind that blows down the eastern slopes of the Rocky Mountains. It is most frequent in the winter months in northern Montana and southern Alberta. There are comparable winds associated with other mountain ranges such as the *foehn* wind of the Alps and the *Santa Ana* of California's Sierra Nevadas.

Circumhorizontal arc A colorful arc of light caused by the refraction of sunlight through the ninety degree prisms of large hexagonal ice crystals. It is parallel to the horizon (hence its name) and formed only when the sun is very high in the sky.

Climate The long-term characteristics of the weather.

Cloud bow A rainbow formed in cloud drops rather than rain drops. The smaller cloud drops cause the bow to appear whiter and broader.

Cloud drop The small drops of water that comprise clouds. A typical cloud drop is 2 to 20 micrometers in radius.

Cloud seeding The introduction into a cloud of substances designed to alter its evolution.

Coalescence With collision, coalescence is one of the processes by which cloud drops grow into rain drops. Two cloud drops, which collide because their different sizes gives them differing terminal velocities, will often coalesce into a single large drop. Repetition of this process will result in rain drops.

Cold front The leading edge of an advancing mass of cold air.

Collision See *coalescence.*

513

Condensation   The process by which water vapor is transformed into a liquid.

Conduction.   The transfer of thermal energy by molecular motions from warmer to cooler material.

Constant pressure chart   A map depicting variables such as temperature, wind, or humidity at a constant pressure.

Continental air mass   An air mass which originated over land and hence is dry.

Contrail   The same as a *condensation trail*.

Condensation trail   The line of cloud that sometimes forms in the wake of a jet aircraft.

Convection   In meteorology, the vertical transfer of air, as in *cumulus convection*.

Convergence   Flowing together, as in horizontally converging currents of air.

Coriolis force   An apparent force that acts upon winds (or other moving objects) and results from the fact that winds are measured with respect to the rotating earth rather than a fixed reference. In the Northern Hemisphere, the Coriolis force acts to the right or the direction of motion.

Corona   A series of rings of colored light seen surrounding the sun or moon; it is caused by the diffraction of light by a cloud of uniformly sized water drops. The corona is much smaller than a halo, having an angular radius usually less than 5 degrees.

Cross section   A vertical slice through atmosphere depicting, for example, temperature, moisture, or wind structure.

Cyclone   An atmospheric circulation in which the winds are rotating in the same direction as the earth. For this to happen, the pressure must be lower than the surrounding regions. The air in such low pressure areas (cyclones) is usually ascending, thus leading to cloudy skies.

Cyclonic   Rotating about a vertical axis in the same direction as the earth's rotation. When viewed from above, cyclonic rotation is counterclockwise in Northern Hemisphere and clockwise in Southern Hemisphere.

Deliquescent   Property by which a surface is wetted due to condensation.

Density   The mass of a substance divided by the volume it occupies.

Dew   Water that condenses directly on objects near the ground.

Dew-point temperature   The temperature to which the air must be cooled (while holding the pressure constant) for the humidity to reach 100 percent and thus for dew to form.

Diabatic process   A process that involves the exchange of heat with the surroundings. Compare with *adiabatic process*.

Diffraction   A process whereby light (or any other radiation), after passing by the edge of an object, bends into the shadow region behind the object.

Diffusion   Spreading out in the horizontal and vertical of a substance; for example, the diffusion of smoke in the air.

Diurnal   Daily, as in a *diurnal cycle*.

Divergence   Flowing apart, as in horizontally diverging currents of air.

Doppler effect   The change of frequency of waves observed when the emitter and the receiver are moving relative to each other.

Doppler radar   A type of radar which makes use of the Doppler effect to determine the component of velocity toward or away from the observer. Used to measure velocity of rain drops.

Downdraft   Downward moving column of air, as in a thunderstorm.

Drizzle   Small (conventionally less than 0.5 mm) water drops that (unlike fog) fall slowly toward the ground.

Dust devil   A whirlwind, either clockwise or counterclockwise, rendered visible by the dust that it picks up. Their diameters can vary from a few meters to tens of meters, and their heights can vary from a few meters to a few hundreds of meters.

Eccentricity   Deviation from circularity, as in eccentricity of an orbit.

Echo   Return signal from a target.

Ecliptic   The circle defined by the apparent annual path of the sun around the earth.

Eddy   An entity of air which usually consists of a circulation about a horizontal or vertical axis.

Emissivity   The ratio of the amount of radiation emitted by an object to the maximum possible radiation (black body radiation) that it can emit.

Equinox   A time of equal day and night; one of two days each year when the sun is directly overhead at the equator.

Evaporation   The process whereby liquid water is converted into water vapor.

Evapotranspiration   Total evaporation from plant surfaces and the ground.

Exosphere   Outermost portion of atmosphere.

Extratropical   Pertaining to phenomena outside the tropics.

Eye   The central region of a hurricane that is often clear with light winds.

Fata Morgana   A type of mirage in which the image is so greatly magnified that the original object is no longer recognizable.

Fog   A cloud that touches the surface of the earth.

Freezing nucleus   A particle that promotes the freezing of supercooled water.

Freezing rain   Rain that freezes upon contact with an object.

Frequency   Number of cycles per time interval in periodic motion.

Front   The dividing line, or transition zone between two air masses of different temperature and humidity.

Frontogenesis   The formation of a front.

Frost   A deposit of ice crystals that forms on an object, such as the ground, by sublimation.

Funnel cloud   Tornado or waterspout cloud, usually used to describe clouds not touching the surface of the earth.

General circulation   The large-scale flow over the earth at different times of year.

Geostrophic wind   The wind that would result if there were a balance between the Coriolis force and the pressure gradient force. The concept is useful because, away from the ground, most natural winds are very nearly geostrophic.

Glory   A series of rings of colored light seen surrounding the head of your shadow when your shadow is cast onto a cloud. It is most common to see from an aircraft as colored rings surrounding the shadow of the plane.

Gradient   The change of some quantity with distance. For example, the pressure gradient is the difference in pressure at two locations divided by the distance between those two locations.

Graupel   Precipitation consisting of white, opaque, approximately round ice particles of 2–5 mm in diameter. They are also called *snow pellets*.

Gravity   The force of attraction exerted by the earth.

Green flash   A small flash of green light sometimes seen at the top of the setting or rising sun.

Greenhouse effect   A misnomer, the greenhouse effect should be known as the *atmospheric effect*.

Hail   Precipitation in the form of balls or irregular lumps of ice. It is always produced by convective clouds, usually thunderstorms. By convention, hail has a diameter of greater than 5 mm, and the smaller particles are called *graupel*.

Hadley cell   A thermal circulation between the tropics and the subtropics. The trade winds are a manifestation of the Hadley cell.

Halo   Any of a wide variety of circles, arcs, or spots of light that occur in the sky and which are caused by the refraction and reflection of sunlight (or moonlight) by ice crystals in the atmosphere.

Haze   Tiny particles of dust, smoke, or small solution droplets that are dispersed through the air. The particles are too small to be seen or felt individually, but they diminish visibility.

Heat   A form of energy associated with the rate of vibration of molecules and transferred between systems by virtue of a difference in temperature.

Heat capacity   The heat energy absorbed by a substance divided by the associated temperature change.

Heat lightning   Ordinary lightning so far away the thunder cannot be heard.

High   An ellipsis for "area of high pressure." See also *anticyclone*.

515

Humidity   Any one of a number of measures of the water vapor content of the air. However, the term is popularly used to be synomous with *relative humidity*.

Hurricane   A severe tropical cyclone with winds exceeding 65 knots.

Hydrologic cycle   The flow and exchange of water in all forms throughout the atmosphere, ground, and sea.

Hydrometer   Solid or liquid water particles in the atmosphere.

Hydrostatic equilibrium   A balance in forces between the vertical pressure gradient force and the force of gravity acting on the air.

Hygrometer   An instrument that measures the amount of water vapor in the air.

Hygroscopic particle   A particle with an affinity for water vapor. Condensation nuclei are most commonly hygroscopic.

Ice crystal   Any one of a number of crystalline forms in which ice can appear such as hexagonal columns, hexagonal plates, dendritic crystals, and ice needles.

Ice pellets   A type of frozen or partially frozen precipitation with a diameter of 5mm or less. Ice pellets usually bounce off of hard ground.

Indian summer   A period of unusually mild, fine weather in autumn.

Inferior mirage   See *mirage*.

Infrared radiation   Electromagnetic radiation with a wavelength longer than visible radiation but shorter than the microwaves (between 0.8 and 1000 micrometers). All significant long-wave or terrestrial radiation lies in this range, and so is infrared radiation.

Instability   See *stability*.

Inversion   An ellipsis for *temperature inversion*. In the troposphere, the temperature normally decreases with height, but in an inversion the temperature increases with height—the normal situation is *inverted*. The term is sometimes used loosely to refer to any stable layer in the atmosphere.

Ion   An electrically charged atom or particle.

Ionosphere   A region of the atmosphere above about 70 km and dense in ions.

Isobar   On a map, a line of equal pressure.

Isotach   On a map, a line of equal wind speed.

Isotherm   On a map, a line of equal temperature.

Jet stream   Relatively narrow bands of high-velocity winds in the atmosphere. The term is usually applied to high winds in the upper troposphere associated with a polar front.

Katabatic wind   A wind blowing down a slope, usually cold; the opposite of *anabatic*.

Kelvin temperature   See *absolute temperature*.

Kinetic energy   The energy possessed by an object as a consequence of its motion.

Knot   A unit of velocity equal to one nautical mile per hour (1.85 km/h). It is common error to speak of "knots per hour" as if it were a measure of velocity.

Lake breeze   A wind blowing from a lake toward the shore.

Land breeze   A wind blowing from land across a coast toward a lake or ocean. See *sea breeze*.

Lapse rate, environmental   The rate of decrease of temperature with height in the atmosphere.

Latent heat   The amount of heat either released or absorbed by a unit mass of a substance when it changes from one state to another, such as when water changes from liquid to vapor.

Lee   Downwind.

Lenticular .   Lens-shaped, as in lenticular cloud.

Lidar   An instrument that uses a laser for determining the distance to and the amount of particles in the atmosphere. *Lidar* stands for *light detection and ranging*.

Light   The visible portion of the electromagnetic spectrum with wavelengths between 0.4 and 0.7 micrometers.

Lightning   A visible electrical discharge produced by thunderstorms.

Long-wave radiation   See *infrared radiation*.

Low   An ellipsis for "area of low pressure." See also *cyclone*.

Maritime air mass   An air mass which originated over water and is therefore moist.

Mesoscale   A horizontal scale ranging from about 1 to 300 kilometers.

Mesosphere   That portion of the atmosphere between the stratosphere and the thermosphere and so extending from about 50 to 70 kilometers. It is characterized by a steady decrease in temperature.

Meteorology   The study of the atmosphere and atmospheric phenomena.

Millibar   A unit of pressure equal to 0.1 kilopascals or about 0.03 inches of mercury.

Mirage   The name applied to the images that are formed when the atmosphere behaves like a lens. This occurs when there are strong temperature gradients (usually vertical temperature gradients associated with a surface such as the ground) and light is caused to bend. An image displaced upwards is called a *superior mirage*, and one displaced downward is called an *inferior mirage*.

Mixed layer   The lowest portion of the atmosphere (usually less than a kilometer deep) in which the air is well mixed in the vertical. There is a substantial exchange of heat by conduction from the earth to the atmosphere.

Mixing ratio   The ratio of the mass of water vapor in a volume of air to the mass of dry air. The mixing ratio is a dimensionless measure of the moisture content of the air.

Moist adiabatic lapse rate   The term should be *wet adiabatic lapse rate*; see *adiabatic lapse rate*.

Monsoon   A wind system that reverses direction between summer and winter.

Mountain wind   A local wind that blows downhill by night. Compare with *valley wind*.

Nautical mile   A distance of 1852 meters (6076 feet).

Numerical weather prediction   Prediction of weather phenomena by solving appropriate equations on a computer.

Obliquity of the ecliptic   Angle between plane of earth's orbit and the plane of the earth's equator, currently 23 1/2°.

Occlusion   Also known as an *occluded front*, the occlusion forms when a cold front overtakes a warm front.

Orographic   An adjective implying of or caused by mountains.

Overruning   A term applied to a layer of air that flows over the top of another layer, such as when warm air ascends above a warm frontal surface and is thus above the cold air mass.

Ozone   A form of molecular oxygen in which the molecule is made up of three atoms of oxygen instead of the usual two.

Parcel   The term used in meteorology to imply a small volume of air (a few meters across) whose properties are uniform. The parcel can therefore be characterized by a single value of the temperature, humidity, or velocity.

Parhelia   Also known as *sun dogs* or *mock suns,* the parhelia are a type of halo which occurs as colored spots of light at an angular distance of about 22 degrees to either side of the sun. They are caused by ice crystals in the form of large hexagonal plates. The singular of parhelia is parhelion.

Perihelion   The point on the orbit of the earth that is closest to the sun.

Pileus   The name applied to a smooth cap cloud when it forms over a cumulus cloud rather than over a mountain.

Planck's law   An expression describing the distribution of radiation, with wavelength, for a black body.

Precipitable water   The depth of water that would result from condensation of all the water vapor in a column of air extending from the surface to the top of the atmosphere.

Precipitation   Any of the various forms of water particles that fall from the atmosphere and reach the ground.

Pressure   A force per unit area. In meteorology, pressure is rarely measured in inches of mercury anymore, but rather in either millibars or kilopascals.

Pressure gradient force   A net force acting on a volume of air and resulting from the difference in pressure on either side of the volume.

Psychrometer   An instrument for determining the wet bulb temperature.

Radar   An instrument using a microwave transmitter and receiver to determine the distance to and

517

amount of particles in the atmosphere. *Radar* stands for *radio detection and ranging*.

Radiation   A form of energy that propagates as a wave.

Radiosonde   A device for taking measurements from just above the ground to an elevation of about 30 km. It consists of an instrument package and a radio transmitter carried aloft by a balloon. Sometimes called *rawinsonde*.

Rain shadow   A region downwind of a mountain where precipitation is reduced because of downward-moving air.

Rawinsonde   See *radiosonde*.

Reflection   The return of a portion of radiation striking an object.

Refraction   The bending of radiation as it passes through an object.

Relative humidity   The ratio of the air's actual vapor pressure to its saturation vapor pressure.

Ridge   In meteorology, a ridge implies an elongated area of relatively high atmospheric pressure.

Rime   A milky, opaque granular deposit of ice formed by the rapid freezing of supercooled water drops as they impinge upon an exposed object.

St. Elmo's fire   A luminous and often audible electrical discharge from sharply pointed objects under conditions of very highly charged electric fields, such as those which accompany thunderstorms.

Santa Ana   See *chinook*.

Saturation   The condition in which the water-vapor pressure is equal to the maximum possible vapor pressure that can be exerted at that temperature.

Scattering   A term applied to the interaction of radiation with objects such that some of the radiation is deflected in new directions. Although scattering is a general term, it is often used in the restricted sense of the interaction of light with very small particles, such as haze or air molecules.

Sea breeze   A wind blowing from an ocean toward land.

Sea level pressure   The atmospheric pressure at sea level. If the station is at sea level, it can be measured directly; but if the station is at some higher (or lower) level, the sea level pressure must be calculated based on the station pressure, temperature, and height above (or below) sea level.

Sea smoke   See *steam fog*.

Shear   See *wind shear*.

Shower   Precipitation from a cumuliform cloud, usually of short duration.

Sidereal day   The time (23 h 56 min 4 s) it takes for the earth to make one complete revolution relative to a given point in space.

Sleet   See *ice pellets*.

Smog   A mixture of smoke and fog

Snow   Precipitation in the form of ice crystals or aggregates of ice crystals.

Solar constant   The rate at which solar radiation is received outside the earth's atmosphere on a surface normal to the incident radiation, and at the earth's mean distance from the sun. The value of the solar constant is about 2 calories per square centimeter per minute (1396 W/m^2).

Solstice   One of two points on the sun's apparent annual path where it reaches its apparent northernmost point (June 21) or southernmost point (December 21).

Sounding   A vertical profile of temperature, moisture, or other variable.

Specific heat   The ratio of the heat absorbed by a unit mass of a substance to the resulting change in temperature.

Spectrum   The complete range of wavelengths of any wave phenomena.

Spiral bands   Bands of cumulus and cumulonimbus clouds in hurricanes that assume a spiral shape.

Stability   A description of a system which is in equilibrium. The stability is determined by what happens when the equilibrium is disturbed by a very small perturbation. If the system returns to its original equilibrium after the disturbance, it is said to be stable; if the system moves farther from the equilibrium state, it is unstable. Finally, if the system remains where the disturbance put it, it is said to be neutral.

Steam fog   A fog produced by the mixing of warm moist air with colder air. It is common over warm bodies of water on cold days (so also is called *sea smoke*), but it is the same mechanism that produces the condensation trail behind jet aircraft and the cloud that forms when you "see your breath" on a cold morning.

518

Storm surge   Rapid rise of sea level as a hurricane makes landfall.

Stratosphere   That region of the atmosphere above the troposphere and below the mesosphere (from 10 to 50 km in height). The lower portion is nearly isothermal but the upper portion increases in temperature with height.

Sublimation   The process whereby ice is converted directly into water vapor (or water vapor is converted directly into ice) without going through the liquid state.

Subsidence   A sinking motion of air.

Subsun   A bright spot of light directly below the sun, as far below the horizon as the sun is above the horizon. It is caused by the reflection of sunlight off the mirror-like faces of large hexagonal plate ice crystals.

Sun pillar   A long column of light extending vertically through the sun and only seen when the sun is near or below the horizon. It is caused by the reflection of sunlight off the mirror-like sides of large hexagonal column ice crystals.

Supercooled water   Water that exists in the liquid state even though its temperature is below the melting temperature (0°C). This is a very common condition for small drops of water, and many clouds at temperatures below 0°C contain supercooled water.

Superior mirage   See *mirage*.

Supernumerary bows   A family of weakly colored arcs of light sometimes seen just inside the rainbow.

Synoptic   A term implying an overall view. In meteorology, *synoptic* was first applied to a presentation of data collected simultaneously from a large area (perhaps greater than 800 km on a side), but it is now applied as an epithet for this large scale itself.

Temperature   A measure of the energy of molecular motion of a substance; measure of the "hotness" or "coldness."

Terminal velocity   The constant velocity reached by a falling object when gravity is balanced by drag on the object.

Thermal   A term used to describe the large (over 100 meters in diameter) buoyant bubbles of warm air that rise from near the ground to produce cumulus clouds.

Thermosphere   A region of the atmosphere above the mesosphere (and so above about 70 kilometers) which is characterized by increasing temperature with height. The ionosphere is included within the thermosphere.

Tornado   A violently rotating column of air extending from the base of a thunderstorm. Intense tornadoes are the most destructive of all storms in the atmosphere, with winds that can exceed 480 km/h.

Trade winds   A system of low-level winds occupying much of the tropics. The trades blow from the northeast towards the equatorial trough in the Northern Hemisphere and from the southeast towards the equatorial trough in the Southern Hemisphere.

Tropical air mass   An air mass originating over low latitudes and therefore warm or hot.

Tropical cyclone   A cyclone originating over low-latitude oceans, generally smaller than extratropical cyclones.

Tropopause   The boundary between the troposphere and stratosphere.

Troposphere   A region of the atmosphere extending from the ground to the base of the stratosphere (up to about 10 km). Almost all of the meteorological systems that give us weather are confined to the troposphere.

Trough   An elongated area of relatively low atmospheric pressure; the opposite of a ridge.

Typhoon   A hurricane occurring in the western Pacific.

Ultraviolet radiation   Electromagnetic radiation with a wavelength shorter than that of light (about 0.4 micrometers) but longer than X rays.

Unstable atmosphere   See *Stability*.

Valley wind   A local wind that blows uphill by day. Compare with *mountain wind*.

Vapor pressure   The pressure exerted by the molecules of water vapor (or any other vapor). It is used as a measure of the moisture content of the air.

519

Veering   A clockwise turning of the wind direction, either with height or time; the opposite of backing.

Virga   Wisps or streaks of water or ice falling out of a cloud but evaporating before reaching the earth's surface.

Visibility   The greatest distance over which an object can be distinguished from its background. The term is often limited to distinguishing a dark object against the background of the sky. Visibility is limited in the atmosphere by the scattering of light by air molecules and haze.

Vorticity   A measure of the spin of a fluid.

Water vapor   The gaseous form of water, found at normal temperatures in the atmosphere.

Wave cyclone   A cyclone that forms along a front; in the early stages the circulation deforms the front, producing a wave-like appearance.

Wet adiabatic lapse rate   See *adiabatic lapse rate*.

Wet-bulb temperature   The temperature a parcel of air would have if it were cooled to saturation by the evaporation of water into it. The wet-bulb temperature is an easily measured quantity which determines the moisture content of the air.

Wind   The motion of the air relative to the surface of the earth.

Wind chill   A measure of the rate of heat loss from a human body, taking into account both the temperature and the wind speed.

Wind shear   The change of wind in a given direction, as in horizontal or vertical wind shear.

Credits

CHAPTER 1

FIGURE 1.1 Aristotle, *Meteorologica*, p. 266.

CHAPTER 2

OPENER Courtesy of Robert Davies-Jones, National Severe Storms Laboratory.
FIGURE 2.2 © Hammond Incorporated #10399.
FIGURE 2.6 Courtesy of John Norman.
FIGURE 2.8 Courtesy of John Norman.
FIGURE 2.9 Courtesy of John Norman.

CHAPTER 3

OPENER © 1979 Alistair B. Fraser.
FIGURE 3.3 After Eric Palmén and Chester W. Newton, *Atmospheric Circulation Systems: Their Structure and Physical Interpretation* (New York: Academic Press, 1969).

CHAPTER 4

OPENER © 1971 Alistair B. Fraser.
FIGURE 4.9 After Robert Fleagle and Joost Businger, *An Introduction to Atmospheric Physics* (New York: Academic Press, 1963).
FIGURE 4.20 From Albert Miller and Jack C. Thompson, *Elements of Meteorology*, 3rd ed. (Columbus OH: Charles E. Merrill Pub. Co., 1979).

CHAPTER 5

OPENER Photograph by C. Hosler.

CHAPTER 6

OPENER © 1980 Alistair B. Fraser.
FIGURE 6.4 From *Patterns and Perspectives in Environmental Science*, National Science Foundation, 1973.
FIGURE 6.13 Photograph by N.M. Reiss.

CHAPTER 7

OPENER © Richard A. Anthes.
FIGURE 7.20 After Cooperative Extension Service, Purdue University.

CHAPTER 8

OPENER © 1973 Alistair B. Fraser.

CHAPTER 9

OPENER Courtesy of Richard A. Anthes.
FIGURE 9.2 Courtesy of Joseph H. Golden, National Severe Storms Laboratory.
FIGURE 9.3 After William Gray, "Global View of the Origin of Tropical Disturbances and Storms," Atmospheric Science Paper No. 14, Colorado State University, 1967.
FIGURE 9.4 After William Gray, *Monthly Weather Review* 96 (1968), pp. 669–700.
FIGURE 9.6 Courtesy of William Shenk, NASA.
FIGURE 9.9 Courtesy of William Shenk, NASA.
FIGURE 9.12 Courtesy of William Shenk, NASA.
FIGURE 9.17 Courtesy of Peter Black, National Hurricane Research Laboratory.

CHAPTER 10

OPENER © 1972 Alistair B. Fraser.
FIGURE 10.4 Photograph by Ronald Holle.
FIGURE 10.5 Photograph by Andrew Watson.
FIGURE 10.6 From Ed Brandes, *Journal of Applied Meteorology*, April, 1977.
FIGURE 10.7 From Albert Miller and Richard A. Anthes, *Meteorology*, 4th ed. (Columbus OH: Charles E. Merrill Pub. Co., 1980).
FIGURE 10.10 Photograph by Robert McAlister.
FIGURE 10.16 Courtesy of T. Fujita.

CHAPTER 11

OPENER   Photograph by Edmund Scientific Co., Barrington NJ.

FIGURE 11.11   From P. Walter Purdom and Stanley H. *Environmental Science* (Columbus OH: Charles E. Merrill Pub. Co., 1980).

FIGURE 11.27   After Clarence A. Woolum, "Notes from a Study of the Microclimatology of the Washington, D.C., Area for the Winter and Spring Seasons," *Weatherwise*, 1964, 17:6.

FIGURE 11.28   After Clarence A. Woolum, "Notes from a Study of the Microclimatology of the Washington, D.C., Area for the Winter and Spring Seasons," *Weatherwise*, 1964, 17:6,

CHAPTER 12

OPENER   © 1980 Richard A. Anthes.

FIGURE 12.2   From *Understanding Climatic Change*, National Academy of Sciences, Washington, D.C., 1975, p. 130.

FIGURE 12.4   From *Climate and Food*, National Academy of Sciences, Washington, D.C., 1976, p. 21.

FIGURE 12.6A   From A. James Wagner, "Weather and Circulation of January 1974," *Monthly Weather Review*, 102:4 (April 1974), pp. 324–31.

FIGURE 12.6B   From A. James Wagner, "Weather and Circulation of January 1977," *Monthly Weather Review*, 105:4 (April 1977), pp. 553–60.

FIGURE 12.7   From A. James Wagner, "Weather and Circulation of January 1974," *Monthly Weather Review*, 102:4 (April 1974), pp. 324–31.

FIGURE 12.8   From A. James Wagner, "Weather and Circulation of January 1977," *Monthly Weather Review*, 105:4 (April 1977), pp. 553–60.

FIGURE 12.9   From "Climate, Weather, Aridity," *Mosaic*, 8:1 (January–February 1977), National Science Foundation, p. 14.

FIGURE 12.10   From "Climate, Weather, Aridity," *Mosaic*, 8:1 (January–February 1977), National Science Foundation, p. 14.

FIGURE 12.11   From Andrew McIntyre, "The Surface of the Ice-age Earth," *Science*, 191:4232 (March 1976), cover figure. © 1976 by the American Association for the Advancement of Science.

FIGURE 12.13   From W.L. Gates, "The Numerical Simulation of Ice-age Climate with a Global General Circulation Model," *Journal of the Atmospheric Sciences*, 33:10 (October 1976), pp. 1844–73.

FIGURE 12.14   From B.J. Mason, "Towards the Understanding and Prediction of Climatic Variations," *Quarterly Journal of the Royal Meteorological Society*, 102:433 (July 1976), pp. 473–98.

FIGURE 12.15   From J.D. Hays, John Imbrie, and N.J. Shackleton, "Variatons in the Earth's Orbit: Pacemaker of the Ice Ages," *Science*, 194:4270 (December 1976), pp. 1121–32. © 1976 by the American Association for the Advancement of Science.

FIGURE 12.16   From V.C. LaMarche, Jr., "Paleoclimatic Inferences from Long Tree-ring Records," *Science*, 183:4129 (March 1974), pp. 1043–48. © 1974 by the American Association for the Advancement of Science.

FIGURE 12.17   From "Energy vs. Productivity: Diminishing Returns," *Mosaic*, 6:3 (May–June 1975), National Science Foundation, p. 9.

FIGURE 12.18   From B.J. Mason, "Towards the Understanding and Prediction of Climatic Variations," *Quarterly Journal of the Royal Meteorological Society*, 102:433 (July 1976), pp. 473–98.

FIGURE 12.19   From "Life at the Desert's Edge," *Mosaic*, 8:1 (January–February 1977), National Science Foundation, p. 24.

FIGURE 12.20   From M.J. Budyko, "The Effect of Solar Radiation Variations on the Climate of the Earth," *Tellus*, 21:5 (1967), pp. 611–19.

FIGURE 12.21   From *Understanding Climatic Change*, National Academy of Sciences, Washington, D.C., 1975, p. 14.

FIGURE 12.23   From J.M. Mitchell, Jr., "The Natural Breakdown of the Present Interglacial and Its Possible Intervention by Human Activities," *Quaternary Research*, 2:3 (November 1972), pp. 436–45.

FIGURE 12.24   From John A. Eddy, "The Maunder Minimum," *Science*, 192:4245 (June 1976), pp. 1189–1202. © 1976 by the American Association for the Advancement of Science.

FIGURE 12.25   From *Understanding Climatic Change*, National Academy of Sciences, Washington, D.C., 1975, p. 187.

CHAPTER 13

OPENER © 1978 Alistair B. Fraser.
FIGURE 13.3 From George Holzworth, "Mixing Heights, Wind Speeds, and Pollution throughout the Contiguous United States," Environmental Protection Agency, Research Triangle Park NC, January 1972.
FIGURE 13.4 From George Holzworth, "Mixing Heights, Wind Speeds, and Pollution throughout the Contiguous United States," Environmental Protection Agency, Research Triangle Park NC, January 1972.
FIGURE 13.5 From George Holzworth, "Meteorological Episodes of Slowest Dilution in the Contiguous United States," Environmental Protection Agency, Research Triangle Park NC, February 1974.
FIGURE 13.8 From John F. Clarke, "Nocturnal Urban Boundary Layer over Cincinnati, Ohio," *Monthly Weather Review*, August 1969.
FIGURE 13.9 From the National Center for Atmospheric Research.
FIGURE 13.10 Courtesy of L. Machta.

CHAPTER 14

OPENER © 1971 Alistair B. Fraser.
FIGURE 14.2 From Edward J. Tarbuck and Frederick K. Lutgens, *Earth Science*, 2nd ed. (Columbus OH: Charles E. Merrill Pub. Co., 1979).

CHAPTER 15

OPENER © 1969 Alistair B. Fraser.

FIGURE 15.4 © Alistair B. Fraser.
FIGURE 15.5 © Alistair B. Fraser.
FIGURE 15.7 © Alistair B. Fraser.
FIGURE 15.13 © Alistair B. Fraser.
FIGURE 15.14 © Alistair B. Fraser.
FIGURE 15.15 © Alistair B. Fraser.

CHAPTER 16

OPENER © 1979 Alistair B. Fraser.
FIGURE 16.1 From E.T. Wilkins, *Quarterly Journal of the Royal Meteorological Society* 80 (1954), pp. 267–71.
FIGURE 16.2 From Paul Kutschenreuter, "Some Effects of Weather on Mortality," Public Health Service Publication No. 999–AP–25, 1967, pp. 81–94, Washington, D.C.
FIGURE 16.3 From *Vital Statistics of the United States*, 1973, U.S. Department of Health, Education, and Welfare.
FIGURE 16.5 From Nicholas Camille Flammarion, *The Atmosphere*. Translated from the French; edited by James Glaisher (New York: Harper Brothers, 1874), p. 439.
FIGURE 16.7 After Linden, "Merchandising Weather," *The Conference Board Business Record* 19 (1962), pp. 15–16.
FIGURE 16.8 From "Wind Energy," *Innovative Systems Conference Proceedings*, May 1979. E1.28: SER 1/TP–245–184.
FIGURE 16.11 From "Solar Energy as a Natural Resource." NS 1.2: 50412.
FIGURE 16.12 From *Energy, Environment, Productivity: Proceedings of the First Symposium in RANN: Research Applied to National Needs*, Washington, D.C., November

18-20, 1973. Available from the National Science Foundation, Superintendent of Documents, U.S. Government Printing Office, Washington, D.C., 20402.
FIGURE 16.14 From "Wind Energy." Hearing before the Subcommittee on Energy of the Committee on Science and Astronautics, U.S. House of Representatives, 93rd Congress, May 21, 1974, U.S. Government Printing Office, Washington, D.C., p. 130.
FIGURE 16.15 From *Resource Energy Conservation* 1, 43.T22/2:2 EN 2/7/v.1.

CHAPTER 17

OPENER Courtesy of Richard A. Anthes.
FIGURE 17.1 After Robert A. Muller, *Weatherwise*, December 1966.
FIGURE 17.2 From "Statistical Probabilities for a White Christmas," U.S. Department of Commerce news release, Washington, D.C., December 15, 1971.

Index

References to figures are printed in **boldface** type. References to plates are noted in *italic* type.